Pathophysiology of Cerebral Energy Metabolism

Pathophysiology of Cerebral Energy Metabolism

Edited by

B. B. Mršulja and Lj. M. Rakić

University of Belgrade, Belgrade, Yugoslavia

and

I. Klatzo and Maria Spatz

U.S. Department of Health, Education, and Welfare
National Institutes of Health, Bethesda, Maryland

SPRINGER SCIENCE+BUSINESS MEDIA, LLC

Library of Congress Cataloging in Publication Data

International Symposium on the Pathophysiology of Cerebral Energy Metabolism,
   University of Belgrade School of Medicine, 1977.

   Pathophysiology of cerebral energy metabolism.
   Includes index.
   1. Cerebral ischemia—Congresses. 2. Epilepsy—Congresses. 3. Energy metabolism—
Congresses. 4. Physiology, Pathological—Congresses. I. Mršulja, B. B. II. Rakić,
Lj. M. III. Klatzo, Igor. V. Spatz, Maria. VI. Title.
RC388.5.I516 1977   616.8′1′07   78-7342
ISBN 978-1-4684-3350-0   ISBN 978-1-4684-3348-7 (eBook)
DOI 10.1007/978-1-4684-3348-7

Proceedings of an International Symposium on the Pathophysiology of Cerebral
Energy Metabolism held at the University of Belgrade School of Medicine,
September 19—22, 1977

© 1979 Springer Science+Business Media New York
Originally published by Plenum Press, New York in 1979
Softcover reprint of the hardcover 1st edition 1979

PREFACE

This monograph contains the proceedings of an
International Symposium on the Pathophysiology of Cerebral Energy Metabolism which was held at the University
of Belgrade School of Medicine, September 19-22, 1977.

The purpose of the symposium was to promote an
interdisciplinary discussion on energy demand in the
brain in pathophysiological states, particularly with
regard to cessation of blood flow, epilepsy, and paradoxical sleep disturbances. Doctors W.F. Caveness,
I. Klatzo, B.B. Mršulja, Lj.M. Rakić, and V. Šušić
served as an advisory board and helped to plan the scientific program. The papers which were presented covered
neurophysiological, neurochemical, morphological, and
clinical approaches to the topics.

The Editors

CONTENTS

INTRODUCTORY REMARKS

J. Ristić

Neuropsychiatric Clinic, Faculty of Medicine
University of Belgrade
11 000 Belgrade, Yugoslavia

The aim of the Symposium on Pathophysiology of Cerebral Energy Metabolism was to present current studies on problems of cerebral ischemia, convulsions, and the mechanisms of sleep, and to allow some synthesis to be reached concerning the basic mechanisms involved. Such studies already have clinical significance, or will acquire such significance in the near future. The powerful tools and insights of the modern biochemistry of the brain furnish both the ways and the means for important developments in this area of research.

The investigation of the metabolism of an organ like the brain, which is practically without any energy reserve, necessarily deals with problems of changes in its constant glucose and oxygen supply. As is well known, oxygen is needed by the brain for the display of two fundamental processes: the unfolding of the oxidative stage of glucose breakdown by phosphorylation in the Krebs's cycle and the syntheses of appropriate neurotransmitters by their enzymes.

The link between intracellular energy-supplying processes and ATP metabolism can be surmised from reports announcing a diminution of sodium-potassium ATPase in the cortical epileptical foci. Then there is a need to distinguish cellular hunger (the falling of the energy metabolism of ATP) from disturbances of intercellular communications (transmissions which lead to clinical

manifestations like the lowering of the adaptation capacity) of learning processes and of consciousness.

The respiration of cells and their energy metabolism
can be maintained much longer, especially in hypothermia,
than the system of intercellular communications. Clin-
icians are extraordinarily interested in both these vital
processes. The disturbances of both - of the oxidative
stage of glucose breakdown and of the synthesis of transmitters - lead in a very short time interval (within
some seconds to a few minutes) to important and severe
biochemical changes in neurons.

The most serious clinical picture is provoked by
the cerebral ischemia, a lack of blood supply, because
it means not only an oxygen deficit but also a shortage
of glucose. Even a slight glucose deficiency provokes
disturbances in transmitter synthesis. In mild hypogly-
cemia the synthesis of ACh is reduced. The disturbances
of consciousness are done by this somewhat roundabout
way of metabolic disturbances. It is well known, for instance, that local brain anemia provokes a diminution of
monoamines in the respective area.

The clinical experiences with transient ischemic
attacks should be correlated with new knowledge of recirculation. Some important contributions in this field
were reported on in this Symposium.

Of utmost importance in the scientific approach to
the clinical work should be a deeper understanding of delayed postischemic changes in both protein and transmitter
metabolism despite the normal energy state. This could
provide an explanation of important changes in glucose
consumption and blood coagulability in remote parts of
the brain.

Consciousness, namely "the awareness of sensation
and of self," results from the proper functioning of
brain cells, which is secured by cerebral metabolism.
Generally, there ought to exist some correlation between
the rate of this metabolism and the preservation of consciousness. There are many exceptions, however, to this
rule. Thus, in sleep, which involves a peculiar state of
quantitative and qualitative consciousness alteration,
there is no diminution of brain metabolism. In hepatic
and diabetic comas, and therefore in states of complete
loss of consciousness, the energy metabolism in the
brain remains fully normal; it is also normal in bar-
bituric and diazepam comas.

An enhancement of the brain energy metabolism can be seen in the situations of fear, stress (e.g. immobilization), schizophrenia, hyperthermia and epileptic status. The diminution of brain metabolism does not signify a lowering of the consciousness level, and, at the same time, an enhancement of the same metabolism does not draw even a slight presence of consciousness as is the case in epileptic status.

Here I quote a famous old neurologist, J. M. Charcot (1825-1893), who wrote "Disease is old and nothing about it has changed. It is we who change as we learn to recognize what was formerly imperceptible."

This symposium will not solve, I surmise, all the important problems dealt with, but it will, I am sure, contribute to a heightening of the level of our understanding of the unknown.

PATHOPHYSIOLOGY OF CEREBRAL ISCHEMIA

K.-A. Hossmann

Max-Planck-Institut für Hirnforschung
Cologne-Merheim, W. Germany

The limiting factor for the reanimation of organisms following cardio-circulatory arrest is, in general, the anoxic vulnerability of the brain. The misrelationship between energy consumption and energy production is the reason for the high sensitivity of the brain. The brain has a high metabolic activity even under rest conditions: cerebral blood flow is about 50 ml/100 g/min, and oxygen consumption more than 3 ml/100 g/min. Brain energy requirements are approximately 8 cal/100 g/min, and are covered almost exclusively by oxidation of glucose which has to be continuously transported from the circulating blood into the brain.

The high energy requirements are in sharp contrast with the low energy reserves of the brain tissue. The total amount of substrates available as primary energy sources or as carbohydrate reserves is equivalent to about 2.5 mmol of energy-rich phosphate bonds ($\sim$P)/100 g brain (37). Since the energy yield of 1 mmol $\sim$P is 8 cal, the cerebral energy reserves are equivalent to about 20 cal/100 g brain. This amount of energy is sufficient for maintaining undisturbed brain function for only 2-3 min.

In reality the depletion of energy reserves is somewhat slower. Acute circulatory arrest elicits a "cortical stress response" (3) which causes a reduction in the energy consumption, and thereby retards the breakdown of the energy-dependent metabolic processes necessary for the structural integrity of the brain. This is illustrated by the fact that the suppression times of central nervous functions - the time from the onset of ischemia

until the function is suppressed - increase with decreasing complexity. Consciousness and EEG activity disappear within 15 sec, synaptically evoked potentials
after 2-4 min and electrically evoked neuronal activity
after 6-8 min (20, 31, 54). At this time the available
energy reserves are used up (37, 38), cell membranes
depolarize (18, 26, 30), and all endergetic biochemical
and biophysical functions of the brain come to a halt
(28).

Up to a certain limit, recovery of central nervous
function may occur when the brain is recirculated with
blood after ischemia. According to Sugar and Gerard (54),
latency of recovery is the interval which elapses after
the end of ischemia until a given function just begins
to reappear, recovery time is the time which elapses until
this function has fully recovered, and revival time is
the longest duration of ischemia that can be followed by
the reappearance of function.

Contrary to the suppression times, the recovery and
revival times increase with increasing complexity of the
functions under investigation. The revival times of the
brain following total ischemia, therefore, may vary considerably depending on the function chosen as the crite-
rium of cerebral viability (Table I). However, even by
monitoring the same function, e.g., the electroencephalogram, considerable differences have been noted suggesting
that recovery of the brain following ischemia does not
depend only on the duration of circulatory arrest but, in
addition, on side factors which may either facilitate or
inhibit the functional recovery process.

The following review summarizes some of these factors.

COMPLETENESS OF ISCHEMIA

The differences in revival times following ischemia
have been, by some authors, related to the possibility of
a remaining blood flow during ischemia. In experiments
performed by Hirsch et al. (24), the latency of recovery
of the electroencephalogram following 10 min of ischemia
decreased from 35-120 min to about 2 min when blood flow
persisted at a rate of only 7% of the normal. When blood
flow was further reduced, the latency of recovery sharply
increased and rapidly approximated that of the total ischemia. The protective role of the slow remaining perfusion was confirmed by Marshall et al. (39), who observed
a distinct prolongation of the revival time at a cerebral
perfusion pressure of only 20 mm Hg.

Table I.   REVIVAL TIMES OF THE BRAIN FOLLOWING COMPLETE ISCHEMIA

Functions	Methods	Species	Revival Times* partial	full recovery	Authors
Neurological performance	pneumatic cuff	dog	6 min	–	Grenell 1946 (13)
	pneumatic cuff	rabbit	–	10 min	Hirsch et al. 1957 (23)
	arterial clamp	monkey	–	13 min	Wolin & Massopust 1972(56)
	intracranial hypertension	dog	–	25 min	Neely and Youmans 1963(44)
	aortic ligat.	monkey	24 min	20 min	Miller and Myers 1970(43)
EEG	isolated head	cat	–	4.5 min	Gänshirt & Zylka 1952(12)
	pneumatic cuff	rabbit	9 min	7 min	Sainio 1974(49)
	intracranial hypertension	rat	18 min	–	Ljunggren et al. 1974(36)
	intracranial hypertension	cat	30 min	–	Van Harreveld 1947(17)
	isolated head	dog	30 min	–	Hinzen et al. 1972(19)
	arterial clamp	monkey	60 min	–	Hossmann & Zimmermann 1974 (33)

*Values refer to the longest time observed.

In contrast, very slow flow rates (of less than 5 ml/ 100 g/min) seem to reduce the chances for successful revival as compared to complete ischemia (28, 33, 48). Several factors may be responsible for this paradoxical situation. One is the fact that less than 15% of the blood glucose can be oxidized by the oxygen contained in the circulating blood. Since at low flow rates energy requirements of the brain cannot be fully covered by oxidative phosphorylation, anaerobic glycolysis is stimulated. This does not contribute to a major improvement of the brain energy state because the energy yield by anaerobic glycolysis is only 5% of that obtained by oxidative metabolism (7). The continuous supply of glucose, however, causes an increase in brain lactate, which greatly exceeds the one observed after complete ischemia. The more severe drop in tissue pH may, in turn, activate catabolic processes and thus reduce the chances of postischemic recovery (48).

Another factor which may complicate brain revival following complete ischemia is that the cerebral vessels remain filled with blood. The continuous ("trickling") fluid supply leads to swelling of capillary endothelial cells which, in conjunction with aggregation of blood particles, may cause secondary microcirculatory disturbances (see below). This could explain the fact that, in earlier experiments, revival time was shortest when the pneumatic cuff method was used for producing cerebral ischemia (Table I). Ljunggren et al. (37) demonstrated that, using this technique, complete ischemia, in fact, is not achieved.

BRAIN TEMPERATURE

During cerebral ischemia brain temperature decreases because heat exchange between blood and brain is interrupted, and heat production by the brain itself ceases as soon as metabolism breaks down. On the surface of the exposed brain the temperature drop may be as much as 1°C during 30 sec ischemia (2). Deeper, the decrease in temperature is much slower, particularly if the brain is not exposed. Temperature recordings at a depth of 2-3 mm below the brain surface revealed an initial drop of 0.2°C/ min at normal room temperature. After 60 min of complete cerebro-circulatory arrest, the brain temperature was $28-30^\circ$C (31). A temperature drop by $7-9^\circ$C causes a decrease in the metabolic rate oxygen by 35-45% (5, 14), reflected by a measurable prolongation of the revival time. In experiments performed by Hirsch et al. (21, the revival time of the EEG increased from 6-10 min at a temperature of $37^{\circ}C$ to about 30 min at a temperature of $28-30^{\circ}C$.

These observations were made under experimental conditions in which brain temperature was lowered <u>prior</u> to ischemia. A temperature decrease <u>after</u> the onset of ischemia, however, does not markedly reduce the energy requirements of the brain because the consumption of the available energy reserves is fast in comparison to the drop in temperature, and because heat production of the brain continues until metabolism ceases. Energy metabolism breaks down completely within less than 10 min (38), and during this time brain temperature does not decrease by more than $2^{\circ}C$. The suppression of all endergetic processes consequently occurs almost under normothermic conditions. It is, therefore, not likely that a drop in temperature during ischemia stimulates postischemic revival by improving the energy state of the brain. This does not exclude the possibility of having a certain influence on catabolic processes; these factors however, have not yet been investigated.

ANAESTHETIC DRUGS

Most anaesthetic drugs cause an inhibition of cerebral metabolism. Barbiturates, most commonly used in experimental ischemia research, cause an increase in the glucose and glycogen concentration of the brain and a decrease in the cerebral metabolic rate of glucose and oxygen by as much as 40% (9). The higher carbohydrate reserves and the reduced metabolic rates are responsible for the fact that under barbiturate anaethesia energy reserves are used up more slowly than in the awake animal (38, 48). There is little evidence, however, that the delayed break-down of energy metabolism is accompanied by a substantial prolongation of the brain revival time. Earlier communications about the protective effect of barbiturate anaesthesia during ischemia have not been confirmed by later investigations (42, 47, 48). An interesting hypothesis which might explain these different opinions has been forwarded by Michenfelder and Theye (42). They believe that anaesthetic drugs have a protective action against oxygen deficiency as long as there is some remaining metabolism, because the discrepancy between energy availability and energy demands of the tissue may be reduced by metabolic-activity reduction. However, as soon as metabolism breaks down completely, anaesthetic drugs become inefficient.

During total ischemia this occurs within 10 min,
i.e, before the limits of cerebral revival are reached
(see below). If this hypothesis is true, barbiturate
anaesthesia should be more protective during hypoxia and
oligemia than during total ischemia.

Barbiturates have in the past also been advocated
for ameliorating brain damage following the ischemic
impact (postischemic barbiturate loading; 4, 25, 41).
The mechanisms responsible for this effect are still
unclear. There is evidently no direct influence on ischemia because barbiturates are administered after and
not during ischemia. It has been discussed that inhibition of free radical reactions (10) or inhibition of
postischemic hypermetabolism (45) could be involved because both processes might be harmful for the structural
integrity of the ischemic brain (30, 48).

POSTISCHEMIC COMPLICATIONS

Cerebral ischemia of equal length is not equally
tolerated in different animals in the same experimental
series (29). This suggests that recovery after ischemia
does not depend on the duration of the ischemia alone,
but on postischemic side factors as well.

One of the most important factors is postischemic
blood recirculation. Recirculation disturbances have been
observed following 7.5 min strangulation ischemia, and
complete failure of reperfusion after 15 min ischemia
(norefolow phenomenon, 1). It is evident that adequate
blood recirculation is an essential condition for recovery. It has therefore been suggested that revival of
the brain is limited by such recirculation disturbances
rather than by the anoxic resistance of the neurons (1).
This has been corroborated by quantitative measurements
of cerebral blood flow demonstrating that electrophysiological recovery after prolonged ischemia depends on a
phase of postischemic hyperemia (29).

Postischemic recirculation disturbances are due to
either postischemic hypotension or postischemic increase
in vascular resistance. Following ischemia, autoregulation
of cerebral blood flow is transiently abolished (15),
causing a decrease in cerebral blood flow at low perfusion
pressure. This is the reason for the fact that the revival
time of the brain is shorter after cardiac arrest than
after isolated cerebro-circulatory arrest, because the
return of blood pressure to normal levels is delayed
during cardiac resuscitation (23).

Postischemic increase in cerebro-vascular resistance results from one or more of the following factors: narrowing of cerebral vessels due to swelling of capillary endothelial cells (6); an increase in blood viscosity, mainly because of aggregation of platelets and erythrocytes (11, 16, 22); local vascular occlusion due to disseminated intravascular coagulopathy (27, 51, 52); and an increase in intracranial pressure due to anoxic brain swelling (33, 40). Anoxic brain swelling is the consequence of an increase in brain osmolality (32) and a decrease in extracellular sodium following ischemic cell membrane depolarization (30). Furthermore, a delayed decrease in blood flow may develop after a few hours due to functional disturbances in flow regulation which appear to be the consequence of a dissociation between an already reestablished autoregulation and a still disturbed CO_2 reactivity (postischemic hypoperfusion, 29, 46).

Narrowing the cerebral vessels, intracranial hypertension and increase in blood viscosity do not obstruct blood flow completely. Increasing the perfusion pressure of the brain by raising systemic blood pressure, osmotherapy for treatment of postischemic brain swelling and postischemic coagulopathy ameliorate postischemic recirculation. By further adjusting the postischemic acid-base and electrolyte balance of the blood to normal in order to obtain optimal oxygen saturation and ion homeiostasis, progressing signs of neurophysiological and biochemical recovery of the brain reappear after prolonged ischemia. Over the past years, partial or complete recovery of oxygen utilization (50), energy metabolism (35, 48), phospholipid metabolism (51), evoked potentials (19), and spontaneous EEG activity (31, 33) have been described following complete normothermic ischemia of 30 min or even one hour. This and evidence of morphological preservation of the brain parenchyma after one hour's ischemia (33) indicate that this period comes close to the actual (theoretical) revival time of the normothermic brain. Failure of recovery after this or shorter periods of ischemia, consequently, would indicate that in such cases the conditions for cerebral resuscitation were not optimal.

CONCLUSIONS

The present review of animal experimental investigations on total cerebral ischemia suggests that revival of the brain does not depend only upon the duration of ischemia, but also upon the prevention of postischemic recirculation disturbances. Such disturbances develop not earlier than 7-8 min after the onset of ischemia, and full neurological revival of the brain within this period is not problematical as long as blood pressure is in the normal range. This time, therefore, can be referred to as the "safe revival time" of the brain. The "theoretical revival time" is limited by anoxic tolerance of the neurons, which presumably is close to one hour. This limit, however, can be reached only under optimal recirculation conditions which, with increasing duration of ischemia, are more and more difficult to achieve. An extension of the brain revival time under clinical conditions will therefore greatly depend on the management of postischemic complicating side effects.

REFERENCES

1. Ames III, A., Wright, R.L., Kowada, M., Thurston, J. M. and Majno, G. (1968): Cerebral ischemia. II. The no-reflow phenomenon. Am. J. Path. 52: 437-453.

2. Anderson, R.E., Waltz, A.G., Yamaguchi, T. and Ostrom, R.D. (1970): Assessment of cerebral circulation (cortical blood flow) with an infrared microscope. Stroke 1: 100-103.

3. Bito, L.Z. and Myers, R.E (1972): On the physiological response of the cerebral cortex to acute stress (reversible asphyxia). J. Physiology 221: 349-370.

4. Bleyaert, A.L., Nemoto, E.M., Stezoski, S.W., Alexander, H. and Safar, P. (1975): Amelioration of post-ischemic encephalopathy by sodium thiopental after 16 minutes of global brain ischemia in monkeys. Physiologist 18: 145.

5. Carlsson, C., Hägerdal, M., and Siesjö, B.K. (1976): The effect of hyperthermia upon oxygen consumption and upon organic phosphates, glycolytic metabolites, citric acid cycle intermediates and associated amino acids in rat cerebral cortex. J. Neurochem. 26: 1001-1006.

6. Chiang, J., Kowada, M., Ames II, A., Wright, R.L. and Majno, G. (1968): Cerebral ischemia. III. Vascular changes. Am. J. Pathol. 52: 455-476.

7. Cohen, P.J. (1972): The metabolic function of oxygen and biochemical lesions of hypoxia. Anesthesiology. 37: 148-177.

8. Cooper, H.K., Zalewska, T., Kawakami, S., Hossmann, K.-A. and Kleinhues, P. (1977): The effect of ischemia and recirculation on protein synthesis in the rat brain. J. Neurochem. 28: 929-934.

9. Cucchiara, R.F. and Michenfelder, J.D. (1973): The effect of interruption of the reticular activating system on metabolism in canine cerebral hemispheres before and after thiopental. Anesthesiology 39: 3-12.

10. Demopoulos, H.B., Flamm, E.S., Seligman, M.L., Jorgensen, E. and Ransohoff, J. (1977): Antioxidant effects of barbiturates in model membranes undergoing free radical damage. Acta Neurologica Scand. 56. Suppl. 64: 152-153.

11. Fischer, E.G. (1973): Impaired perfusion following cerebro-vascular stasis. A review. Arch. Neurol. 29: 361-366.

12. Gänshirt, H. und Zylka, W. (1952): Die Erholungszeit am Wärmeblütergehirn nach kompletter Ischämie. Arch. Psychiat. Nervenkr. 189: 23-36.

13. Grenell, R.G. (1946): Central nervous system resistance. I. The effects of temporary arrest of cerebral circulation for periods of two to ten minutes. J. Neuropath. Exp. Neurol. 5: 131-154.

14. Hägerdal, M., Harp, J., Nilsson, L., and Siesjö, B.K. (1975): The effect of induced hypothermia upon oxygen consumption in the rat brain. J. Neurochem. 24: 311-316.

15. Häggendal, E., Löfgren, J., Nilsson, N.J. and Zwetnow, N.N. (1970): Prolonged cerebral hyperemia after periods of increased cerebrospinal fluid pressure in dogs. Acta physiol. Scand. 79: 272-279.

16. Hallenbeck, J.M. (1977): Prevention of postischemic impairment of microvascular perfusion. Neurology 27: 3-10.

17. Van Harreveld, A. (1947): The electroencephalogram after prolonged brain asphyxiation. J. Neurophysiol. 10: 361-370.

18. Heilbrun, M.P. and Goldring, S. (1968): Steady potential and pathological correlates of cerebrovascular occlusion of dog. Arch. Neurol. 19: 410-420.

19. Hinzen, D.H., Müller, U., Sobotka, P., Gebert, E.,
    Lang, R. and Hirsch, H. (1972): Metabolism and func-
    tion of dog´s brain recovering from longtime ische-
    mia. Amer. J. Physiol. 223: 1158-1164.

20. Hirsch, H., Bange, F., Pulver, G. and Steffens, J.
    (1960): Evoked responses of the cat´s visual cortex
    to optic tract stimulation at temperatures between
    39⁰ and 15⁰C. Electroencephal. Clin. Neurophysiol.
    12: 679-684.

21. Hirsch, H., Bolte, A., Schaudig, A. und Tönnis, D.
    (1957): Über die Wiederbelebung des Gehirnes bei
    Hypothermie. Pflügers Arch. ges. Physiol. 265: 328-
    336.

22. Hirsch, H., Breuer, M., Künzel, H.P., Marx, E. und
    Sachweh, D. (1964): Über die Bildung von Thrombozyten-
    aggregaten und die Änderung des Hämatokrits durch
    komplette Gehirnischämie. Dtsch. Z. Nervenheilk. 168:
    58-66.

23. Hirsch, H., Euler, K.H. und Schneider, M. (1957):
    Über die Erholung und Wiederbelebung des Gehirnes
    nach Ischämie bei Normothermie. Pflügers Arch. ges.
    Physiol. 265: 281-313.

24. Hirsch, H., Koch, D., Krenkel, W. und Schneider, M.
    (1955): Die Erholungslatenz des Warmeblütergehirns
    bei Ischämie und die Bedeutung des Restkreislaufes.
    Pflügers Arch. ges. Physiol. 261: 392-401.

25. Hoff, J.T., Smith, A.L., Hankinson, H.L. and Nielsen,
    S.L. (1975): Barbiturate protection from cerebral
    infarction in primates. Stroke 6: 28-33.

26. Hossmann, K.-A. (1971): Cortical steady potential,
    impedance and excitability changes during and after
    total ischemia of cat brain. Experimental Neurology
    32: 163-175.

27. Hossmann, K.-A. and Hossmann, V. (1977): Coagulopathy
    following experimental cerebral ischemia. Stroke 8:
    249-254.

28. Hossmann, K.-A. and Kleihues, P. (1973): Reversibility
    of ischemic brain damage. Arch. Neurol. 29: 375-384.

29. Hossmann, K.-A., Lechtape-Grüter, H. and Hossmann, V. (1973): The role of cerebral blood flow for the recovery of the brain after prolonged ischemia. Z. Neurol. 204: 281-299.

30. Hossmann, K.-A., Sakai, S. and Zimmermann, V. (1976): Cation activities in reversible ischemia of the cat brain. Stroke 8: 77-81.

31. Hossmann, K.-A., and Sato, K. (1971): Effect of ischemia on the function of the sensorimotor cortex in cat. Electroencephal. Clin. Neurophysiol. 30: 535-545.

32. Hossmann, K.-A. and Takagi, S. (1976): Osmolality of brain in cerebral ischemia. Exp. Neurology 51: 124-131.

33. Hossmann, K.-A. and Zimmermann, V. (1974): Resuscitation of the monkey brain after 1 h complete ischemia. I. Physiological and morphological observations. Brain Res. 81: 59-74.

34. Kleinhues, P. and Hossmann, K.-A. (1973): Regional incorporation of L-(3-^{3}H)-tyrosine into cat brain proteins after 1 hour of complete ischemia. Acta Neuropath. 25: 313-324.

35. Kleinhues, P., Kobayashi, K. and Hossmann, K.-A. (1974): Purine nucleotide metabolism in the cat brain after one hour of complete ischemia. J. Neurochem. 23: 417-425.

36. Ljunggren, B., Ratcheson, R.A. and Siesjö, B.K. (1974): Cerebral metabolic state following complete compression ischemia. Brain Res. 73: 291-307.

37. Ljunggren, B., Schutz, H. and Siesjö, B.K. (1974): Changes in energy state and acid-base parameters of the rat brain during complete compression ischemia. Brain Res. 73: 277-289.

38. Lowry, O.H., Passonneau, J.V., Hasselberger, F.X. and Schulz, D.W. (1964): Effect of ischemia on known substrates and cofactors of the glycolytic pathway in brain. J. Biol. Chem. 239: 18-30.

39. Marshall, L.F., Durity, F., Lounsbury, R., Graham, D. I., Welsh, F. and Langfitt, T.W. (1975): Experimental cerebral oligemia and ischemia produced by intracranial hypertension. I. Pathophysiology, electroenceph-

alography, cerebral blood flow, blood brain barrier
and neurological function. J. Neurosurg. 43: 308-317.

40. Matakas, F., Cervos-Navarro, J. and Schneider, H.
    (1973): Experimental brain death. I. Morphology and
    fine structure of the brain. J. Neurol. Neurosurg.
    Psychiatr. 36: 497-508.

41. Michenfelder, J.D., Milde, J.H. and Sundt, T.M.
    (1976): Cerebral protection by barbiturate anaesthesia
    use after middle artery occlusion in Java monkeys.
    Arch. Neurol. 33: 345-350.

42. Michenfelder, J.D., and Theye, R. (1973): Cerebral
    protection by thiopental during hypoxia. Anesthesiolo-
    gy 39: 510-517.

43. Miller, J.R. and Myers, R.E. (1970): Neurological ef-
    fects of systemic circulatory arrest in the monkey.
    Neurology 20: 715-724.

44. Neely, W.A. and Youmans, J.R. (1963): Anoxia of canine
    brain without damage. JAMA 183: 1085-1087.

45. Nemoto, E.M., Kofke, W.A., Kessler, P., Hossmann,
    K.-A., Stezoski, S.W. and Safar, P. (1977): Studies
    on the pathogenesis of ischemic brain damage and the
    mechanism of its amelioration by thiopental. Acta
    Neurol. Scand. 56, Suppl. 64: 142-145.

46. Nemoto, E.M., Snyder, J.V., Carroll, R.G. and Morita,
    H. (1975): Global ischemia in dogs: Cerebrovascular
    CO_2 reactivity and autoregulation. Stroke 6: 425-431.

47. Nilsson, L. (1971): The influence of barbiturate an-
    aesthesia upon the energy state and upon acid-base
    parameters of the brain in arterial hypotension and
    asphyxia. Acta Neurol. Scand. 47: 233-253.

48. Nordström, C.-H., Rehncrona, S. and Siesjö, B.K.
    Effects of phenobarbital in cerebral ischemia. 2.
    Restitution of cerebral energy state as well as of
    glycolytic metabolites, citric acid cycle interme-
    diates and associated amino acids after pronounced,
    incomplete ischemia. Stroke (in press).

49. Sainio, K. (1974): Computer analysis of rabbit EEG
    after cerebral ischemia. Electroencephal. Clin.
    Neurophysiol. 36: 471-479.

50. Sobotka, P., Gebert, E. and Lang, R. (1972): The uti-
    lization of oxygen in the brain after complete ische-
    mia. Physiol. Bohemoslav. 21: 436.

51. Sobotka, P. and Hinzen, D.H. (1973): Effect of com-
    plete cerebral ischemia on brain phospholipid metabo-
    lism. Act. Nervosa Superior 15: 28.

52. Strumza, M.-V., Migne, J. et Maupin, B. (1970):
    Effets rapides de l'hypoxie sur certains facteurs de
    la coagulation sanguine et de la fibrinolyse. C.R.
    Soc. Biol. 164: 962-966.

53. Sturm, K.W., Wenzel, E., Tamaska, L. and Holzhüter,
    H. (1974): Comparative study of neuropathological
    findings and coagulation parameters in disseminated
    intravascular coagulation. In: Pathology of Cerebral
    Microcirculation (J. Cervos-Navarro, Ed.), De Gruyter
    Berlin-New York, 419-424.

54. Sugar, O. and Gerard, R.W. (1938): Anoxia and brain
    potentials. J. Neurophysiol. 1: 558-572.

55. Wilhjelm, B. (1966): Protective action of anaesthe-
    tics against anoxia. Acta Amaesth. Sand. Suppl. 25:
    318-321.

56. Wolin, L.R. and Massopust, L.C. Jr. (1972): Behavioral
    effects of arrest of cerebral circulation in the
    Rhesus monkey. Exp. Neurol. 34: 323-330.

THE INTERPRETATION OF ULTRASTRUCTURAL ABNORMALITIES

IN CEREBRAL ISCHEMIA

J.H. Garcia, A.S. Lossinsky, K. Donger
and F.C. Kauffman

Department of Pathology and Department of
    Pharmacology and Experimental Therapeutics
University of Maryland,School of Medicine
Baltimore, Maryland 21201, U.S.A.

INTRODUCTION

It has been customary to equate the effects of pure
lack of oxygen (hypoxia, anoxia) with those of ischemia.
Significant differences between them have been determined
at the cellular level; these have been attributed, among
other reasons, to the fact that ischemia is characterized
by major compartmental fluid shifts, which appear to be
minimal in hypoxic cells (6). Experimental models of brain
ischemia, or impaired intracranial blood flow, range from
those in which the entire arterial circulation is interrupted either temporarly (10) or permanently (12), to
situations involving interruptions of both arterial and
venous flow (1) to models of single-artery occlusion (18).
In terms of applicability to human brain injuries, two
mechanisms of ischemia are particularly frequent: *temporary global* ischemia (secondary to hypotension and cardiac
arrest) and *regional* ischemia (due to occlusion of a major artery). In addition to the differences attributable
to variations in the mechanism of ischemia (global vs.
regional), it has become increasingly apparent that the
consequences of ischemic injury very considerably depend
on whether or not ischemia is followed by reperfusion.
In several models of *global* ischemia, secondary to hypotension, the extent of brain damage (evaluated by histology) was inversely related to the level to which the
blood pressure was brought at the end of the hypotensive
episode (5), and in a model of *regional* cerebral ischemia,

the extent and pattern of tissue damage was considerably
altered by the *time* when arterial reperfusion was insti-
tuted (15).

Marked contrasts in the degree and pattern of structural alterations were noted when comparisons were made
between absolute, complete, irreversible *global* brain
ischemia at 37°C (12) and *regional* brain ischemia of
comparable duration (7).

This portion of our studies addresses itself to
these questions: (a) what is the nature and sequence of
the several structural alterations that occur in the
brain as a consequence of regional ischemia?, and (b)
what is the significance of these alterations in terms
of changes in perfusion? We attempt to provide some
answers with the combined use of ultrastructural and
biochemical methods.

DESCRIPTION OF MODELS OF ISCHEMIA

We have studied,in detail, the structural alterations
that develop in a cerebral hemisphere after inducting
regional ischemia either by extrinsic occlusion, in the
cat (11), or by intrinsic occlusion, in the monkey (19)
of the same vessel: middle cerebral artery (MCA). After
completing the experimental occlusion all animals, including the controls, were killed by the same methods
at intervals ranging between 17 minutes and 7 hours (7,
8). The method of fixation in all instances consisted of
intracardiac perfusion with aldehydes, to which various
tracers were added in order to ascertain the adequacy of
the perfusion (7, 17).

After completion of the aldehyde fixation, adjacent
tissue samples were collected from areas specified below
and distributed as follows: (a) embedded in paraffin for
subsequent evaluation by light microscopy, (b) embedded
in epoxy resins for subsequent evaluation by electron
microscopy, and (c) frozen in liquid nitrogen and stored
at -80°C for subsequent biochemical analysis. Pyruvate,
lactate, alanine, aspartate and glutamate, in $HClO_4$ ex-
tracts, were assayed by direct fluorometric enzymatic
procedures (3).

Cerebral areas sampled, in the above manner, were:
body of the caudate nucleus, putamen, corona radiata,
internal capsule, globus pallidus, insular cortex, superior temporal gyrus and inferior frontal gyrus. Tissue

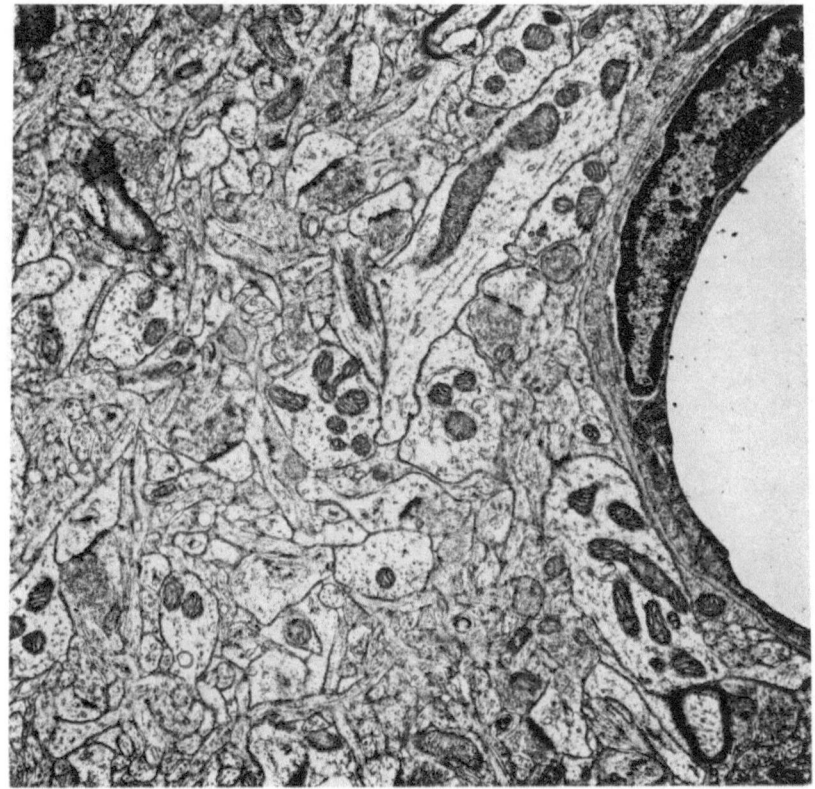

Fig. 1. Sham-operated Rhesus monkey caudate nucleus; note
        intact capillary wall and homogenous appearance
        of neuropil (orig. magnif. 6,000 X).

samples designated *ipsilateral* were derived from the hemisphere whose MCA had been occluded either via a transorbital approach or by a silastic embolus. Tissue samples
designated *control* were derived from sham-operated animals.
Evaluation of samples from cerebral hemisphere contralateral to the side of the arterial occlusion were also
conducted in all experiments.

SUMMARY OF ULTRASTRUCTURAL OBSERVATIONS

    As discussed previously, all animals with MCA occlu-
sion - of the type used in these experiments - develop
cerebral-blood-flow modifications and neurologic deficits
consistent with the condition known in human neurology

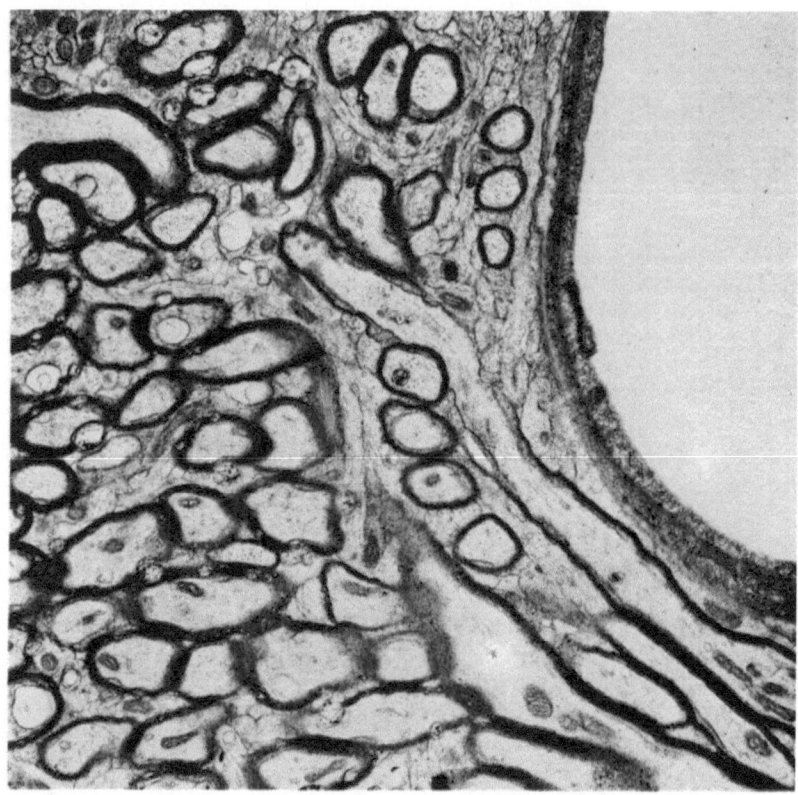

Fig. 2. Control tissue: white matter fibers adjacent to
        the putamen. Occasional ripples in myelin sheaths
        are attributed to defects in aldehyde fixation
        (orig. magnif. 5,000 X).

as cerebral *infarction* (7, 15). In the experiments reported here, those animals surviving long enough to permit a
neurologic evaluation (some were killed before awakening
from anesthesia) showed consistent contralateral motor
deficit, rotational gait, conjugate-eye deviation, et
cetera, as previously reported in detail (15). All sham-
operated animals were free of any neurologic deficit.

     In all *control* samples examined, the ultrastructural
features (Figs 1 & 2) were identical to those described
as normal for laboratory animals which had not been subjected to any experimental maneuver prior to death by

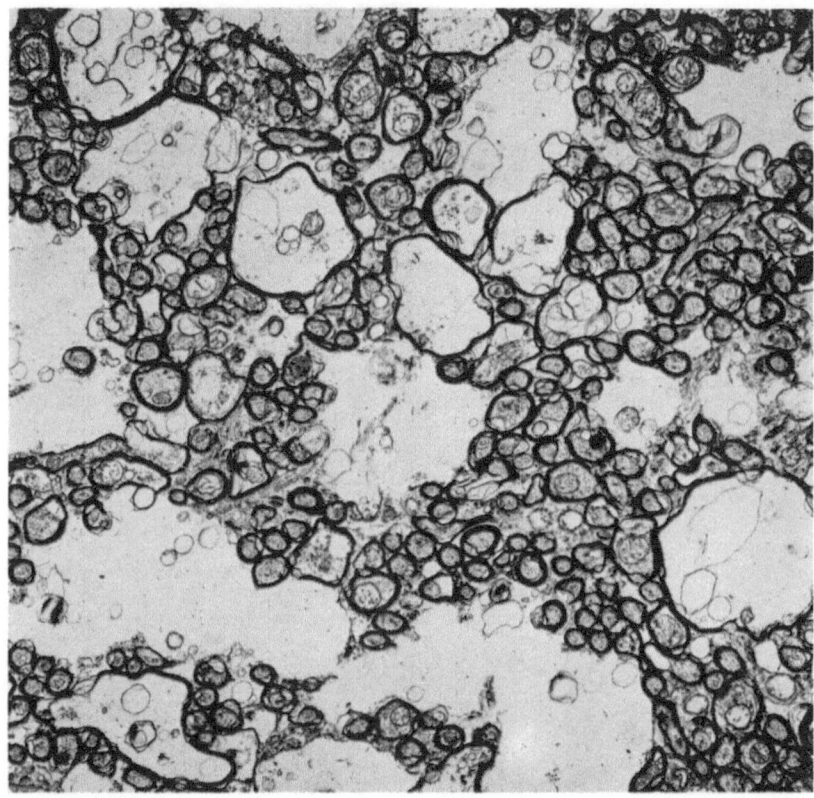

Fig. 3. MCA occlusion of 7 hour duration. Internal capsule:
some of the larger spaces probably correspond to
swollen astrocytic processes; many large myelin-
ated fibers show massive axonal swelling. No myelin splitting is visible here (orig. magnif.
3,000 X).

aldehyde perfusion (21).

    In the *ipsilateral* samples, three patterns of alter-
ations were noted:
        (a) massive enlargement and increased electronlucency
            of astrocytic compartment, axonal compartment and
            myelin sheath (Figs. 3 & 4); these changes are
            designated *edematous*,

        (b) breakdown of capillary walls with occlusion of
            the lumen by cell debris, escape of tracers

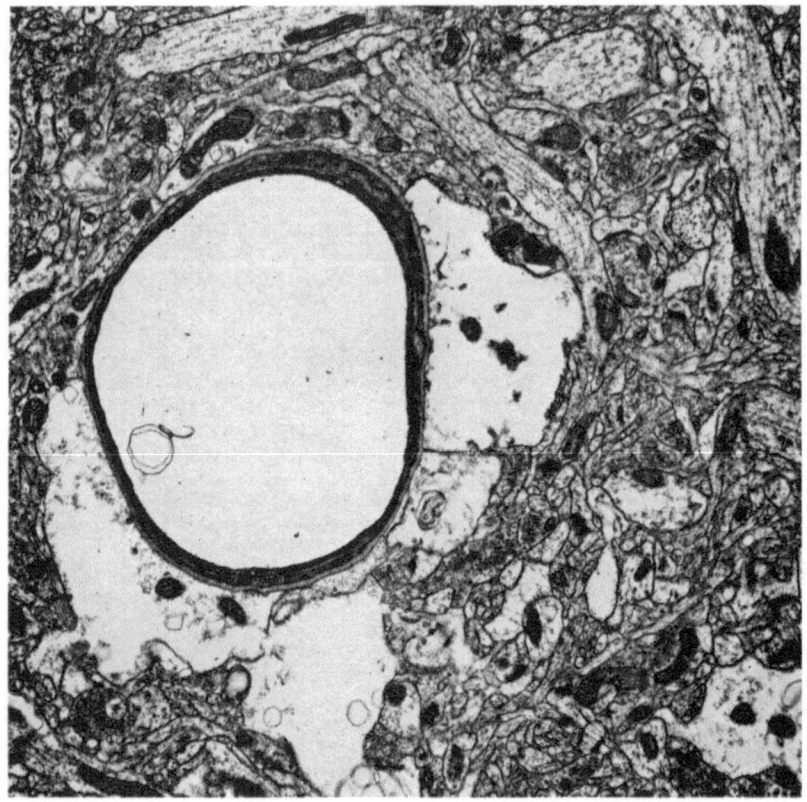

Fig. 4. MCA occlusion of 5 hour duration. Superior tempo-
        ral gyrus: swelling is confined to astrocytic
        processes. Note breakdown of astrocytic plasma
        membrane (orig. magnif. 5,000 X).

        administered post-mortem into the vascular com-
        partment and breakdown of surrounding cellular
        elements (Fig. 5); these changes are designated
        necrotic,and

    (c) various alterations of neuronal volume, neuronal
        organelles, and focal areas of massive synaptic
        swelling, particularly noticeable in terminal
        boutons (Fig. 6) without evidence of lethal cell
        injury; these changes are designated *ischemic*.

    The topographic distribution of the above changes
was random but diffuse; nevertheless, certain changes

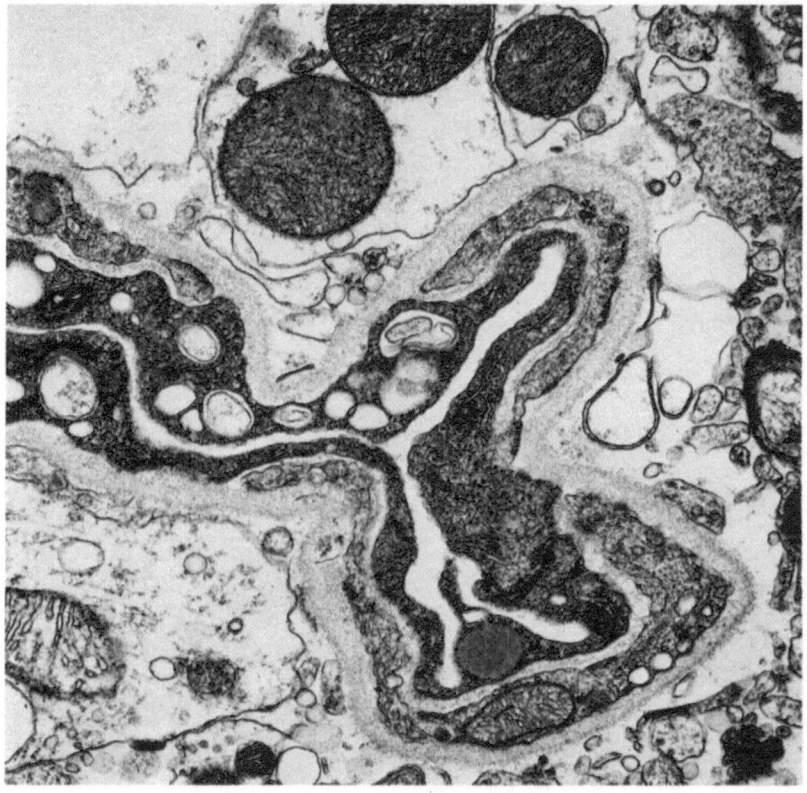

Fig. 5. MCA occlusion (5 hr.). Superior temporal gyrus:
         collapsed capillary lumen, "dark" endothelial
         cells, thickened basement membrane and three
         mitochondria, one of which contains flocculent
         densities (orig. magnif. 20,000 X).

predominated in some areas. Thus, myelinic and axonal
alterations were more prominent in the corona radiata
and the internal capsule; foci of necrotic alterations
were more readily found in the putamen and these foci
became larger with increasingly longer ischemic periods.
Sublethal alterations in neuronal perikarya became obvious first in the insular cortex, whereas dendritic changes were common both in the caudate nucleus and the
cerebral neocortex. Additional details on the range of
the ultrastructural abnormalities encountered in regional
ischemia are provided in separate publications (7, 9).

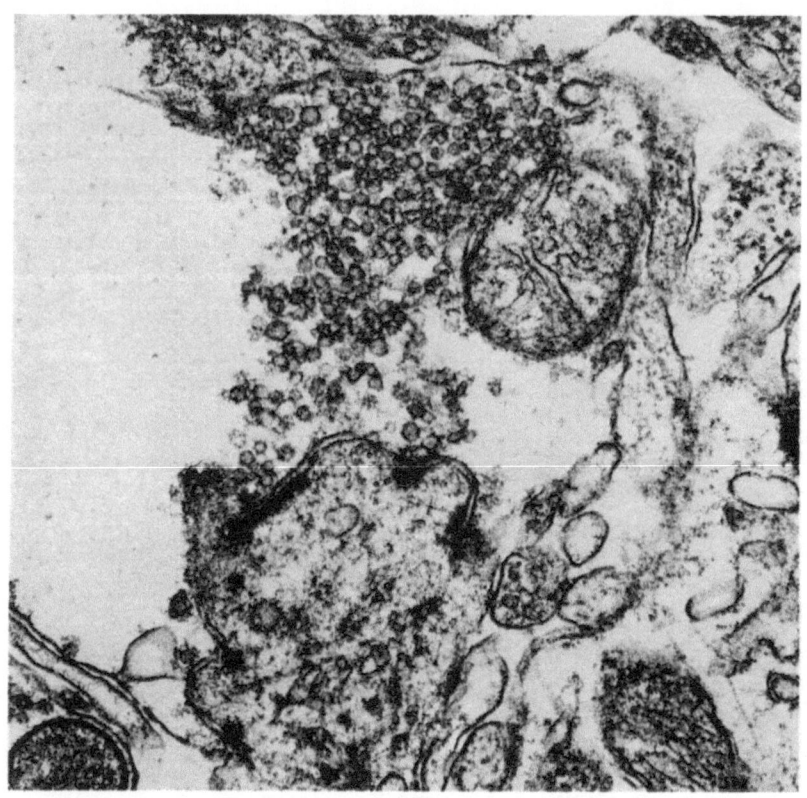

Fig. 6. MCA occlusion (6 hr.) Caudate nucleus: massive
        swelling of terminal bouton with partial disin-
        tegration of mitochondria and synaptic vesicles
        (orig. magnif. 25,000 X).

SUMMARY OF BIOCHEMICAL OBSERVATIONS

All alterations in the *ipsilateral* tissue samples
are expressed in terms of percentage changes of the *control* samples. A detailed description of all biochemical
derangements and of the methods used are included in a
separate communication, which is being readied for publication (2).

Three sets of biochemical changes were noted:

(a) Areas with marked reduction in glutamate and
    aspartate levels, particularly evident in the

Table I

CONTROL LEVELS OF AMINO ACIDS IN CAUDATE NUCLEUS AND
INSULAR CORTEX OF MONKEY BRAIN

AMINO ACID	CAUDATE NUCLEUS	INSULAR CORTEX
	m moles . Kg wet weight^{-1}	
Glutamate	9.54 ± 0.97	9.53 ± 0.17
Aspartate	1.43 ± 0.23	1.50 ± 0.27
Alanine	0.33 ± 0.03	0.23 ± 0.04
Glutamine	3.01 ± 0.19	1.09 ± 0.28
N-acetyl-Aspartate	8.06 ± 0.82	10.04 ± 0.67
$\frac{\text{(Alanine)}}{\text{(Glutamate)}}$ X 100	3.87 ± 0.50	3.12 ± 0.16

Amino acid concentrations from two sham-operated control
monkeys perfused with aldehydes are given. Each value is
the average of 12 samples. These means were used to cal-
culate the percent change given in Table II. Values are
means ± S.E.M.

striatum, which increased with longer ischemic
periods. Since the stability of these amino acids
in aldehyde-fixed tissues had been previously
established (3), their reduction was attributed
to dilutional effects or *edema*.

(b) Areas with marked increases in the alanine/glu-
tamate ratio. Since it was shown that the magnitude of this figure is indirectly related to
tissue *perfusability* (by either blood or fixa-
tive) (3), we attribute these biochemical changes to the effect of blocked or collapsed capillaries in areas of tissue *necrosis*.

(c) Areas with increased glutamine content. Since
ATP is required for glutamine synthesis (4, 16,
22), its rise in ischemic cortex and white matter
is interpreted as evidence for active ammonia

Table II

SELECTED AMINO ACIDS IN CAUDATE NUCLEUS AND INSULAR
CORTEX OF MONKEY BRAIN 24 HOURS AFTER MCA OCCLUSION

AMINO ACID	CAUDATE NUCLEUS		INSULAR CORTEX	
	Contralat.	Ipsilat.	Contralat.	Ipsil.
		% CONTROL		
Glutamate	109 ± 5	20 ± 1	83 ± 3	131 ± 1
Aspartate	110 ± 4	24 ± 3	105 ± 5	95 ± 10
Alanine	73 ± 3	255 ± 27	93 ± 18	376 ± 84
Glutamine	90 ± 3	14 ± 3	97 ± 7	471 ± 69
N-acetyl-Aspartate	84 ± 3	10 ± 1	103 ± 23	73 ± 3
$\frac{(Alanine)}{(Glutamate)} \times 100$	60 ± 2	1160 ± 41	83 ± 13	231 ± 80

Amino acids are expressed as percent of control concentration for the two areas given in Table I. Each value is the
average of 3 replicate samples from areas supplied by
branches of the occluded cerebral artery (ipsilateral)
and patent artery (contralateral). Values are means $\pm$ SEM.

detoxification and the presence of reasonably
high ATP levels. We attribute this increase in
glutamine to recovery from reversible *ischemic*
alterations at 24 hours after occlusion.

As in the case of the ultrastructural alterations,
biochemical derangements increase as a function of ischemic time, i.e., they are more pronounced at 24 hours than
at 5 hours, and at any given time period, biochemical
changes vary in severity according to the topographic
distribution. After six hours of MCA occlusion alanine
increases are much higher in striatum than in any of the
other areas studied. Control values for five amino acids
and the ratio of alanine:glutamate in the caudate nucleus
and insular cortex are given in Table I. At six hours,
alanine was increased in both the caudate nucleus and
insular cortex. Changes in alanine are reflected by 6-fold

increase in the alanine:glutamate ratio in the caudate
nucleus and a 2-fold increase in this ratio in the insular
cortex. The occurrence of edema in the caudate nucleus is
indicated by proportional decreases in aspartate and
glutamate. Twenty-four hours after occlusion, alanine was
increased more than 3-fold in both the insular cortex and
caudate nucleus (Table II); however the alanine:glutamate
ratio increased 19 fold in the caudate nucleus and only
3-fold in the insular cortex. The large difference in the
ratio in these two areas, at 24 hours, corresponded with
light and electron microscopic evidence indicating severe
compromise of perfusion in basal ganglia region but not
in the cortex. Large differences between cortical and
striatal areas were also reflected in the concentration
of glutamine and N-acetyl-aspartate (Table II).

The above data are in correspondence with the *in vivo*
studies of cerebral-blood-flow, after MCA occlusion, which
showed a heterogenous pattern of multi-focal alterations
that included: hyperperfusion, hypoperfusion and some areas of normal perfusion (24).

SUMMARY AND CONCLUSIONS

Trump and Arstila (1975, 23) have developed a model
of cell injury that permits the structural and biochemical
evaluation of lethal cell injury in preparations subjected
to simulated ischemia. On the basis of such schema, it
appears that one of the earliest structural markers of
irreversible cell injury is a mitochondrial abnormality,
which has been designated *flocculent density*.

Our analysis of ultrastructural abnormalities in neuronal perikarya of the cerebral neocortex, that up to
2 1/2 hours after MCA occlusion, there are still numerous
neuronal soma with either sublethal injury, i.e., potentially reversible changes, or neuronal soma that are
structurally intact (7). Moreover, the release of the MCA
occlusion in cats, prior to six hours, resulted in
either no visible tissue injury or focal alterations
that were both smaller or different from those observed
in a classic cerebral infarction (15).

The viability of neurons situated in the territory
of an occluded MCA was tested by measuring the rate at
which the ischemic tissue could synthesize dopamine (a
function which is thought to be exclusively neuronal).
It was found that five hours of MCA occlusion led to
minimal alterations of tyrosine uptake and dopamine synthesis (20).

Based on the above evidence, we hypothesize that immediately after occluding a MCA the progressive and increasing set of changes may follow this approximate sequence:

(1) alterations in microcirculation lead to multifocal changes in ATP content, which lead to

(2) massive swelling of astrocytes and possibly swelling of ECS, via breakdown of astrocytic membranes,

(3) the above abnormalities are accompanied by volumetric changes in neuronal soma and especially in mitochondria and rough endoplasmic reticulum cisternae.

(4) simultaneously, there is a multifocal but diffuse alteration of brain circuitry which is particularly apparent in the terminal boutons and the neurotransmitter vesicles.

We wish to emphasize that although there may be an occasional necrotic neuron, after a few minutes of regional ischemia, the majority of neurons - in an area of MCA occlusion - are still viable several hours after the occlusion, as judged by ultrastructural and biochemical criteria. The classical dictum that 4-6 minutes of ischemia means neuronal death does not apply to all neurons located in the territory of an occluded large cerebral artery.

ACKNOWLEDGEMENTS

Financial Support: USPHS Grant NS 06779

The secretarial collaboration of Mrs. Liz Tinnell is acknowledged with gratitude.

REFERENCES

1. Ames, A., Wright, R.L., Kowada, M., Thurston, J.M. and Majno, G. (1968): Cerebral ischemia. II. The no-reflow phenomenon. Am. J. Pathol. 52: 437-454

2. Conger, K.A., Garcia, J.H., Lossinsky, A.S., Bielefend, J.L., Fuld, R.A. and Kauffman, F.C. (1977): Amino acids in aldehyde-paraformaldehyde perfused monkey brain after middle cerebral artery occlusion. (In preparation)

3. Conger, K.A., Garcia, J.H., Lossinsky, A.S. and Kauffman, F.C. (1977): The effect of aldehyde fixation on selected substrates for energy metabolism and amino acids in mouse brain. J. Histochem. Cytochem. (Submitted).

4. Elliot, W.H. (1948): Adenosinetriphosphate in glutamine synthesis. Nature, 161: 128-129.

5. Gamache, F.W. and Myers, R.E. (1975): Effects of hypotension on rhesus monkeys. Arch. Neurol. 32: 374-380.

6. Garcia, J.H. (1976): The cellular pathology of hypoxic injuries: Ultrastructure. In: Oxygen and Physiological Function ", Ed. Frans F. Jöbsis, Publisher - Professional Information Library, Dallas, 277-284.

7. Garcia, J.H., Kalimo, H., Kamijyo, Y. and Trump, B.F. (1977): Cellular events during early regional cerebral ischemia. Virchows Arch. B (In Press).

8. Garcia, J.H. and Kamijyo, Y. (1974): Cerebral infarction. Evolution of histopathological changes after occlusion of a middle cerebral artery in primates. J. Neuropath. Exp. Neurol. 33: 409-421.

9. Garcia, J.H., Lossinsky, A.S. and Jamaris, J. (1977): Ultrastructural investigations of regional ischemia in the rhesus monkey (Research in Progress, Unpublished Observations).

10. Hossman, K.A. and Kleihues, P (1973): Reversibility of ischemic brain damage. Arch. Neurol. 29: 375-384.

11. Hudgins, W.R. and Garcia, J.H. (1970): Transorbital approach to the middle cerebral artery of the squirrel monkey: A technique for experimental cerebral infarction applicable to ultrastructural studies. Stroke 1: 107.

12. Kalimo, H., Garcia, J.H., Kamijyo, Y., Tanaka, J. and Trump, B.F. (1977): Electron microscopy of feline cerebral cortex after complete ischemia. Virch. Arch. B. (In Press).

13. Kalimo, H., Garcia, J.H., Kamijyo, Y., Tanaka, J., Viloria, J.E., Valigorsky, J.M., Jones, R.T., Kim, K. M., Mergner, W.J., Pendergrass, R.E. and Trump,B.F.

(1974): Cellular and subcellular alterations of human CNS: Studies utilizing *in situ* perfusion-fixation in immediate autopsies. Arch. Pathol. 97: 352-359.

14. Kamijyo, Y. and Garcia, J.H. (1975): Carotid arterial supply of the feline brain. Applications to the study of regional cerebral ischemia. Stroke 6: 361-369.

15. Kamijyo, Y., Garcia, J.H. and Cooper,J. (1977): Temporary middle cerebral artery occlusion: A model of hemorrhagic and subcortical infarction. J. Neuropath. Exp. Neurol. 36: 338-350.

16. Leuthardt, F. and Bujard, E. (1947): Über Glutamin-bildung im Leberhomogenat. Helv. med. acta 14: 274-278.

17. Lossinsky, A.S. and Garcia, J.H. (1977): Vascular perfusion of the central nervous system for light and electron microscopy (In Preparation).

18. Molinari, G.F. (1970): Experimental cerebral infarction I. Selective segmental occlusion of intracranial arteries in the dog. Stroke 1: 224-231.

19. Molinari, G.F., Moseley, J.I. and Laurent, J.P.(1974): Segmental middle cerebral artery occlusion in primates An experimental method requiring minimal surgery and anesthesia. Stroke 5: 334-339.

20. Pence, R.S. and Garcia, J.H. (1976): Tyrosine uptake and metabolism during regional cerebral ischemia. Circulation, 54: Suppl. II 103.

21. Peters, A., Palay, S.L. and Webster,H.deF. (1976): The fine structure of the nervous system: The neurons and supporting cells. W.B. Saunders. Philadelphia.

22. Speck, J.F. (1947): The enzymatic synthesis of glutamine. J. Biolog. Chem. 168: 403-404.

23. Trump, B.F. and Arstila, A.U. (1975): Cell membranes and disease processes. In: Pathobiology of Cell Membranes. Vol. I Academic Press, Inc., New York, 1-103.

24. Yamaguchi, T., Waltz, A.G. and Okazaki, H.(1971): Hyperemia in experimental cerebral infarction: Correlation of histopathology and regional blood flow. Neurol. 21: 565-578.

SHORT-TERM UNILATERAL ISCHEMIA IN GERBILS:

A REEVALUATION

N. Murakami, W.D. Lust, A.B. Wheaton, and
J.V. Passonneau

Laboratory of Neurochemistry
National Institute of Neurological and
    Communicative Disorders and Stroke
National Institutes of Health
Bethesda, Maryland 20014, U.S.A.

Our previous studies on unilateral and bilateral ischemia in the gerbil have been primarly concerned with the sites directly affected by the loss of circulation. In these regions, the changes in brain metabolites, including energy reserves and cyclic nucleotides, have been pronounced (4, 8). In earlier studies with unilateral ischemia, there were little or no observable changes in these metabolites in the contralateral cortex (8, 11). However, Levy and Duffy (6) have recently demonstrated that there are changes not only in the contralateral cortex of affected gerbils but also in the ipsilateral cortex of animals not exhibiting positive neurological signs of ischemia, a condition which was completely ignored in our previous studies. It appears that the biochemical effects of an ischemic episode are variable depending upon the region of the brain investigated and perhaps on the neurological responses elicited. The intermediate range of responses in those areas not directly affected by the carotid artery occlusion is the subject of this paper.

By using unilateral ligation of the common carotid artery of the gerbil, the more subtle changes in brain metabolites could be studied in the contralateral cortex of symptom-positive gerbils as well as in the cortex of those gerbils behaviorally unaffected by the occlusion.

In addition, exposure to a 5% oxygen atmosphere during
the last five minutes of the ten minute period of occlusion enhanced the incidence of the number of animals exhibiting positive signs of ischemia. The biochemical response was then compared on the basis of the neurological
signs.

MATERIAL AND METHODS

   Animals. Mongolian gerbils obtained from Chickline
Co. (Vineland, NJ) weighing 50-70 gms were fed ad libitum.
The right common carotid artery was exposed and looped
with surgical threads. The animals were allowed more
than 2 hours to recover from the surgery and then were
treated as follows:
Experiment I. The right common carotid artery was ligated
with surgical thread for a 10 minute period (Figs.1 and 2).
Experiment II. The procedure was the same as Experiment I,
except that animals were exposed to 5% oxygen during the
last 5 minutes of occlusion. Intact animals exposed to
the low oxygen atmosphere serve as a basis of comparison
for the effects of hypoxia alone.

   For the studies on recovery, Heifetz aneurym clips
were used to occlude the artery for 10 min and then were
removed for five minutes (Figs. 3 and 4). Sham-operated
animals served as controls.

   Neurological signs. Experiment I. On the basis of
neurological signs, according to Kahn (2) the animals
were grouped as follows:

   (-) no behavioral effects were exhibited.

   (±) exhibited some of the signs described for the
       third group, but certainly not as well marked.

   (+) in this group, there was pronounced abnormal
       motor behavior including circular movements
       and seizures.

Experiment II. The neurological signs were assessed
1) during the first five minutes following unilateral
occlusion 2) during the second five minutes while being
exposed to 5% oxygen atmosphere and 3) during the third
five minutes when the occlusion was removed and the gerbils were exposed to air. On the basis of neurological
signs, the animals were grouped into the following categories: (-,-), (-,+) and (+,+) for unilateral occlusion

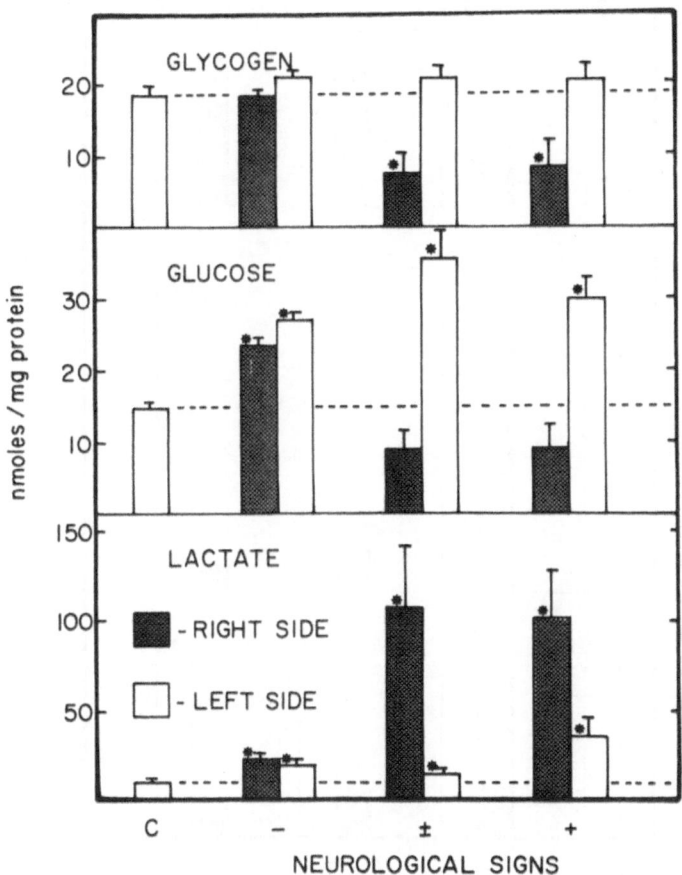

Fig. 1. The concentrations of glycogen, glucose and lactate in the gerbil cerebral cortex following
unilateral ligation of the common carotid artery.
C denotes control values which are also indicated by the dashed lines. The right side (hatched
bars) is the cerebral cortex to which the circulation has been occluded for 10 min and the left
side (open bar) is the contralateral cortex. The
symbols represent those animals which exhibited
the following neurological signs: (-) none, (+)
mild, and (+) pronounced (see text). The values
represent the mean ± SEM for the following number of animals: (-) 21, (±) 6, and (+) 10. The
asterisks indicate significant differences from
C values, $p < 0.05$.

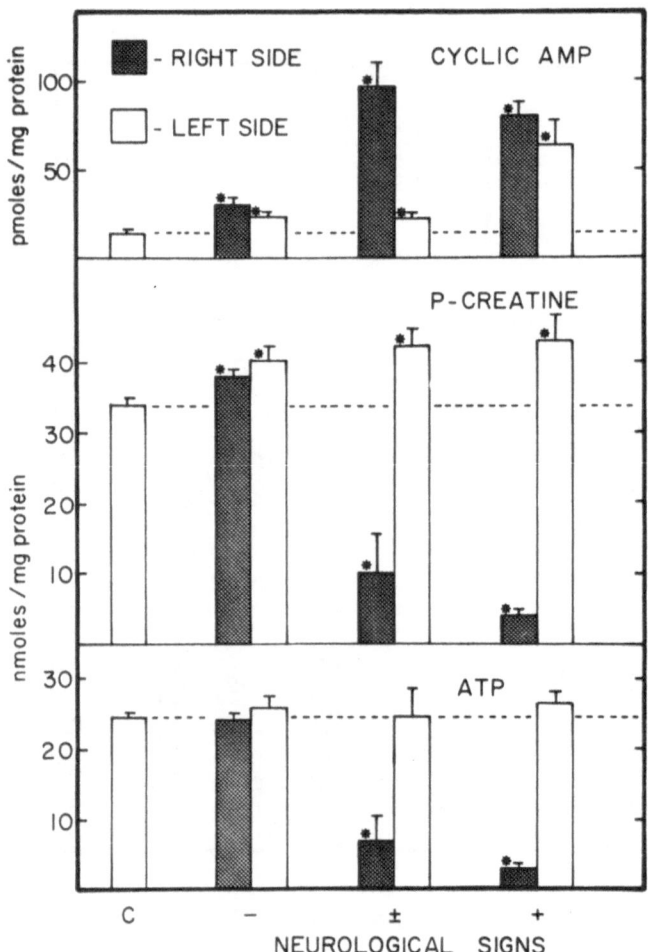

Fig. 2. The concentrations of cyclic AMP, P-creatine,
and ATP in gerbil brain following unilateral
ischemia. The animals and symbols are as presented in Fig. 1.

with hypoxia and (-,-,-) (-,+,-),(-,+,+) and (+,+,+) for
unilateral occlusion with hypoxia and recovery. The metabolite concentrations in each group were measured.

Preparation of tissues. The gerbils were frozen in liquid
nitrogen and were subsequently stored at -70°C until the
dissections were made. The outer 1-2 mm of the cerebral
cortex were removed in a cryostat maintained at -20°C.

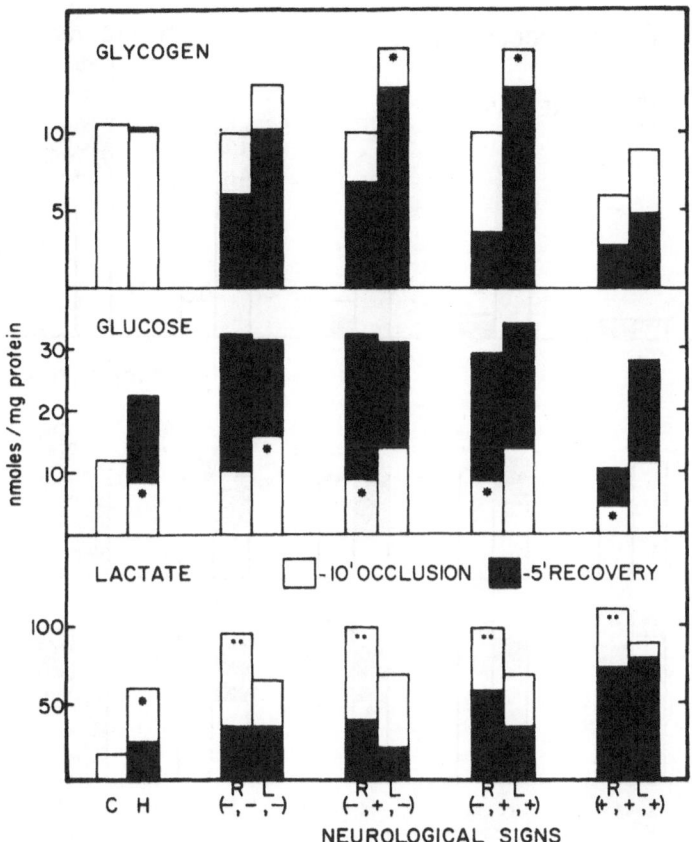

Fig. 3. The concentrations of glycogen, glucose and lactate in gerbil brain from control animals, (C);
from animals exposed to 5% O_2 for 5 min, (H);
from animals following 10 min of unilateral occlusion including exposure to 5% O_2 during the
last 5 min (white bars); and from animals in
which the preceding treatments were followed by
5 min of recirculation in normal atmosphere
(shaded bars). The sequence of neurological
signs in the group without a recovery period
(white bars) is represented by the first 2 out
of the 3 symbols and in the group with recovery
period (shaded bars) by all 3 symbols. The total
number of animals was 103, with 4 to 20 animals
in each group. Single asteriks indicate significant differences from C values, double asterisks indicate significant differences from H
values. Neurological signs are explained in
Materials and Methods, Experiment II.

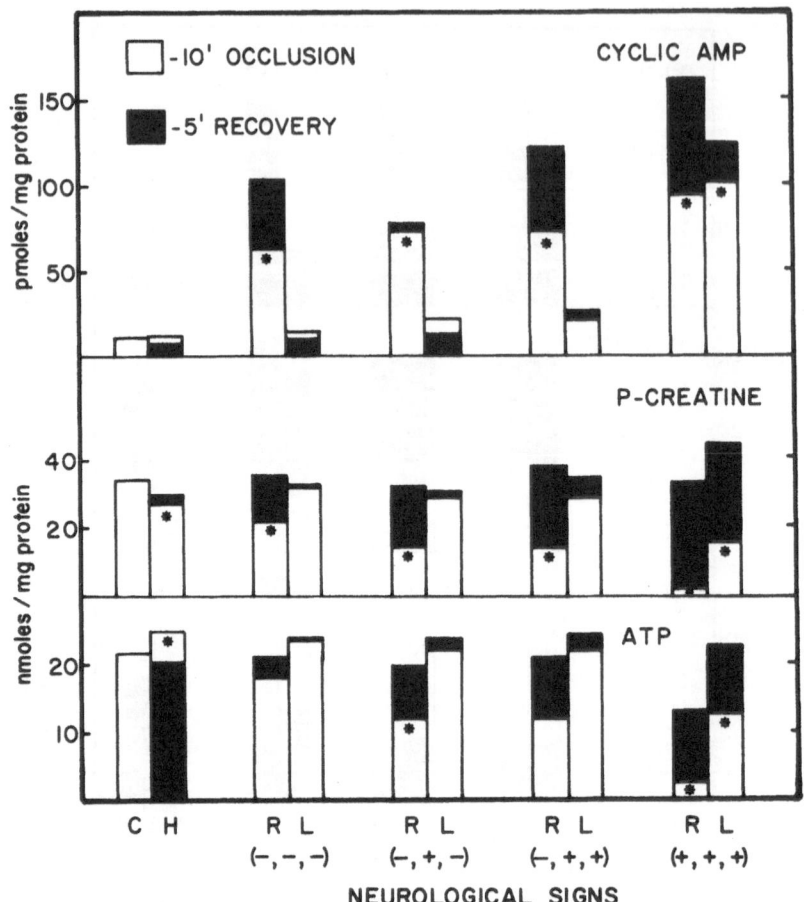

Fig. 4. The concentrations of cyclic AMP, P-creatine
and ATP in the cerebral cortex of the animals
of experiments described in Fig. 3, and Materials and Methods, Experiment II.

        The tissue was equilibrated with 0.2 ml of 0.1 N HCl
in methanol at -50°C for 1 hour after which 1 ml of 0.017
N HCl was added to the tube and the samples were homogenized. A 0.1 aliquot of the homogenate was removed for
glycogen and glucose determination and to the remaining
homogenate a 0.1 ml of 3 N PCA containing 10 mM EGTA was
added. The tubes were centrifuged, the supernatant was
removed and neutralized with one-tenth volume of 3 M
potassium bicarbonate. To the pellet, 1 ml of 1 N NaOH
was added to dissolve the protein for analysis. Cyclic

nucleotides were measured according to the method of
Steiner et al. (14). Glycogen and glucose were analyzed
according to the method of Passonneau and Lauderdale (12).
All other metabolites were determined with enzymic fluorometric methods as described by Lowry and Passonneau (7).

RESULTS

I. Unilateral ischemia

Thirty-eight gerbils had their right common carotid
arteries occluded for 10 minutes and of that group 16
animals or 42% exhibited some neurological signs of
ischemia, a value in close agreement with other studies
(6). There were 22 animals in the (-) group, 6 in the
(±) group and 10 in the (+) group. A mild seizure state
occurred in 70% of the gerbils in the (+) group. These
were characterized by the presence of intermittent tre-
mors and myoclonic jerks associated with an opisthotonic
posture. This behavioral pattern was minimal or absent
in the other 2 groups.

The results for the cerebral metabolites when grouped
on the basis of neurological signs are shown in Figs. 1
and 2. The glycogen levels decreased only on the ipsilateral side of those animals exhibiting some signs of
ischemia (+ and ±). The concentration of glucose was
significantly elevated on the contralateral cortex of
all three groups as well as in the ipsilateral cortex
of the unaffected group. While the lactate levels were
increased on both sides in all three groups, the major
changes of 10-fold or greater only occurred in the ipsilateral cortex of those animals with positive signs of
ischemia. In the contralateral cortex of animals exhibiting positive signs, lactate increased almost 4-fold;
whereas, in the other groups the increase was about 2-
fold or less. These results indicate that the metabolite
changes in the animals which exhibit even minor neurological signs (±) resemble those of the (+) group more
than those of the (-) group.

The interpretation above is confirmed by the fall
in both ATP and P-creatine in the ipsilateral cortex
of animals exhibiting positive signs (± and +, Fig. 2).
ATP levels were not affected in any of the groups in
the contralateral cortex nor in the ipsilateral cortex
of symptom-negative gerbils. The levels of P-creatine
were elevated in the contralateral cortex of all three
groups and in the ipsilateral cortex of the negative
group. The levels of cyclic AMP were significantly in-

creased on both sides of the cortex in all three groups, but the magnitude of the response was greater in the groups exhibiting positive signs (Fig. 2). As with lactate, the increase of cyclic AMP in the contralateral cortex of the positive group was markedly higher than in that of the other two groups, and wa·s almost as great as that for the ipsilateral cortex.

Thus, as shown by others, occlusion of the common carotid artery does result in changes in supposedly non-ischemic regions of the brain. Even in animals not exhibiting neurological signs, there were changes in lactate, cyclic AMP, P-creatine and glucose. The major differences between the (+) and (±) groups were observed not in the ipsilateral cortex, but rather in the contralateral cortex. The changes in lactate and cyclic AMP in the contralateral cortex of the (+) group but not the (+) group resemble those following an ischemic episode.

II. Unilateral ischemia with 5% oxygen.

One major drawback in using unilateral ischemia is the relatively low incidence of positive ischemic episodes. From the data presented earlier in this paper, it would appear that the biochemical changes in the contralateral cortex of animals exhibiting positive signs are greater than those in the ispilateral cortex of unaffected gerbils. In order to extend this observation, we applied a procedure similar to the anoxic-ischemic model of Levine to the gerbil (5). In the first five minutes following the occlusion of the common carotid artery, the neurological signs of ischemia were assessed and only 14 of the 60 animals exhibited positive signs of ischemia. At the end of 5 minutes, the animals were placed in a chamber with a 5% oxygen atmosphere and the neurological signs were again noted. The number of positive animals increased from 14 (23%) to 45 (75%) under this reduced oxygen atmosphere. The onset of signs occurred at approximately 3 minutes after the animals were placed in the chamber. During recovery when the animals were placed in air, the percent of animals which showed positive signs decreased only slightly to 67%. Reduced oxygen atmosphere thus dramatically increased the incidence of positive symptoms.

In contrast to Experiment I, the incidence and severity of seizures were far greater in Experiment II. Clonic seizures occurred in 20 percent of the gerbils exhibiting positive neurological signs of ischemia in the first 5 min after occlusion, while none occurred in Experiment I.

Once the gerbils were exposed to 5% oxygen, over 75% of
the (+,+) animals exhibited either tonic or clonic convulsions or both. In the (-,+) group, seizures were evident in 50% of the gerbils during the hypoxic exposure.
Seizures have been demonstrated to affect many of the
metabolites measured (1, 3, 9). Consequently, gerbils
were never frozen during a convulsion in an attempt to
minimize the influnece of seizures on cortical metabolism.

The effect of treatment with 5 min of only 5% oxygen
on the various brain metabolites is shown in Figs. 3 and
4 H. There was a striking 3-fold increase in cortical
lactate levels, a decrease in glucose to 70% of control,
and much smaller changes in glycogen, P-creatine, ATP
and cyclic AMP. The energy status of the tissue based
on the levels of the high-energy intermediates, ATP, and
P-creatine, was not severly compromised by reduced oxygen
availability for 5 min.

The metabolic effects of 10 min of unilateral ischemia with an inclusive five minute period of 5% oxygen are
also shown in Figs. 3 and 4. The only significant change
in glycogen levels was a 30% increase observed in the
contralateral cortex of the (-,+,-) and (-,+,+) groups.
The glucose concentrations in the ipsilateral cortex of
all groups except for the (-,-,-) group were depressed
compared to the control cerebral cortex; the levels in
the (+,+,+) group were also significantly lower than those in the H group (p < 0.001). Since the lactate levels
were significantly increased in the 5% oxygen group, the
other groups which were exposed to similar conditions
were compared statistically to the H (not the C) group.
While there was no effect in the contralateral cortex in
any of the groups, the ipsilateral cortical levels of
lactate were significantly elevated in all groups.

The levels of P-creatine were reduced in the ipsilateral cerebral cortex of all the groups, but were decreased in the contralateral cortex of only the (+,+,+)
group.The ipsilateral side of the (-,+,-) and (-,+,+)
groups and both sides of the (+,+,+) group were significantly lower than the values for the 5% oxygen group
as well as controls. The ATP concentrations were depressed on the ipsilateral side of all groups except the
(-,-,-) group. On the contralateral side, only the (+,+,
+) group was significantly lower than the control values.
Cyclic AMP levels were increased in the ipsilateral cortex in all the groups, and also in the contralateral
cortexof the (+,+,+) group. These result suggest that

the energy metabolism in the contralateral cortex of the
(+,+,+) group is compromised during unilateral ischemia
and exposure to 5% oxygen atmosphere. The greater than
two-fold increase of cyclic AMP in the contralateral cortex of both (-,+,+) (+,+,+) groups contrasts with our previous findings (8). This discrepancy probably can be best
explained on the basis of the anesthetic schedule used in
the previous experiments; the animals were still under
the influence of anesthesia at the time of occlusion.
All of the contralateral changes shown here, but not observed in previous studies, may similarly reflect that
in the earlier studies the animals still showed effects
of anesthesia.

Recovery.

A group of animals were also allowed to recover for
five minutes after unilateral ischemia; the clip was removed and animals exposed to air. The top of the dark
bars in Figs. 3 and 4 represent the values for the given
metabolite after 5 min of recirculation. The glycogen levels decreased on both sides of the cerebral cortex during the recovery period in all but the 5% oxygen group.
In contrast, glucose concentrations increased in all experimental groups including those exposed to the 5% oxygen only. The large post-ischemic rise in glucose in all
other groups did not occur in the ipsilateral cortex of
the (+,+,+) group. Lactate levels decreased in all groups
during recirculation, but the restoration process on both
sides of the cerebral cortex in the (+,+,+) group appeared to be delayed. The recovery of ATP was retarded only
in the ipsilateral cortex of the (+,+,+) group. The values
for all other groups were essentially back to control by
5 min of recirculation.

It has been previously reported that cyclic AMP increased during recirculation over the already elevated
levels during ischemia (4, 11). In the present study, the
post-ischemic increase in cyclic AMP occurred not only in
the ipsilateral cortex of affected gerbils, but also in
the ipsilateral cortex of unaffected animals and the
contralateral cortex of the (+,+,+) group.

DISCUSSION

An ischemic episode essentially depletes the brain
of its energy stores, including glycogen, glucose, P-creatine and ATP. In previous studies on unilateral ischemia
we concluded that the changes in these energy metabolites

were minimal in the contralateral cortex of affected gerbils. In addition, the cortical levels in animals which
did not exhibit positive signs of ischemia were not examined. Levy and Duffy (6), however, were able to show
that certain metabolites from cerebral cortex of apparently unaffected animals,and of the contralateral cortex of clinically affected gerbils did indeed change.
Our inability to see any changes in the contralateral
cortex of affected animals in previous studies can probably be attributed, in part, to the use of a barbiturate
to anesthetize the animals for surgery. McGraw (10) has
reported thatbarbiturates have a protective effect in
gerbils against an ischemic insult. Barbiturates were not
used in the present study to eliminate the possible influence of the drug on the ischemic events. The animals were
also allowed 2 hours of recovery between surgery and occlusion to assess the early onset of the neurological
signs without ambiguity. With these modifications of our
previous procedures, the changes not previously observed
in cerebral metabolites were unmasked.

When the results were assessed on the basis of the
observed neurological signs, the following observations
were made. First, even in those animals which did not
exhibit signs of ischemia, there were significant increases in lactate, glucose, P-creatine and cyclic AMP on
both sides of the cortex. Secondly, the metabolite pattern of the ipsilateral cortex of the (+) group was
quite similar to that of the (+) group. On major difference between the (+) and the (+) groups was the biochemical
response in the contralateral cortex of the (+) group.
These changes in metabolites suggest that there is biochemical justification for grouping these animals on the
basis of neurological signs.

Exposure of the gerbils to 5 min of 5% oxygen atmosphere without occlusion caused an increase in lactate and
a decrease in P-creatine and glucose. Otherwise, it would
appear that the energy metabolism was not severly compromised by this procedure. However, in animals with carotid
occlusion, the exposure to 5% O_2 increased the incidence
of positive ischemic signs from 23 to 75% in this experiment. It appears that even in those animals with no
manifestation of ischemia, the margin of safety is reduced by the unilateral occlusion such that the lowering
of the oxygen tension in the atmosphere triggers an ischemic episode.

The changes in metabolites after 10 min. of unilateral occlusion were essentially the same with or without

exposure to 5% oxygen (Experiment I versus Experiment II).
In the group of animals which developed positive signs of
ischemia during treatment with 5% oxygen, the onset of
signs quite predictably occurred at approximately 3 min
(197 + 7 sec) from the start of hypoxia. If the ischemic
onset is coincidental to that of the neurological signs,
only the final 2 min of occlusion represent the ischemic
duration in these animals. Thus, the duration of the actual ischemic episode may explain some of the quantitative
differences between the (-,+) and the (+,+) groups. For
example, the greater changes in the contralateral cortex
of the (+,+,+) group may be a result of the total length
of ischemia exceeding that of the other groups.

A major consideration in the interpretation of the
data is the presenceof seizures in a number of the gerbils examined. During tonic extension, the energy demands
have been shown to exceed energy production (1), the net
effect of which is a reduction of brain energy reserves
qualitatively similar to that observed during ischemia.
In Experiment I, the problem with seizures was minimal,
but in Experiment II it was a real concern. The exposure
of the gerbils to 5 min of 5% oxygen during the second
5 min after occlusion increased in the contralateral cortex of the (+,+,+) group could be attributed to an effect
of convulsions. However, quantitatively many of the changes on the contralateral cortex were also evident in the
nonconvulsing (+) gerbils in Experiment I. The evidence
suggests that since the animals were not frozen during a
convulsive episode, the metabolic effects of the seizure
added only minimally to the effects of ischemia.

The dramatic changes in cerebral metabolites in the
ipsilateral cortex of affected animals are undoubtedly
due to the cessation of blood flow to this region of the
brain. Less clearly defined are the causes of the metabolite changes either in the ipsilateral cortex of behavioral unaffected gerbils or in the contralateral cortex
of affected gerbils. In the case of the gerbils without
neurological signs, reduction of blood flow, albeit insufficient to cause ischemic signs, may be partially
responsible for the alterations in the metabolites. In
contrast, the effect on the contralateral cortex of affected animals is probably secondary to the ischemic episode.
The nature of the effect whether neurally mediated or not
is presently unclear, but deserves additional investigation as part of the phenomena of ischemia. Thus, these
areas of the brain which exhibit intermediate metabolite
responses may be of some significance to understanding
the pathology of ischemia and its sequelae.

REFERENCES

1. Ferrendelli, J.A. and McDougal, D.B. (1971): The effect of electroshock on regional CNS energy reserves in mice J. Neurochem. 18: 1197-1208.

2. Kahn, K. (1972): The natural course of experimental cerebral infarction in the gerbil. Neurology, 22: 510-515.

3. King, L.J., Lowry, O.H., Passonneau, J.V. and Venson,V. (1967): Effects of convulsants on energy reserves in the cerebral cortex. J. Neurochem. 14: 599-611.

4. Kobayashi, M., Lust, W.D. and Passonneau, J.V.(1977): Concentrations of energy metabolites and cyclic nucleotides during and after bilateral ischemia in the gerbil cerebral cortex. J. Neurochem. 29: 53-59.

5. Levine, S. (1960): Anoxic-ischemic encephalopathy in rats. Amer. J. Path. 36: 1-15.

6. Levy, D.E. and Duffy, T.E. (1977): Cerebral energy metabolism during transient ischemia and recovery in the gerbil. J. Neurochem. 28: 63-70.

7. Lowry, O.H. and Passonneau, J.V. (1972): In A Flexible System of Enzymatic Analysis. Academic Press, New York 151-156.

8. Lust, W.D., Mršulja, B.B., Mršulja, B.J., Passoneau, J.V. and Klatzo, I. (1975): Putative neurotransmitters and cyclic nucleotides in prolonged ischemia of the cerebral cortex. Brain Research, 98: 394-399.

9. Lust, W.D. and Passonneau, J.V. (1976): Cyclic nucleotides in murine brain: Effect of hypothermia on cyclic AMP, glycogen phosphorylase, glycogen synthase and metabolites following maximal electroshock or decapitation. J. Neurochem. 26: 11-16.

10. McGraw, C.P. (1977): Experimental cerebral infarctation effects of phenobarbital in mongolian gerbils. Arch. Neurol. 34: 334-336.

11. Mršulja, B.B., Lust, W.D., Mršulja, B.J., Passonneau, J.V. and Klatzo, I. (1976): Post-ischemic changes in certain metabolites following prolonged ischemia in the gerbil cerebral cortex. J. Neurochem. 26: 1099-1103.

12. Passonneau, J.V. and Lauderdale, V.R. (1974): A
    comparison of three methods of glycogen measurement
    in tissues. Anal. Biochem. 60: 405-512.

13. Smith, A.L. and Wollman, H. (1972): Cerebral blood
    flow and metabolism: Effects of anesthesia drugs and
    techniques. Anesthesiology. 36: 378-400.

14. Steiner, A.L., Wehmann, R.E., Parker, C.W. and Kipnis,
    D.M. (1972): Radioimmunoassay for the measurement of
    cyclic nucleotides. In Advances in Cyclic Nucleotide
    Research. (Greengard, P. and Robinson, G.A., eds.).
    Raven Press, New York, 2: 51-61.

SOME NEW ASPECTS OF THE PATHOCHEMISTRY OF THE POST-ISCHEMIC PERIOD

B.B. Mršulja

Laboratory for Neurochemistry
Institute of Biochemistry
Faculty of Medicine
Belgrade, Yugoslavia

The susceptibility of Mongolian gerbils (Meriones unguiculatus) has provided a suitable model for the investigation of not only prolonged ischemia but also the recovery process after such ischemic insult. The characteristics of metabolite restoration during recirculation indicate that the recovery process is more than a simple reversal of the events which occurred during ischemia. Recirculation appeared to trigger its own series of events including reported changes in certain metabolites and enzymes (15, 16, 22). Although glutamate remained unchanged during ischemia, it has been shown to decrease during postischemia (15). Also, cyclic AMP levels increased an additional 3- to 6-fold during subsequent recirculation (15). While the restoration of P-creatine was rapid and complete, recovery of ATP appeared to occur in stages. The initial increase of ATP was rapid, but complete restoration quite slow, apparently requiring the replenishment of the total adenylate pool. The levels of ATP, P-creatinine, glucose and glycogen decreased during ischemia, but were never totally depleted (13, 15). Furthermore, once these metabolites reached a new steady-state level, they remained unchanged for up to 6 hours of ischemia. But the restoration of high-energy phosphate depends on the intensity (duration) of ischemia. Therefore, the biochemical events during recirculation reflect the severity of the ischemic insult more accurately than those occurring during ischemia. The time required for metabolite restoration during recirculation appeared to be proportional to the length of the ischemic insult.

Recently the changes during the recovery period of certain enzymes which were unchanged in ischemia were shown. Schwartz et al. (22) have been able to show that both Na^+- K^+-ATPase and adenylate cyclase are unchanged during ischemia but decrease in activity during recovery. Kleinhues et al. (11) reported increased activity of two enzymes, ornithine decarboxylase and S-adenosyl-methionine decarboxylase, during the late recovery phase. These changes unique to the postischemic period may be useful in determining the necessary steps for the restoration of function after ischemia. Furthermore, the existence of the "maturation phenomenon" (5) as well as the altered responsiveness of brain tissue to the second ischemic insult (18) indicate that the postischemic period is not just the recovery from ischemia but a completely different pathophysiologic situation.

This paper describes (1) biochemical aspects of the maturation phenomenon; (2) the response of brain tissue to repeated ischemic insult; and the deleterious effect of increased systemic blood pressure on the recovery of brain metabolites in ischemically affected tissue.

METHODS

All the experiments were carried out on Mongolian gerbils. Unilateral ischemia was produced as described elsewhere (15). Symptom-positive animals (9) were the only ones used for the evaluation of the ischemia effect in the brain. For assessing the initial ischemic insult effect, gerbils were frozen in liquid nitrogen following the ischemic periods as well as at different periods of postischemia. At 1, 5 and 20 h and 1 and 2 weeks after an initial 60-min occlusion some animals were made ischemic for a second period of either 5 or 60 min. and then frozen. The systemic blood pressure (SBP) was raised by repeated subcutaneous injection of 1:10,000 diluted epinephrine, and the SBP was monitored at 10-min. intervals with a transducer connected to an arterial catheter.

Brain metabolites were measured according to established fluorimetric (12, 20) and spectrophotofluorimetric (1) methods.

MATURATION PHENOMENON

Ito et al. (5) examined the changes in brain structures visible by light microscopy in animals subjected to brief periods of occlusion followed by variable periods

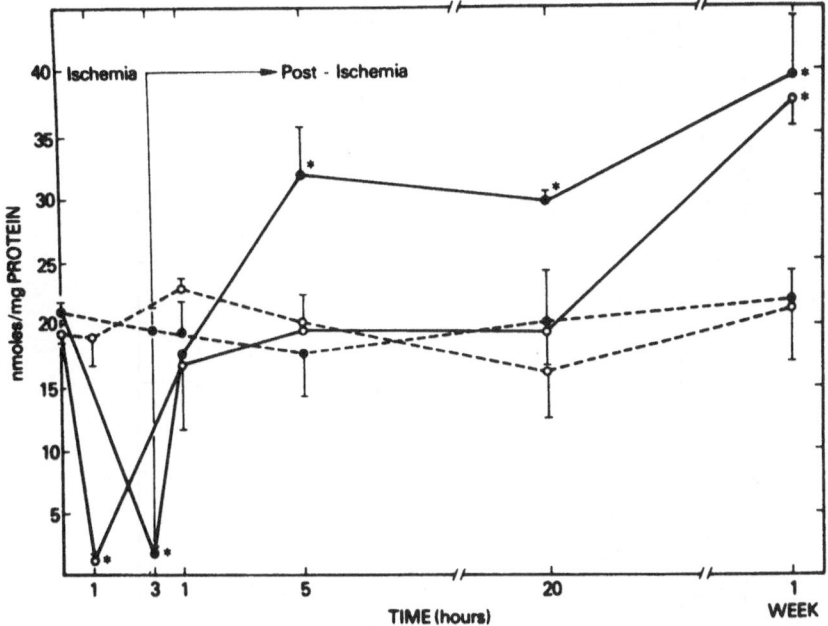

Fig. 1. Glycogen levels at various intervals following
1 (o) or 3 (●) h of ischemia.

of restored carotid blood flow. These studies revealed
that lesions developed not only during ischemia but even
after restoration of the circulation; lesions developed
during ischemia continued to "mature" and progress after
hemispheric blood flow was reestablished. The two factors
which determine lesion formation are the intensity (duration) of the ischemic episode and the duration of the
postischemic period. Also, studies of the behavior of
the blood-brain barrier following different episodes of
ischemia showed the same phenomenon; earlier barrier
breakdown was observed in gerbils exposed to longer occlusion periods (6).

Glycogen accumulation in the central nervous system
is a common occurrence following a variety of brain injuries. Our results show that the restoration of glycogen
during unilateral ischemia in gerbils is followed by re-
accumulation of this polysaccharide during postischemia
to concentrations greater than those in controls (16).
Cortical glycogen levels decreased after both 1 and 3 h
of unilateral ischemia (Fig. 1). After 1 h of recirculation, the levels of glycogen were restored to the control

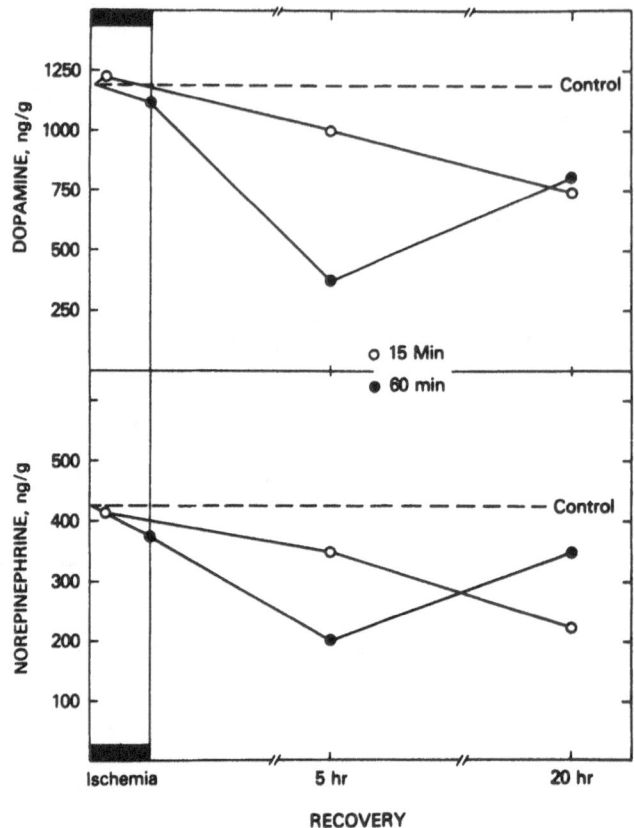

Fig. 2. Dopamine and norepinephrine levels at various
        times following 15 or 60 min. of ischemia.

values in both groups. Subsequently, glycogen increased
above normal levels after 1 week of blood restoration in
the 1-h ischemic group, and after 5 h in the 3-h ischemic
group. Thus, the onset of glycogen accumulation appeared
to be dependent upon the intensity of the ischemic insult.

Long-term ischemia is followed by the reduction of
both dopamine and norepinephrine (17). In contrast, the
data presented here show that short-term ischemia is
characterized by the unchanged levels of catecholamines,
and that reduction of monoamines occurs in the postischemic period (Fig. 2). In 15-min. ischemic animals, the
catecholamine values showed the lowest level after 20 h
of recirculation, whereas in 60-min. ischemia, the same
was observed after a 5-h interval.

These data provide additional biochemical evidence
supporting earlier findings on the existence of the maturation phenomenon following ischemic insult; the phenomenon was also confirmed by the histochemical studies in
hippocampus (19). The main feature of this phenomenon is
the fact that the rate of maturation of ischemic injury
is directly related to the intensity of ischemic insult,
a lesser intensity resulting in delayed lesion development.

The importance of the maturation phenomenon is still
intriguing. Apparently a considerable neuronal recovery
from ischemia is possible (5); the majority of gerbils
sacrificed 1 week after 1-h ischemia showed remarkably
well-preserved cortical neurons. An indication of neuronal ability to recover from ischemia was also provided
by histochemical observations in gerbils (19). However,
a release of an occlusion and a reestablishment of blood
circulation in the previously deprived areas does not
stop the further developments in ischemic injury. The
final outcome of an ischemic insult depends on the intensity of injury. Histochemical studies gave evidence
that glycogen accumulates in the collateral zone to the
central (ischemic) zone (14). Therefore, it can be speculated that the maturation phenomenon may also be operative
in permanent occlusion, and that the rate of the process
is related to a range of zonal gradients in the intensity
of injury from the focus of maximal impaired blood circulation to the histologically normal brain tissue. It
is also possible that in clinical situations negative
isotope scanning following short-term transient ischemia
becomes positive after some latency period, being directly
related to the maturation phenomenon.

BRAIN TOLERANCE TO SECOND ISCHEMIA

Repetitive ischemic episodes are characteristic of
certain pathological conditions affecting the brain circulation. In the gerbil, release of the occluded artery
permits circulation to be restored. Since our previous
investigations of these animals have demonstrated that
following release of a single unilateral carotid occlusion, a range of biochemical events continues to occur
long after reestablishment of circulation (15, 22), the
question of full biochemical recovery of brain tissue
following 1-h ischemia was still intriguing. We thought
that a repeated ischemic insult produced at the time when
brain energy metabolites are essentially restored would
help in obtaining such an answer. However, the study has
demonstrated that the biochemical reactivity of the brain

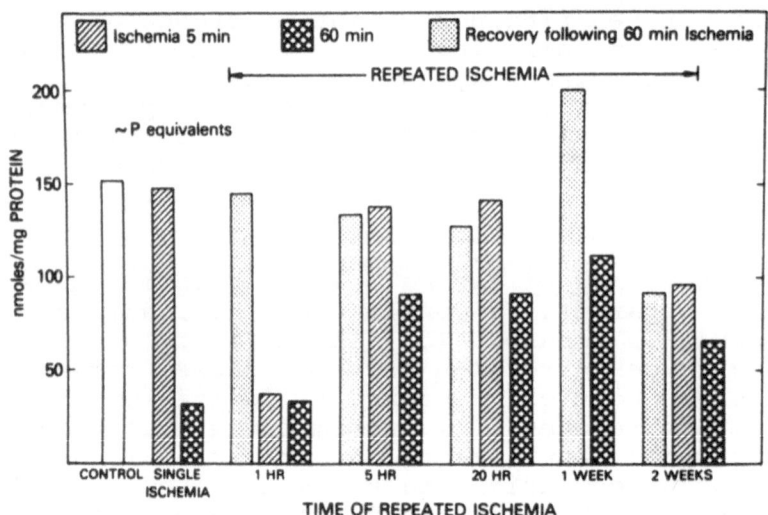

Fig. 3. High-energy phosphate equivalents (glucose + P-
        creatine + 1.4 ATP + 2.9 glycogen) after single
        and repeated periods of ischemia and varying
        periods of recovery.

after an initial 1-hr ischemic episode was altered when
the ischemic insult was repeated (Fig. 3). After 1 h of
recovery, a second period of ischemia introduced much
greater changes than those observed after a similar single
insult; conversely, after 5 and 20 h, and 1 or 2 weeks
of recovery, a second period of ischemia evoked a diminished response. Besides ATP, P-creatine, glucose and glycogen, lactate, GABA and cyclic AMP showed the same behavior in repetitive insult (18). Two distinct postischemic periods have been recognized in the study. After
the relatively short interval of recirculation, the effect of a second carotid occlusion was evident even after
5 min. of ischemia (Fig. 3). On the other hand, after 5
h to 2 weeks of recirculation, a second period of ischemia had relatively little effect on the concentration of
brain metabolites. The periods of hyper- and hyporeactiv-
ity correspond to both the metabolic rate in postischemia
(Fig. 4) and the neurological symptoms (Fig. 5). At a
time of recirculation, when the metabolic rate in the
brain is higher than normal, the brain shows an increased
sensitivity to a second period of ischemia; the signs
of neurological disorders, originally described by Kahn
(9), were evident during the 1 h of initial ischemia, as

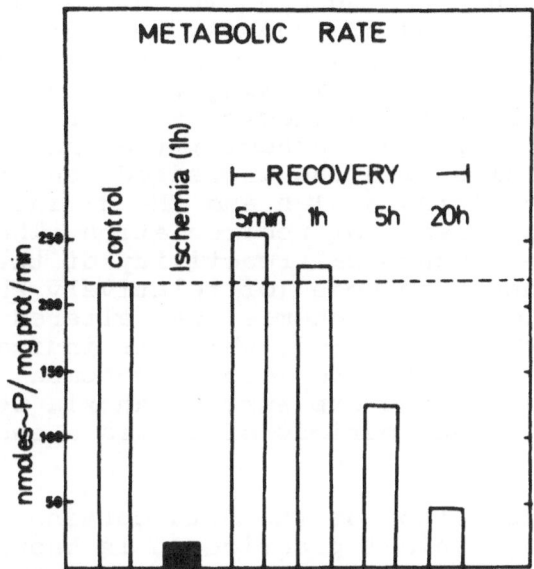

Fig. 4. Metabolic rate of cerebral cortex in control,
ischemic and postischemic gerbils at varying
periods of recovery.

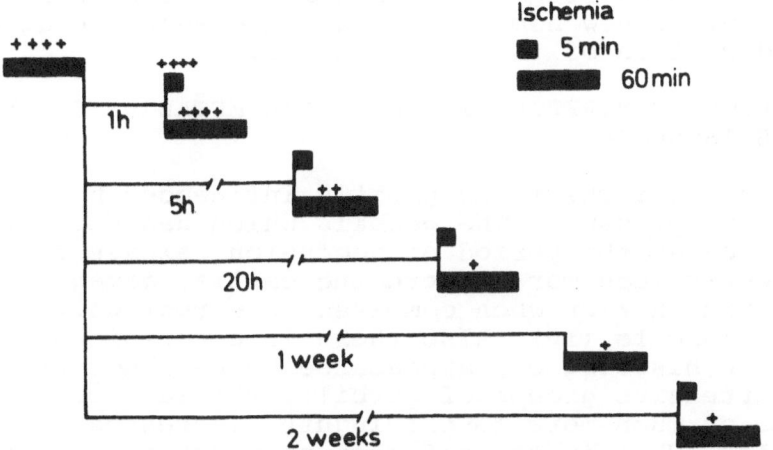

Fig. 5. Appearance of neurological symptoms in single
and repeated periods of ischemia at varying
periods of recovery.

well as at 1 h of postischemic period during the second
ischemia, of both 5 and 60 min. The neurological symptoms
during a second period of ischemia were diminished after
longer periods of recovery, as were the metabolite rate
and biochemical changes following the second insult. Thus
the neurological symptoms show close correspondence to
the biochemical signs of ischemia; the changes in metabolic rate coincide with the increased sensitivity to a
second ischemic period at 1 h and the diminished response
at later time intervals of recirculation. The data demonstrate that the biochemical reactivity of the brain after
an initial ischemic episode (of relatively short duration
in the gerbil model of ischemia) was altered when the
ischemic insult was repeated, which is indicative of the
development of tolerance to second ischemia. Similarly,
rats which survived one exposure to anoxia were able to
withstand much longer periods of anoxia a second time
(2).

The interpretation of the data obtained is still
speculative. In clinical practice it is known that patients with a partially occluded carotid artery may withstand the ischemic attack much more easily. Also the "es-
cape phenomenon" is known in neurophysiology; diminished
response to the same stimuli had been recorded when repeated several times (21). Two explanations of the hypo-
reactive period can be offered: (1) the development of
tachyphylaxis on release of biogenic amine(s) and/or (2)
development of new homeostasis as a consequence of metabolic adaptation of the brain tissue.

EFFECT OF HYPERTENSION ON BRAIN METABOLITES IN
POSTISCHEMIA

Cerebral ischemia is greatly influenced by the systemic blood pressure. The gerbils which had been hypertensive during the period of occlusion release revealed
consequently much more severe and earlier damage of the
blood-brain barrier when compared to normotensive postischemic animals (10). Also there is a conspicuous difference in histological appearance between normotensive
and hypertensive groups of gerbils, the latter showing
evidence of much more severe injury; increased cerebral
blood flow (CBF) in the affected postischemic hemisphere
of hypertensive animals was evident (7).

In ischemia the levels of ATP, P-creatine, glucose
and glycogen are reduced, while the concentrations of
lactate, pyruvate and GABA are increased; glutamate is
unchanged during 1-h unilateral ischemia in gerbils

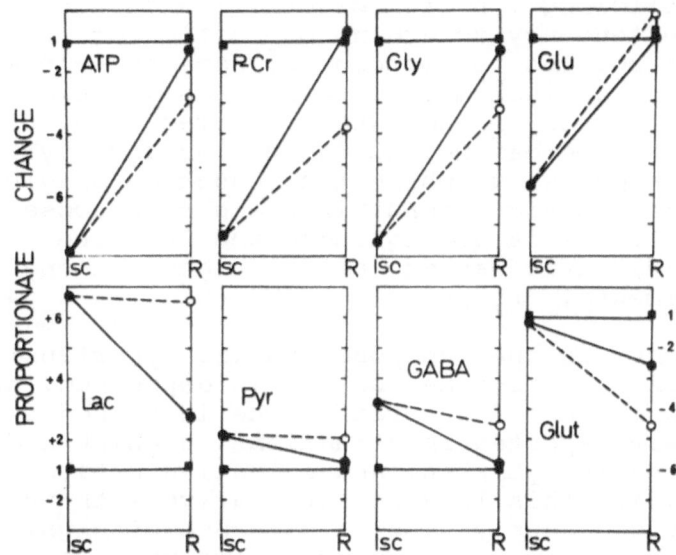

Fig. 6. Effect of increased SBP on the levels of ATP, P-creatine (P-Cr), glycogen (Gly), glucose (Glu), lactate (Lac), pyruvate (Pyr), GABA and glutamate (Glut) in postischemia. (■) Sham-operated gerbils; (●) ischemic gerbils; (o) ischemic gerbils with increased SBP; (Isc) ischemia (1 h); (R) recovery (1 h).

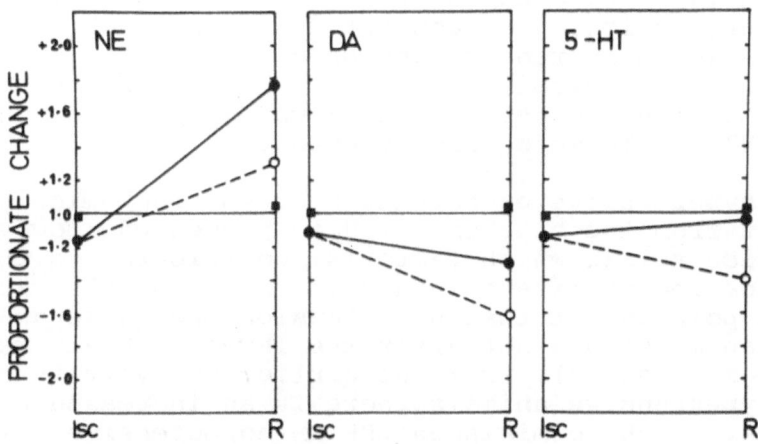

Fig. 7. Effect of increased SBP on the levels of norepinephrine (NE), dopamine (DA) and serotonin (5-HT). Abbreviations and symbols as in Fig. 6.

(Fig. 6). Following 1 h of recirculation in normotensive animals, the tendency toward normalization of the levels of these metabolites is impressive; however, glutamate is reduced in postischemia. In hypertensive animals this tendency to recovery is somewhat affected (Fig. 6). Recovery of ATP, P-creatine, glycogen, lactate, pyruvate and GABA is suppressed, while the reduction of glutamate is potentiated. A significant increase of glucose was found in postischemia during increased SBP. The lactate/pyruvate (L/P) ratio was one of the highest among all these experimental groups.

It is obvious that in postischemic hypertensive gerbils brain oxidative metabolism is depressed, supported by an increased L/P ratio. The increased L/P ratio was previously reported in situations in which hydrogen is accumulated in cytoplasm and oxidative metabolism is impaired (3, 4); this is also correlative with the deficit in tissue oxygen supply. The persistance of significantly elevated lactate levels in hypertensive postischemic animals appears to implicate this metabolite as playing a role in the pathomechanism of ischemic injury; more pronounced cellular degeneration and neuropilic vacuolization was found in hypertensive as compared to normotensive postischemic gerbils (7).

The behavior of biogenic amines in the postischemic period following 1-h ischemia is somewhat peculiar. Norepinephrine (NE), dopamine (DA) and serotonin (5-HT) were slightly reduced during 1-h ischemia, but following the recovery period, 5-HT content returned to the control values, DA was more reduced and NE concentration was enhanced over the control level (Fig. 7). In hypertensive animals postischemic rebound of NE was suppressed, while DA and 5-HT levels were more affected.

The hypertensive exacerbation of postischemic injury is evident from the biochemical data presented. However, the question arises which factor(s) contributed to the intensification of tissue injury in these animals. It should be pointed out that hypertension per se (sham-operated animals) did not alter the level of brain metabolites (Figs. 6 and 7). From the earlier study it is clear that in hypertensive animals there is an increased CBF, in contrast to the diminished CBF in normotensive animals during the recovery period from 1-h ischemia (7). One may speculate on the correlation among biochemical changes (particularly those of biogenic amines and lactate), disturbances of cerebral autoregulation known to occur in hypertension (8), and CBF disturbances. No matter what the basic mechanisms are, our results indicate that hypertension had a deleterious effect on the recovery process following cerebral ischemia.

COMMENT

It is obvious that the postischemic period is more complex than has been thought. The recovery from ischemia is not just a reversal of the events which occurred during ischemia. Release of an occlusion and reestablishment of blood circulation in the previously deprived areas does not stop the further development of ischemic injury. Two phenomena occurring during such postischemic periods are the maturation phenomenon and altered reactivity of the brain in repetitive ischemia, probably deciding the final outcome of an ischemic insult. The nature of these phenomena remains unexplained. However, in clinical situations, the phenomena can be relevant for the timing of cerebrovascular surgery after preceding ischemic insult, as well as in the treatment and prognosis of patients with repetitive ischemic attacks or patients with increased systemic blood pressure.

REFERENCES

1. Cox, R.H., Jr. and Perbach, J.L., Jr. (1973): A sensitive, rapid and simple method for the simultaneous spectrophotofluorimetric determination of norepinephrine, dopamine, 5-hydroxytryptamine and 5-hydroxyindolacetic acid in discrete areas of brain. J. Neurochem. 20: 1777-1780.

2. Dahl, N.A. and Balfour, W.M. (1974): Prolonged anoxic survival due to anoxia pre-exposure: brain ATP, lactate, and pyruvate. Amer. J. Physiol. 207: 452-456.

3. Forsander, A.O., Raiha, N., Salaspuro, M. and Maenpaa, P. (1965): Influence of ethanol on the liver metabolism of fed and starved rats. Biochem. J. 94: 259-265.

4. Hohorst, H-J., Kreutz, F.H. and Bucher, T.: (1959: Über Metabolitgehälte und Metabolitkonzentrationen in der Leber der Ratte. Biochem. Z. 332: 18-49.

5. Ito, U., Spatz, M., Walker, J.T., Jr. and Klatzo, I. (1975): Experimental cerebral ischemia in Mongolian gerbils. I. Light microscopic observations. Acta Neuropathol. (Berlin) 32: 209-223.

6. Ito, U., Go, K.G., Walker, J.T., Jr., Spatz, M. and Klatzo, I. (1976): Experimental cerebral ischemia in Mongolian gerbils. III. Behavior of blood-brain barrier. Acta Neuropathol. (Berlin) 34, 1-6.

7. Ito, U., Mršulja, B.B., Fujimoto, T., Klatzo, I. and Spatz, M. (1976): Effects of hypertension on the postischemic brain following temporary ischemia in gerbils, In: Pathophysiological, Biochemical and Morphological Aspects of Cerebral Ischemia and Arterial Hypertension. Ed. M. Mossakowski, Polish Academy of Sciences, Warsaw, 73-78.

8. Johansson, B., Li, C.-L., Olsson, Y. and Klatzo, I. (1970): The effect of acute arterial hypertension on the blood-brain barrier to protein tracer, Acta Neuropathol. (Berlin) 16: 117-124.

9. Kahn, K.M.D. (1972): The natural course of experimental cerebral infarction in the gerbil. Neurology (Minneapolis) 22: 510-515.

10. Klatzo, I. (1975): Pathophysiologic aspects of cerebral ischemia. In: The Nervous System. Vol. 1. The Basic Neurosciences. Ed. D.W. Tower, Raven Press, New York, 313-322.

11. Kleihues, P., Hossmann, K.A., Pegg, A.E., Kobayashi, K. and Zimmermann, V. (1975): Resuscitation of the monkey brain after one hour complete ischemia. III. Indications of methabolic recovery. Brain Res. 95: 61-73.

12. Lowry, O.H. and Passonneau, J.V. (1972): A Flexible System of Enzymatic Analysis, Academic Press, New York, 146-218.

13. Lust, W.D., Mršulja, B.B., Mršulja, B.J., Passonneau, J.V. and Klatzo, I. (1975): Putative neurotransmitters and cyclic nucleotides in prolonged ischemia of the cerebral cortex. Brain Res. 98: 394-399.

14. Mršulja, B.B., Mršulja, B.J., Ito, U., Walker, J.T. Jr., Spatz, M. and Klatzo, I. (1975): Experimental cerebral ischemia in Mongolian gerbils. II. Changes in carbohydrates. Acta Neuropathol. (Berlin) 33: 91-103.

15. Mršulja, B.B., Lust, W.D., Mršulja, B.J., Passonneau,
    J.V. and Klatzo, I. (1976): Post-ischemic changes in
    certain metabolites following prolonged ischemia in
    the gerbil cerebral cortex. J. Neurochem. 26: 1099-
    1103.

16. Mršulja, B.B., Lust, W.D., Mršulja, B.J., Passonneau,
    J.V. and Klatzo, I. (1976): Brain glycogen following
    experimental cerebral ischemia in gerbils (Meriones
    unguiculatus). Experientia 32: 732-734.

17. Mršulja, B.B., Mršulja, B.J., Spatz, M. and Klatzo,
    I. (1976): Catecholamines in brain ischemia - effects
    of alpha-methyl-p-tyrosine and paraglyne. Brain Res.
    104: 373-378.

18. Mršulja, B.B., Lust, W.D., Mršulja, B.J. and Passoneau,
    J.V. (1977): Effect of repeated cerebral ischemia on
    metabolites and metabolic rate in gerbil cortex.
    Brain Res. 119: 480-486.

19. Mršulja, B.J., Spatz, M. and Klatzo, I. (1978):
    Cytochemistry of hippocampus following cerebral
    ischemia. In: Pathophysiology of Cerebral Energy
    Metabolism. Eds. B.B. Mrsulja, Lj.M. Rakic and I.
    Klatzo, Plenum Press, New York (in press).

20. Passonneau, J.V. and Lauderdale, A.R. (1974): A com-
    parison of three methods of glycogen measurement in
    tissues. Analyt. Biochem. 60: 405-412.

21. Rakić, L.M. (1965): Cortical inhibition and subcorti-
    cal inhibitory influence. Israel J. Med. Sci. 1: 1376-
    1383.

22. Schwartz, J.P., Mršulja, B.B., Mršulja, B.J.,
    Passonneau, J.V. and Klatzo, I. (1976): Alterations
    of cyclic nucleotide-related enzymes and of ATPase
    during unilateral ischemia and recirculation in gerbil
    cortex. J. Neurochem. 27: 101-107.

BLOOD FLOW, OXYGEN AND ELECTRICAL DYNAMICS

IN CEREBRAL ISCHEMIA

R. Cohn, B.J. Sussman, I. Klatzo
and J.F. Turman

Howard University College of Medicine
Washington, D.C.
                          and
National Institutes of Health
Bethesda, Maryland 20014, U.S.A.

Blood flow and oxygen availability are obvious basic requirements for normal brain function. We have measured blood flow and oxygen partial pressures of brain tissue during experimental ischemia in the gerbil. These O_2-tension and blood-flow changes have been correlated with contemporaneous changes in the concatenated EEG and in the visual evoked cortical responses (VECR).

METHOD

Blood flow was measured by temperature changes transduced by 1 mm thermistors placed in the epidural space through trephine holes in the skull. The thermistors were fixed on a stereotaxic frame. The temperature changes were graphically recorded continuously with DC instrumentation. Calibrations were made prior to each set of measurements.

Oxygen-partial-pressure measurements were made by a polarization technique (5).The active sensor consisted of a gold plated metal alloy coated with an oxygen permeable membrane, supplied by IBC. The reference electrode was a 1/2 mm silver-silver chloride wire placed in the subcutaneous tissue. The microsensor amplifiers were current-measuring instruments manufactured by IBC.

The chemical action of this system at the active electrode is: $O_2 + 2H_2O + 4e^- \rightarrow 4OH^-$; at the reference electrode the reaction is: $4 Ag^+ + 4Cl^- \rightarrow 4 AgCl + 4e^-$. In general, the oxygen and temperature measurements were obtained on the same graph simultaneously. It is well known that oxygen pressure measurements are temperature dependent. Experimentally we have shown in our system that there was an approximate 6% change with each centigrade degree change in temperature. This variation with temperature did not appreciably alter the basic oxygen curves.

The EEG, both concatenated and VECR, were recorded from electrodes placed in homologous frontal and occipital regions over each cerebral hemisphere. These electrodes were placed approximately 2 mm lateral to the midsagittal plane. The electrodes consisted of various types such as lead spheroids somewhat less than 1mm in diameter placed in the epidural space through trephine holes (2), or brass pores, or gold plated steel screws which were fixed by the precision tapping of the appropriately drilled skull. The amplifier band-pass was essentially flat from 1 1/2 to 2500 cps; the time constants were 0.3 to 0.1 seconds. Recordings were made with pen and ink and cathod ray oscillographs. The summated recordings (VECR) were made with a CAT 1000, and with a PDP /8e Digital computer, using basic averaging software.

The animals were anesthetized with intraperitoneal nembutal, 0.03 to 0.04 cc of 50 mg/cc. Under anesthesia the carotids were exposed and a fine silk suture was placed around the vessels to allow surfacing of the arteries for subsequent clipping. Special care was taken not to occlude the arteries during the suture placement. Heifitz clips were used to occlude the vessels. The animals were fixed in a head holder with ear plugs and muzzle retainer; they were maintained in such a way that little or no movement artifact occured with vessel manipulation. The EEG electrodes were placed while the animal was in the head holder. The heads were remarkably uniform in the 70 or more gerbils studied.

RESULTS

Blood flow and oxygen availability measurements tended to have similar characteristics following unilateral carotid occlusion, as seen in Fig. 1. The immediate consequence of unilateral carotid occlusion was a rapid decrease in brain oxygen tension and blood flow ipsilateral to the side of occlusion; the contralateral O_2

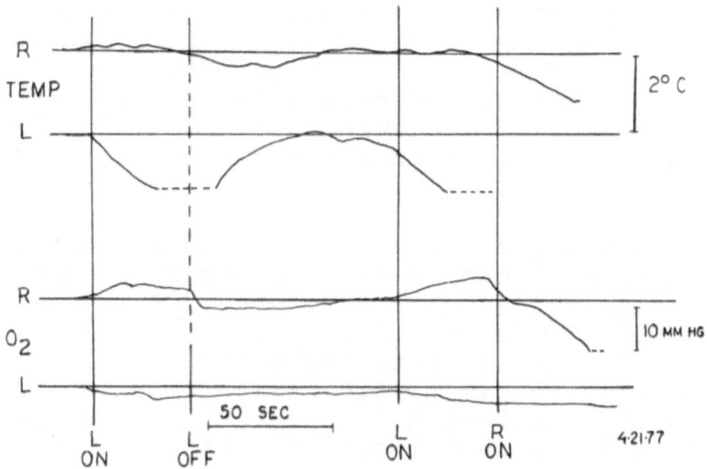

Fig. 1. Upper two lines represent blood flow. Lower two lines represent brain-tissue oxygen. Note the immediate ipsilateral decrease in blood flow and oxygen availability following ipsilateral common-carotid-artery occlusion and the contralateral increased blood flow and O_2.

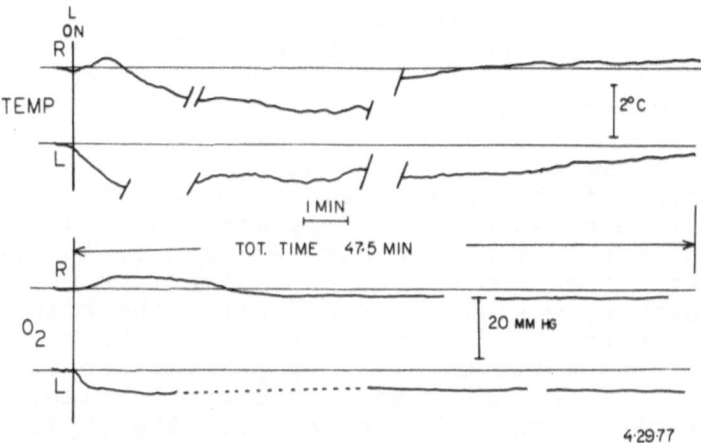

Fig. 2. Longer interval of common carotid artery occlusion with some recovery (and even hyperperfusion) over the contralateral hemisphere (top line).

tension and blood flow generally showed variable amplitude transient increases. These fundamental responses
could be rapidly recycled as shown in the figure. A
similar phenomenon was observed in Figure 2. In the
contralateral blood flow (top line of the figure) a
hyperperfusion was observed during the period of recovery.
Occasionally the O_2 partial pressure and blood flow did
not show contralateral enhancement with unilateral carotid
occlusion.

The acute EEG changes (concatenated and VECR)
following unilateral common-carotid occlusion are shown
in Figure 3: (A) and (B) are controls; (C) and (D) represent changes at 1 and 2 minutes respectively following
the artery occlusion. In both instances the amplitudes
of the waves greatly diminished bilaterally. Within 6
minutes after the occlusion (E) the EEG showed prominent
respiratory activity and almost no fast activity; the
VECR 9 minutes post-clipping (F) showed only a very low
amplitude of response on the right; the left was hardly
discernible. At 11 minutes post-occlusion (G) the right
brain concatenated EEG showed some recovery of the relatively fast activity; the left (occluded side) showed
only respiratory activity and electrical noise. Following
the removal of the clip there was fair recovery of the
VECR within 8 minutes, particularly on the right unoccluded side (H) of the figure. In (I), 62 minutes after the
release of the clip the concatenated EEG had recovered
much of its fast activity, but was somewhat lower in
amplitude when compared with the control; however the
VECR as shown in I, recovered its essential control
contours and amplitude.

Several sequences of O_2 tension changes with clipping
and releasing of the unilateral and bilateral cerebral
artery flow were induced. Usually with the right carotid
clipped the ipsilateral response was rapid; the transient
contralateral enhancement response sometimes was not
marked. Recovery was always rapid after the clip removal.
At times when the left clip was released the right O_2
tension dropped while the left O_2 partial pressure recovered to somewhat more than its control value. With
the clip replaced on the right carotid the ipsilateral
O_2 tissue tension dropped; with the clip now on the left
carotid too (bilateral occlusion) the left O_2 partial
pressure often dropped sharply. Removal of the left clip
frequently showed no change in O_2 tension, but with the
right clip also off, the right brain O_2 tension rapidly
recovered to control values; slowly and steadily the O_2

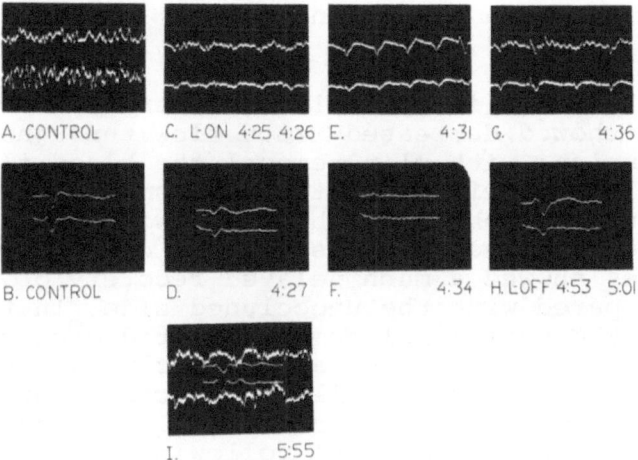

Fig. 3. Concatenated and VECR activities - A and B: Controls. C: Bilateral responses with unilateral
occlusion (note the respiratory loss of electrical activity during occlusion, but some recovery
of the contralateral activity in the concatenated
EEG). G, H and I: Recovery after clip removal.
All VECRs were summated 25 times. Each sweep was
500 msec. The concatenated EEG sequences were
4 seconds in duration.

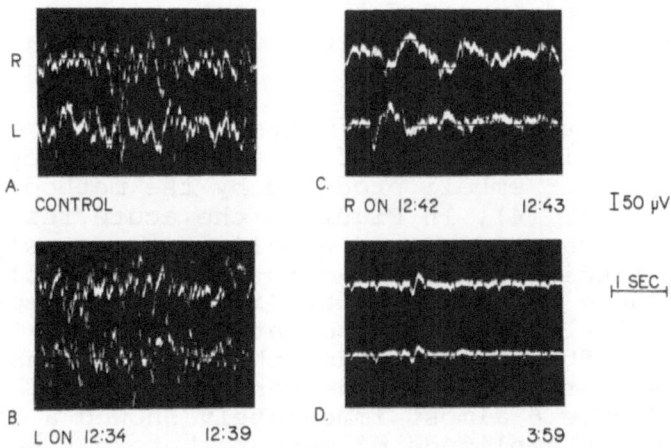

Fig. 4. Concatenated EEG. Effect of prolonged bilateral
occlusion. A: Control. B: Left artery occlusion
with no change. C: Bilateral occlusion with loss
of fast component of EEG. D: Essentially no electric output for over 3 hours of recording. Each
set of waves was 4 seconds in duration.

pressure of the brain tissue on the left generally reached
its baseline level.

In nearly all animals unilateral carotid occlusion
for one hour showed decreased blood flow throughout the
time of occlusion; with clip removal the blood flow recovered slowly, but with a continuous, smooth trajectory.
In a like manner we observed that following unilateral
occlusion and subsequent release of the clip the occluded
side sometimes showed a much delayed recovery of blood
flow when compared with the unoccluded side. In the
concatenated EEG unilateral common-carotid occlusion in
the *sensitive animal, and bilateral occlusion in all
animals, resulted in essentially no electric output
during the entire interval of occlusion. In Figure 4,
(A) represents the control, (B) following unilateral
occlusion (no change, insensitive animal), (C), 1 minute
after bilateral occlusion, and (D) no recovery with clips
on for more than 3 hours of recording.

During these carotid-artery-occlusion-induced-ische-
mia experiments we became concerned whether common-carotid-
artery occlusion generated only effects local to the brain.
We thus recorded peritoneal body temperature simultaneously with the brain-flow studies. It was observed that
despite brain-temperature changes, the body temperature
was not affected. The minimal decrease in body temperature
that was observed appeared to be part of a general cooling
process and not immediately related to the great-vessel
occlusion.

Another series of experiments was carried out to
study the relation of blood flow and EEG changes that
occurred with air emboli produced by the method of
Nishimoto et al. (4). In Figure 5 the acute response with
left-internal-carotid-artery embolization grossly recapit-
ulated the findings in acute-carotid-artery occlusion in
the neck. There was an immediate ipsilateral decrease in
blood flow; the latter enhancement was present for
approximately 90 seconds. Under these conditions of
embolic vascular occlusion the concatenated EEG, as seen
in (B) of Figure 6 almost immediately showed a bilateral
decrease in the amplitude of output, but this was much
more intense over the side embolized. Within 5 minutes
the occluded left side was dominated by respiratory action,

---

* A sensitive animal is defined as one that responds
  with marked homolateral electrical, clinical and
  neuropathological changes following the occlusion
  of one common carotid artery.

with no evident faster activity. Six minutes later (D) the
right-sided EEG had returned to its approximate control
type, but the left side continued to show a decreased
amplitude of potentials. This decreased amplitude of
activity persisted for nearly 2 hours (F). In other
animals even 24 hours after unilateral embolization the
amplitude of the ipsilateral concatenated EEG remained
of reduced amplitude when compared with the opposite side.

The blood-flow responses to unilateral air embolus
sometimes showed recovery of the involved side within 15
minutes after the embolization.

These concatenated EEG findings occurred irrespective
of whether the right or left internal carotid artery was
embolized. The bilateral immediate electrical response,
with greater involvement ipsilaterally was an almost
universal finding. In most animals the contralateral
recovery was full within 30 minutes or 1 hour. The insul-
ted brain generally continued to show residual slowing
together with decreased faster activity.

Because the external carotid arteries are such prominent vessels in the gerbil, as in other rodents, we
felt it important to determine whether unilateral occlusion of these extracranial vessels, as a result of menin-
geal embarrassment, would result in differences in
cerebral blood flow and/or O_2 partial pressure changes.
When either the right or left external carotid was occluded; the cerebral oxygen tension was not effected; but in
these same animals, when the internal carotid artery was
occluded, the cerebral effects were quite obvious.

To control the cerebral effects of local body cooling,
ice was applied around the neck. The intraperitoneal body
temperature always decreased first; this was followed
within one or two minutes by bilateral almost synchronous
decreased cerebral blood flow.

To establish clinically and electrically whether
posterior collateral circulation played a role in the
observed material we performed chronic carotid-artery
occlusions. One animal had the right common carotid
artery ligated with double ligatures ten days prior to
the acute experiments. Within 2 minutes following the
clipping of the left carotid artery no VECR was seen
on the left, and only a low-amplitude response was observed on the right. Within 4 minutes no concatenated
EEG was evident; the animal continued spontaneous
respirations for nearly 9 minutes after the final

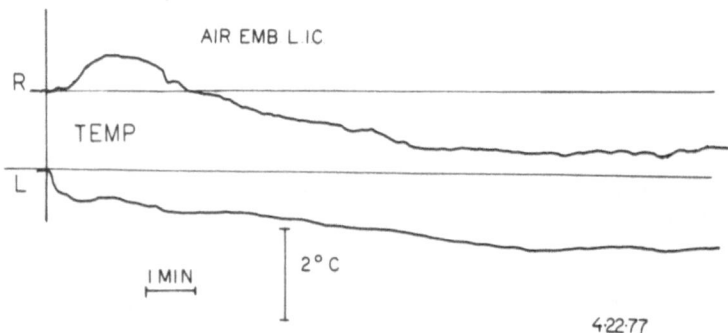

Fig. 5. Effect on blood flow of unilateral internal-
       carotid air embolus.

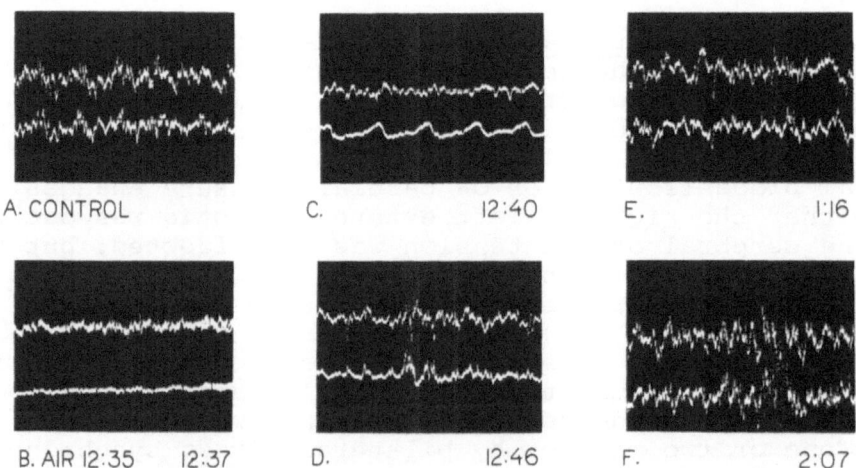

Fig. 6. Concatenated EEG with left-internal-carotid em-
       bolus. A: Control. B. Bilateral effect within
       2 kin. C, D, E and F: Progressive recovery during
       subsequent 92 minutes of recording. Note resp-
       ratory artifact primarily in the ipsilateral hemi-
       sphere. Note progressive recovery within 90 min-
       utes post-embolus.

occlusion. In another chronic experiment we occluded the
left carotid 11 days prior to the acute procedure
Within 2 minutes after occlusion of the right carotid,
no VECR could be elicited. Following removal of the clip,
within 23 minutes a VECR was obtained from the left cerebral hemisphere and 50 minutes later a fairly well con-
toured VECR was evident.

Because of the variable times of onset of marked
electrical change following carotid-artery occlusions
it was deemed important for all basic studies to determine the rapidity of electrical decay when the brain was
actuely deprived of all blood supply. We initially attempted to accomplish this by unilateral vagal stimulation;
this produced slowing of the heart rate but did not stop
the blood flow. We then resorted to decapitation. We were
fortunate that in some animals the operation could be
achieved with little or no movement artifact. When the
electrical data were represented by the area swept out
by the concatenated EEG (integration) it was observed
that immediately following decapitation there was markedly decreased electrical activity; this was followed
within one-half second by a high voltage slow output
which persisted for approximately 1 1/2 seconds.
Following this burst there was a progressive decrease in
electrical activity. At 15 seconds after the cessation of
all blood flow, there was no electrical activity of brain
origin.

Contrary to some reports (1) we have almost never
observed true seizures following unilateral or bilateral
common-carotid-artery occlusion under our experimental
conditions. Certainly the movement patterns seen never
had the character of the focal or general discharges seen
following the application of minute amounts of crystalline
strychnine to the cortex in these same animals. In that
the electrical output rapidly approached zero with common-
carotid-artery occlusion the cortical cells thereby lost
their ability to discharge. The clinical movement patterns
observed under these conditions must be of lower segmental
type (brainstem and/or cord discharges).

DISCUSSION

The acute bilateral hemispheric responses in the
oxygen availability/blood flow experiments and the electrical concomitants, strongly indicate that even in sensitive animals complete isolation of the effects of unilateral cerebral ischemia produced by carotid-artery occlusion is very exceptional. The bilateral effects ob-

served may be the consequence of shunting blood through
the heart into the intact cerebral vascular channels.
Under these conditions the apparent cerebral interaction
would be in fact an extracerebral phenomenon. Other explanations for the observed bilateral blood flow and electrical effects, particularly those that long outlast the
immediate responses, might be that unilateral vascular
occlusion alters the membrane (DC characteristics); these
DC potentials might easily be transmitted across the
corpus callosum and through direct anatomical passages
via the large subcortical nuclear masses, especially the
large massa intermedia in the rodent, to signal the contralateral hemisphere that the ipsilateral hemisphere is
in trouble.

The depressed electrical activity, particularly the
concatenated EEG, far outlasts that ordinarily anticipated
in spreading depression (3). A more direct transtissue
mediation might be entertained through pH and other more
purely chemical substances such as carbohydrate, catecholamine and polypeptide elements. The interhemispheric
interaction can thus be mediated directly or by secondary
neurotransmitters. Irrespective of which dynamics is
operational the two hemispheres appear to be in close
contiguity physiologically as well as anatomically.

In some experiments it is apparent that the right
brain controls the blood flow of the left side. This
may be due in part to the simple fact that in all these
experiments the right side was initially occluded. Whether
this is the explanation will be subjected to experimental
investigation.

In general we have observed that the occluded side
had a longer recovery time for return of blood flow and
oxygen partial pressure than that of the unoccluded side
when there were strong bilateral responses. These findings
closely parallel the concomitant electrical activities
and are probably related to cerebral edema.

In the decapitation data the electrical activity
became essentially nonexistent within 15 seconds. That
the effect of unilateral-carotid artery occlusion was
manifest only at around 1 to 5 minutes post-occlusion
indicates that there are emergency reservoirs for oxygen
and other metabolic necessities when one or both common
carotid arteries are occluded.

The fact that despite recovery of cerebral blood
flow there is a longer delay in recovery of the electrical

activity must be an expression of impaired neuronal function, such as membrane integrity and disturbed metabolism
resulting from the initial insult. Thus even when the
basic requirements of blood flow and subsequent oxygen
availability are met the ability to take oxygen becomes
defective and oxidative-dependent mechanisms must thereby suffer.

The major index of irreversible cell death of a
large population of neurons, at least as a transient
phenomenon, is a reduced amplitude of output in the
concatenated EEG. In the VECR the expression of such
disturbance is manifest in both amplitude and contours
of the summated potentials. The latency changes noted
and described under these conditions are also highly
dependent on amplitudes and contours of the degraded
waveforms.

That the brain impedance changes remarkably, and
early, in the occlusive process is evident by the presence of a prominent respiratory rhythm and the EKG
artifact in the records. Prior to tissue disturbance,
even during hyperventilation and increased systemic blood
pressure, the repiratory activity does not have the
pervasive character following cerebral tissue damage.
Thus when tissue impedance of the brain is altered by
disturbed membrane function the potentials of extracerebral origin are shunted into the cerebral recording
network.

Although temperature measurements represent a gross
and complex integration of heat-production and heat-
radiation losses, and thus are not amenable to state-
ments of local-blood flow conditions in our experiments,
there is no evidence that there was a significant "no-
reflow" phenomenon operating. All the temperature recordings had a smooth continuous trajectory; they showed
no plateaus or other quantal characteristics.

CONCLUSIONS

1. The general effect of unilateral common-carotid-
artery occlusion, even in sensitive animals, is at least
transiently manifest in each cerebral hemisphere.

2. Acute carotid-artery occlusion, whether by embolus
or clipping, evokes similar initial blood-flow, oxygen-
availability and electrical changes.

3. The external carotid arteries, despite their large

size and extensive extracranial distribution, do not
contribute to the oxygen, blood-flow or electrical
changes observed during common-carotid-artery occlusion.

    4. Collateral circulation is not acquired in the
gerbil even after eleven days of unilateral common-
carotid-artery ligation.

    5. Cortical seizure activities are not a general
phenomenon in unilateral or bilateral occlusion of the
common carotid artery.

    REFERENCES

1. Brierly, J.B., Brown, A.W. and Levy, D.E. (1976):
   Dynamics of brain edema. Edited by Hannah Pappius and
   William Feindel. Springer-Verlag, Berlin and Heidelberg.

2. Cohn, R. (1974): Epidural electrode for animals with
   thin skulls. EEG Clin. Neurophysiol. 36: 671-672.

3. Leao, A.A.P. (1944): Spreading depression activity in
   the cerebral cortex. J. Neurophysiol. 7: 359-390.

4. Nishimoto, K., Spatz, M. and Klatzo, I. (1977): Pathophysiological correlations in the blood-brain barrier
   damage due to air embolism. International Erwin Riesch
   Symposium on the Pathology of Cerebrospinal Microcirculation. Sept. 7-10, 1977, Berlin, Raven press.
   (in press).

5. Sussman, B.J., Barber, J.B. and Goald, H. (1974): Experimental intracerebral hematoma. J. Neurosurg. 41:
   177-186.

CYTOCHEMISTRY OF HIPPOCAMPUS FOLLOWING CEREBRAL ISCHEMIA

B.J. Mršulja, M. Spatz and I. Klatzo

Institute for Biological Research
Belgrade, Yugoslavia
        and
National Institute of Neurological and
   Communicative Disorders and Stroke
National Institutes of Health
Bethesda, Maryland 20014, U.S.A.

The feature of selective vulnerability to ischemia displayed by certain topistic units and most clearly observed in the hippocampus, the brain region particularly sensitive to ischemia (12, 14), was the important finding brought out by the light microscopic observation on gerbils. The most striking lesion was the "reactive change", confined to the H3 sector of the hippocampus (26). This change occurred only in the animals subjected to relatively slight ischemic insult and fully developed only upon recirculation. The neurons of the H3 sector were characterized by a peripheral shift of the nucleus whereas the voluminous cytoplasm showed a central chromatolysis. The reversibility of these changes was evident from the almost normal appearance of the H3 sector in the animals sacrificed 1 week after the reestablishment of blood supply to the brain. It is thus evident that "reactive change" represents a cellular reaction to ischemia in which the neurons are capable of full recovery from an ischemic injury.

The present investigations have been focused on the evaluation of histo- and cytochemical changes of Ammon's horn and aimed at supplementing information for a better understanding of the pathophysiological mechanism involved in ischemic lesion.

MATERIAL AND METHODS

Regional cerebral ischemia was produced in anesthetized (sodium pentobarbital, 3 mg/100 gm i.p.) Mongolian
gerbils by unilateral common carotid artery clipping for
15 min. Animals with neurological signs of infarction
(13, 17) were sacrificed either immediately following the
termination of the ischemic period or after 1,5, 12 hrs,
4 days, 1 week and 1 and 2 months of recirculation.

Control and sensitive animals, described by Ito et
al. (12) were submitted to histological investigation.
The brains were carefully exposed, quickly removed, immediately frozen in dry ice, mounted on holders and stored in cryostat at -20°C. Storage was never longer than
24 hrs and did not lead to any demonstrable loss of enzymatic activity. Cryostat sections, 16 µ in thickness,
were mounted on nonalbuminised coverslip, dried at room
temperature for several min and than stored temporarily
in the refrigerator at 4°C. Period of time between cutting and incubation was not more than 3-6 hrs. All the
enzymatic reactions were performed by simultaneous incubation of sections from the experimental and control
animals in the same medium.

The activities of the following enzymes were studied:
(1) respiratory enzymes: Succinic (SDH), lactic (LDH),
glutamic (GDH) and glucose-6-phosphate (G-6-P-DH) dehydrogenase were demonstrated by nitro-BT-methods of Faber
as modified by Hess et al. (11). (2) Cytochrome oxydase
(COX) was studied according to Burstone´s p-Amonodiphe-
nylamine method (5). In addition to p-Aminodiphenylamine,
the incubation medium contained one drop/50 ml of 8-amino-
1,2,3,4-tetraquinoline. Cytochrome C in final concentration of 5 mg per 50 ml of incubation medium was added;
control reactions were performed by admixing sodium cya-
nate (10^{-3}M, final concentration) to the incubation medium. (3) Acid monophosphatase activity was assessed by
Burstone´s methods (4, 6). Sections for acid monophosphatase activity were prefixed in cold acetone. Formol-calcium perfused and post-fixed frozen sections were also
used in addition to fresh frozen sections for both acid
monophosphatase and thiamine pyrophosphatase (TPP-ase).
Frozen sections 10 µ thick, were cut on a freezing micro-
tome and for 2 hrs incubated at 37°C in the Novikoff
and Goldfisher´s medium (24), containing 5% sucrose for
TPP-ase and in Gomori´s medium modified for acid monophosphatase activity by Novikoff (23). Glycerophosphate
as a substrate was substituted for 5´-cytidilic acid.

The controls employed for all enzymatic reactions were: a) incubation of slides preheated for 10 min at 85°C and b) incubation without substrate.

RESULTS

Pyramidal neurons of unoperated and from the right side of the hippocampus of operated gerbils are refered as the controls.

At the end of ischemic insult and during several hrs following 15 min occlusion, changes neither in enzymatic activities in any hippocampal layer nor in H3 sector´s neurons of Ammon´s horn were found.

The earliest detectable increase of SDH, LDH, GDH, G-6-P-DH, COX and acid monophosphatase(AcPh) activities were seen in the H3 sector´s neurons 12 hrs after recirculation. The neurons responded by an increased number of small individual or clumped granules and by the presence of the AcPh activity, were presumed to be lysosomes.

In the group of gerbils with 20 hrs release following 15 min occlusion, most of the animals, in the H3 sector of the hippocampus, showed a striking and constant histochemical picture. The activity of almost all enzymes investigated reached its peak in H3 sector´s neurons at 20 hrs release in comparison with the previous and the successive groups. A distinct increased activities of above mentioned enzymes were found in the central region of the H3 neuronal voluminous cytoplasm, while nuclei were quite excentric in position and remained histochemically negative (Figs. 1 and 2) as compared to the cells from the H3 sector of the hippocampus in controls. The granular histochemical reaction of AcPh, characteristic for normal condition (Figs. 3A and 3B) showed a diffuse character in H3 neurons of the experimental animals (Figs. 4A and 4B).

Morphological studies showed that central chromatolysis and peripheral shift of the nucleus which is a characteristic of "reactive change", were observed in H3 sector´s neurons of the hippocampus 20 hrs after the recirculation following relatively slight ischemic insult. (Figs. 5 and 6). However, at 20 hrs of recirculation, neurons of the H2 sectorappeared to be morphologically intact (Fig. 5) and histochemically unchanged as compared to the appropriate region of controls (Fig. 1).

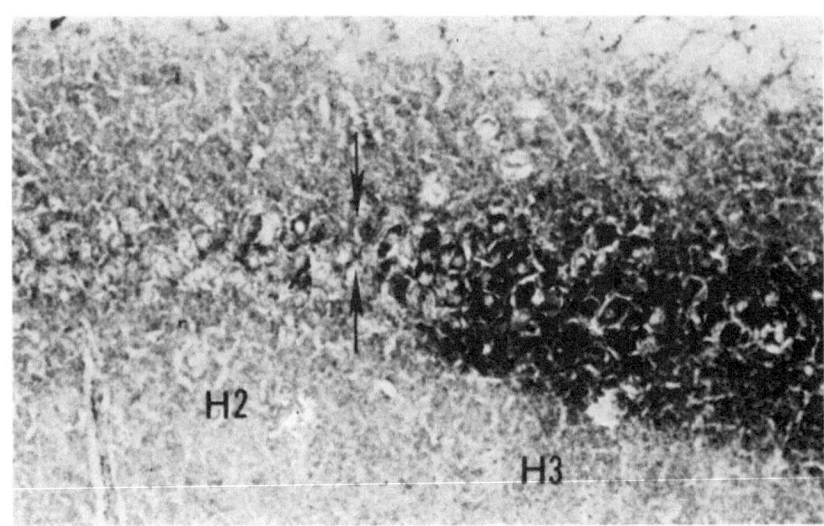

Fig. 1. The left hippocampus in a gerbil subjected to
15 min occlusion and 20 hrs release; the H3 sector's neurons undergoing a "reactive change" reveal dark formazan deposits in cytoplasm.
Succinic dehydrogenase X220.

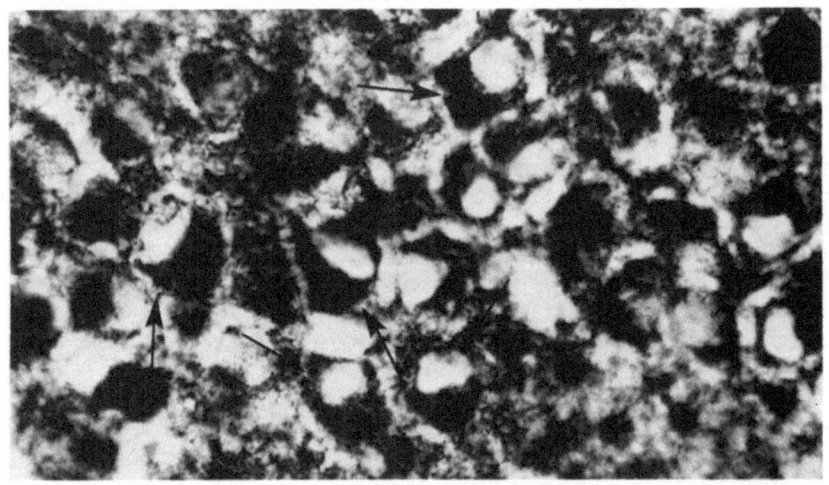

Fig. 2. A higher magnification of the same H3 neurons as
in Fig. 1. Succinic dehydrogenase activity is
diffuse and sharply increased, concentrated in a
neuronal voluminous cytoplasm (arrows) while
nuclei are peripherally dislocated and histochemically negative. X680.

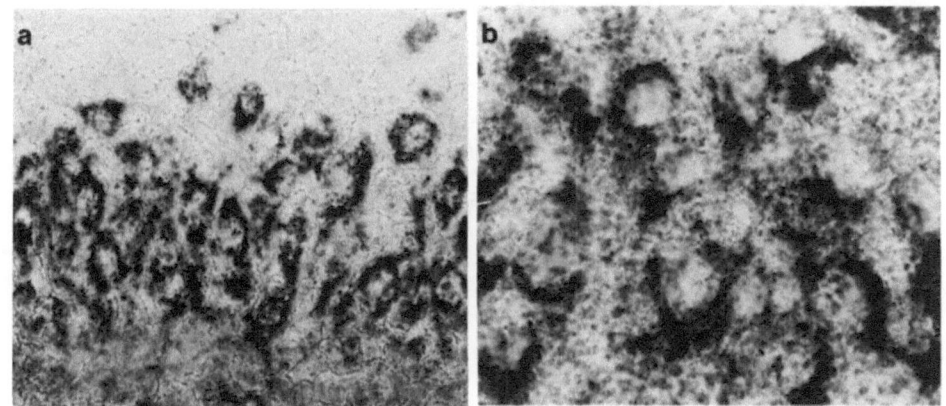

Fig. 3. Normal acid monophosphatase activity (azodye
        coupling method) granular in type; note even
        cytoplasmic distribution of the reaction product
        in H3 neurons of control hippocampus.   3a X350.
        3b X880.

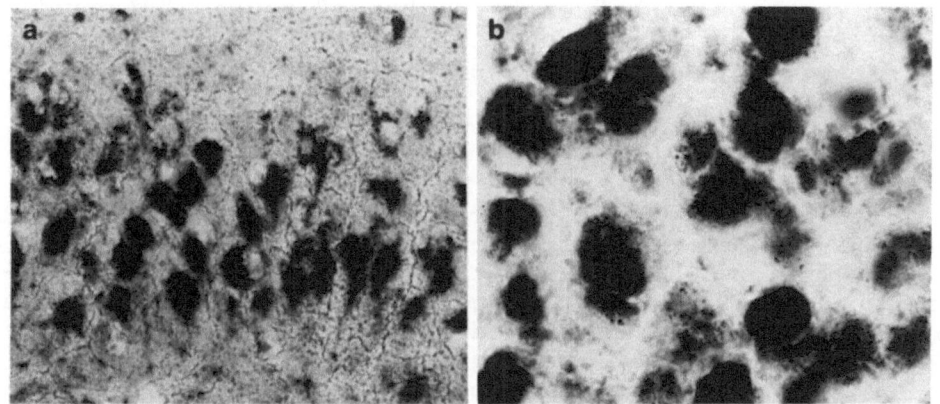

Fig. 4a. Diffuse increase in intraneuronal acid monophos-
         phatase activity is evident in H3 neurons in the
         gerbil sacrificed at 20 hrs of recirculation
         following 15 min occlusion. X350

     4b. Notice completely enclosed a region of cytoplasm
         by prominent diffuse increase of acid monophos-
         phatase reaction product in contrast to the
         histochemically negative nuclei in a damaged
         H3 sector's neurons. X880

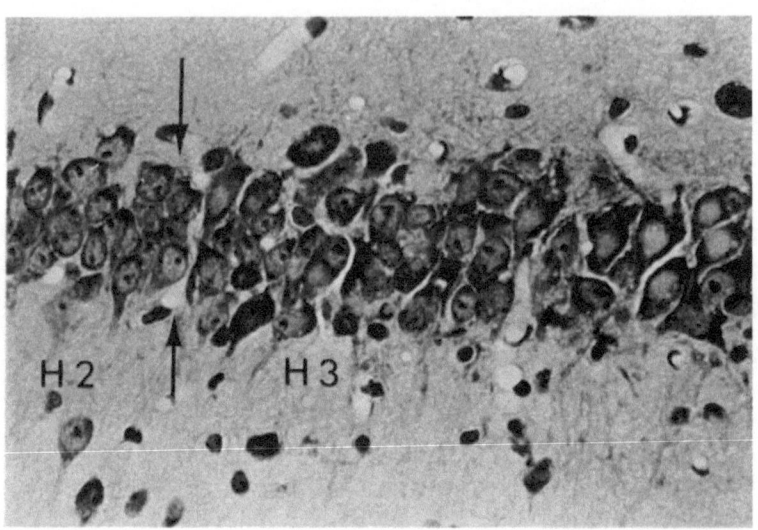

Fig. 5. Border area between H2 and H3 of the left hippocampus in a gerbil subjected to 15 min left carotid occlusion and sacrificed 20 hrs later. H3
neurons with "reactive change" reveal "glassy",
pale cytoplasm and peripheral shift of the nucleus containing enlarged nucleolus. Cresyl violet
X270.

The neurons of H2 sector showed dual behavior of TPP-
ase activity. In controls, coma-shaped reaction product
scattered evenly throughout the cytoplasm (Fig. 7B), became changed in this pathological condition; some of the
H3 sector's neurons displayed increased TPP-ase activity
(Fig. 7A), while the others did not show any TPP-ase activity having the typical appearance of "reactive change"
(Fig. 7C).

The fourth day of recirculation expressed transient
phase of fully developed histochemical changes in H2 sector described below in gerbils with 15 min occlusion and
a week of recirculation.

One week after the ischemic insult, all of the investigated enzymes in the H3 sector were increased as compared to controls, but to a lesser degree than after 20 hrs
and 4 days of clip release. Here small foci of disappeared
neurons have been found (Fig. 8C and 12). However, the
rest of the H3 sector's neurons showed evenly distributed

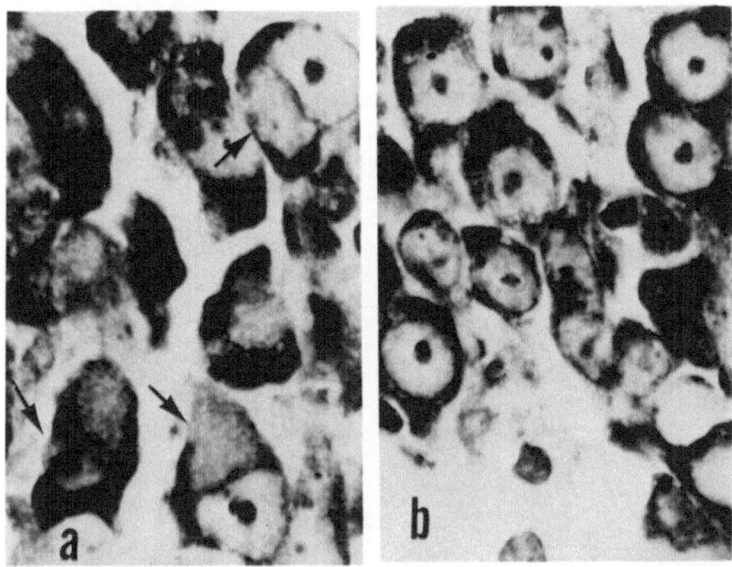

Fig. 6. A gerbil with 15 min ischemia and sacrificed after 20 hrs of recirculation. The H3 neurons with "reactive change" (arrows) on the left side. Cresyl violet. Both photographs X680.

enzymatic products comparable to the appropriate region of controls and/or neurons with still increased enzymatic activities (Fig. 8C). H3 sector's neurons displayed dual pattern of TPP-ase activity; there was no TPP-ase activity in a number of cells while neurons with increased TPP-ase activity, observed after 20 hrs of recirculation, showed further enzyme activity intensification. In addition, H2 sector of hippocampus, a week after recirculation, was almost completely disrupted; only and occasional nerve cell could be found among the numerous micro-astroglial elements (Fig. 9). These morphologically distinctive traits observed are excellent support for histochemical findings. Focal differences in enzymatic activities in the damaged H2 area were observed. A marked increase of enzyme activity in areas with very intensive microglial proliferation and hyperactive astrocytes was adjacent to the tissue with absent reaction due to loss of neurons (Figs. 8A and 10A). Distinct decrease of all dehydrogenases was observed in the neuropil within all layers of correspondent H2 sector (stratum oriens, radiatum and lacunosum-moleculare) while AcPh was diffusely and distinctly increased in the same localization

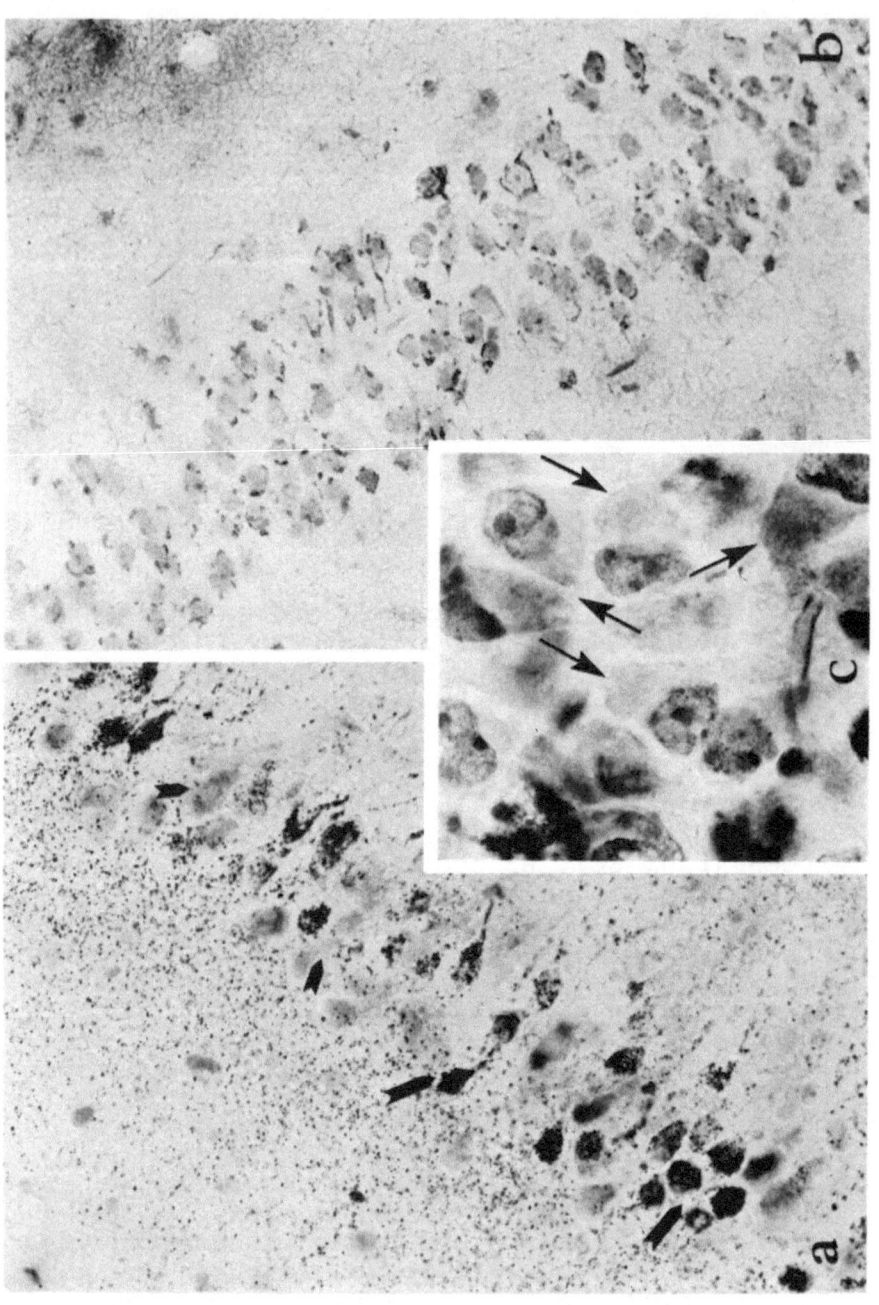

Fig. 7

Fig. 7. Golgi reticulum pattern of H3 neurons in section
incubated for TPP-ase activity. Formol-calcium
fixed frozen sections. a. Gerbil sacrificed after
20 hrs release of 15 min occlusion. Based on the
TPP-ase reaction product, H3 neurons could be
differentiated in two groups; neurons with prominent content of TPP-ase reaction product (long
arrows) and neurons showing no TPP-ase activity
(short arrows), (enlarged in inset c. X880), and
with all characteristics of "reactive change".
X350. b. TPP-ase activity outlining the structures of the Golgi apparatus in the H3 neurons of
control hippocampus. X350.

(Fig. 8a). A prominent increase in acid monophosphatase
(Fig. 8A) and TPP-ase activities was noticed in microglial and astroglial elements spread out through all
layers, placed above and below the H3 sector of the
hippocampus.

Gerbils sacrificed 1 and 2 months after release of
15 min occlusion revealed no significant differences in
histochemical alterations within affected hippocampus.
A multifocal disappearance of neurons in H2 sector was
replaced by numerous enzymatic hyperactive micro- and
astroglial elements, showing on dehydrogenase (Fig. 10A)
TPP-ase and acid phosphatase (Fig. 12) preparations distinct increase of the above-mentioned enzymes. Inverse
activity between dehydrogenases and acid phosphatase was
restricted to the neuropil of the H2 sector; all layers
were deeply deprived of dehydrogenases (Fig. 10A) and
at the same time, showed marked increase of acid phosphatase activity (Fig. 12). The H3 sector's neurons were
mostly characterized by the normal activity of the enzymes, except that dehydrogenases appeared to be still
increased (Fig. 11A). In addition to this, focal loss of
neurons was evident in the affected hippocampal H3 sector as an absence of enzyme activity (Figs. 11A and 12).

DISCUSSION

Necrosis limited to the H2 (Sommer) sector has long
been known to be the only one of the variations in the
extent of neuronal loss that may be encountered in
Ammon's horn. As a result of the variations, Spielmeyer
was firmly convinced that a vascular factor determines
origin and extent of the neuronal loss (31). On the

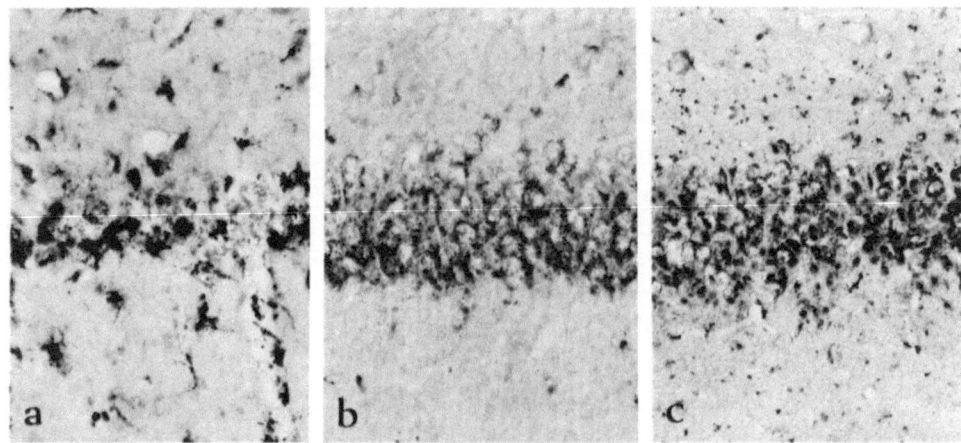

Fig. 8. Acid monophosphatase activity in the hippocampus
of the gerbil sacrificed 1 week after 15 min
ischemia. Formol-calcium fixed frozen section.
a. H2 sector of hippocampus. Hypertrophic astrocytes with localization in both pyramidal layer
and throughout the neuropil, particularly around
blood vessels, reveal abundant enzymatic activity in their cytoplasm and processes. X350. b.
Acid monophosphatase activity in pyramidal cells
of contralateral hippocampus, appeared in the
form of a fine-granular reaction evenly distributed in the neuronal cytoplasm. X220. c. H3 sector
of hippocampus. Most of the neurons show acid monophosphatase reaction product similar to the
control, b., although occasionally there are activated roundish microglial elements. X220.

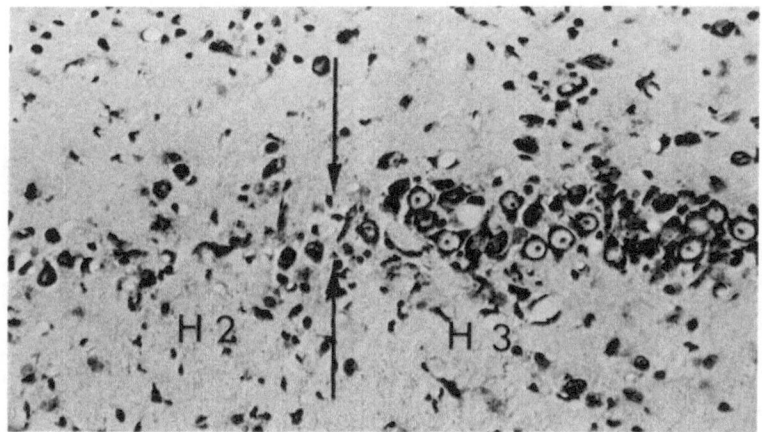

Fig. 9. Border area between H2 and H3 in the left hippocampus of the gerbil subjected to 15 min occlusion and sacrificed 1 week later. H2 sector
shows almost complete disappearance of neurons.
The majority of H3 neurons appear normal. Cresyl
violet X176.

other hand, Friede demonstrated certain metabolic pecu-
liarities concerning the hippocampal cortex, which appear
to be supplemental factors of the selective vulnerability
to oxygen deficiency (9).

The most conspicuous difference between the H2 and
H3 sectors' neurons of hippocampus, observed after reestablishment of the circulation, was related to the rate
of the histochemical response and the final outcome;
there was considerable recovery of enzymatic activities
within the H3 sector's neurons while the H2 sector's
neurons revealed the widespread disintegration.

A characteristic feature of ischemic hippocampal injury is that histochemical abnormalities do not become
apparent until several hours afterwards. These observations suggest that the delayed changes are secondary to
lesions induced at the time of blood flow cessation.
Hence, neurons could retain enough of the enzymatic apparatus for their temporary functioning (32).

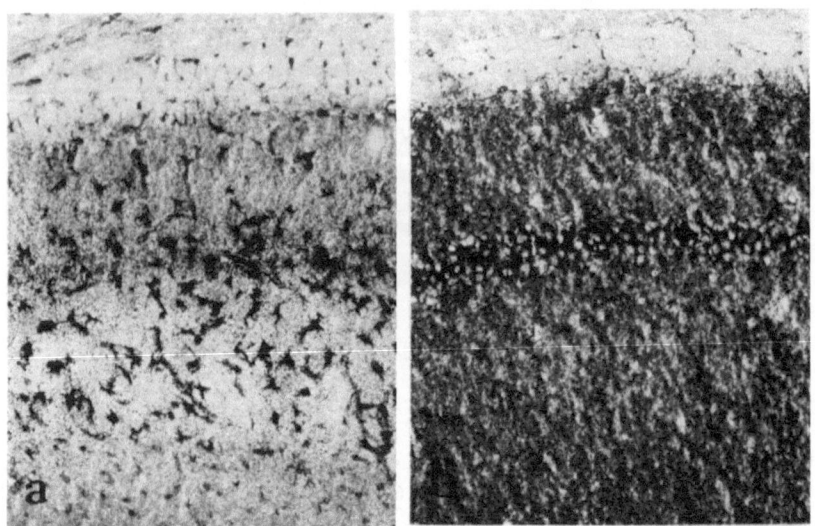

Fig. 10. Area of H2 sector of hippocampus. a. Left (ope-
         rated) side. b. Right (unoperated) side. There
         is conspicious increase of glucose-6-P-dehydro-
         genase within glial elements (mostly astrocytes)
         against the relatively clear background staining
         of the neuropil. Note disappearance of pyramidal
         layer. Both photographs X140.

     The behaviour of AcPh demonstrated no evidence of a
general liberation of lysosomal enzyme into the cytoplasm
during the early development of infarction. Either the
amount of enzyme release was insufficient for early histochemical detection or occurred between 12 and 20 hrs following the reestablishment of the circulation. Significant increase of AcPh, observed within H3 sector's neurons at 20 hrs of recirculation, appeared to be markedly
decreased at 1 week of postischemia with a further decrease tendency. The histochemical reaction observed for
AcPh may be the evidence of an increase in the number of
lysosomes (1, 3, 22) or else an exponent of activated
catabolic processes in the nervous tissue subjected to
temporary ischemia. Our data do not, however, exclude the
possibility of local or in situ activation of AcPh.

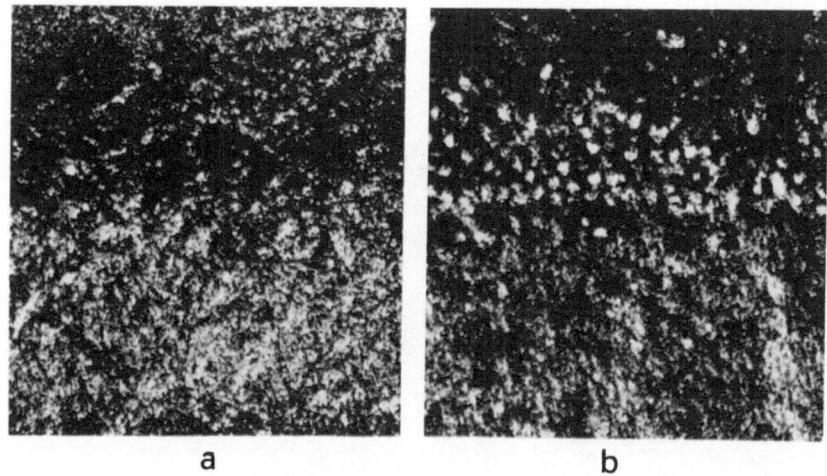

Fig. 11. H3 sector of hippocampus. a. operated side.
         b. unoperated side. There is still an increased
         enzyme activity in perikarya of H3 sector as
         compared to the control side. Neuronal processes
         in H3 sector of ischemic hippocampus stand out
         by their faint accentuation of enzyme activity.
         Both photographs X220.

        In relation to the H2 and H3 sectors of hippocampus,
the activity of the enzymes went in two different direc-
tions; nerve cells of H3 sector showed considerably fast
and distinct increase of the enzymes activity while H2
sector's neurons revealed a delay response and depression
of the enzymatic activities which appeared to be aggra-
vated by increasing recirculation times.

        The H2 sector of the hippocampus was severely and
irreversibly damaged in our experiment. An evident loss
of AcPh from the lysosomes, attributed to the irreversible
neuronal damage of the H2 sector, was also observed in
advanced cerebral ischemia (21). Our finding related to
the hippocampal H2 sector supports the study of Spataro

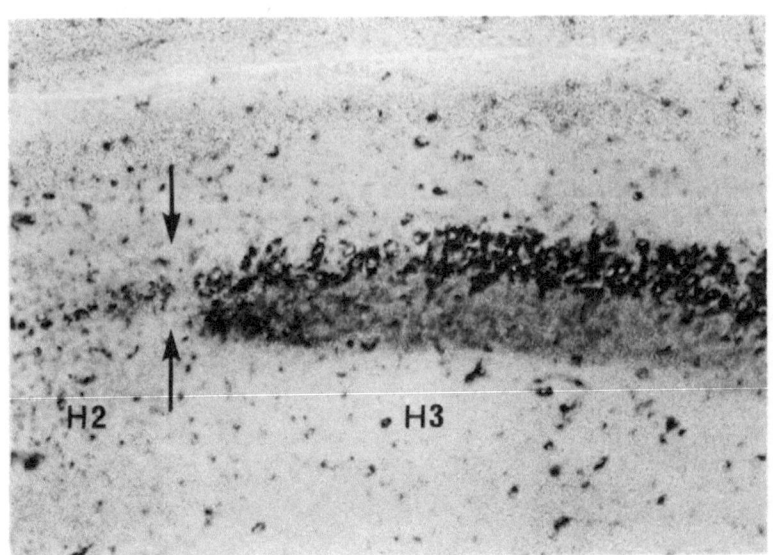

Fig. 12. Acid monophosphatase. Fresh frozen section. The
         left hippocampus in a gerbil subjected to 15 min
         occlusion and 30 days release. Complete disappe-
         arance of perykaria in H2 sector. In the H3
         sector, most of the neurons revealed acid mono-
         phosphatase reaction product resembling control,
         although occasionally there is some reduction in
         the number of the neurons.X130.

et al. (30), who demonstrated that the lysosomes are markedly decreased in anoxic ischemic encephalopathy. Thus,
the lysosomes appear to be less susceptible to ischemia
than the other neuronal components, and the alterations
that did develop occurred at a time when cell damage was
already severe and irreversible. Our findings support the
results of Clendenon et al. (7) and Little et al. (18)
and suggest that lysosomes do not play an important role
in the development of the early ischemic changes ultimately progressing to necrosis. Besides H2 sector, the
other layers located above and below the H2 sector of
the pyramidal band were histochemically altered. Neuropil, as a high vulnerable compartment of the brain tissue, revealed an opposite histochemical response. There
was an abundant AcPh activity within the neuropil due to
the lysosomes' content. At the same time, marked loss of

dehydrogenases from the altered neuropil was in contrast
to homogenous distribution of enzyme activities in controls. Another characteristic feature of both the H2
sector and the correspondent layers of ischemic hippocampus was the extensive glial reaction, including astrocytic hypertrophy and reactive microglial changes, displaying prominent increases in formazan deposits. This
is in agreement with hypertrophic astrocytes and reactive
microglial cells in white matter seen after different
types of edema associated with brain injury (8, 20, 27).

Another group of histochemical differences as compared to the control state consists of the changes in the
activity of oxidative-reductive enzymes, manifested in an
enhancement of G-6-P-DH, SDH, LDH and GDH activity. Enhanced activity of SDH, an enzyme considered to be localized exclusively in mitochondria (25, 28) appears to be
due to the disrupted mitochondria. This is in accordance
with the ultrastructural study on the reactive change
which revealed that mitochondriae, located in the central
part of the perikaryon, varied widely in their sizes and
shapes, some showing bizarre, even triangular contours
(3). Also, damage to the mitochondria may cause an increase in the mitochondrial surface area, thus facilita-
ting the contact between the substrate and the enzyme
(15). As a consequence of this process a spurious increase in enzyme activity may appear. On the other hand,
the increase in succinate dehydrogenase may be explained,
at least in part, as an evidence of oxygen metabolism
activation occurring in the postischemic periods (10).
In addition, there was faint reduction in the staining of
the neuropil, within the layers situated below the H3
sector, observed even at the 30th day of recirculation.
Hence, dendritic processes, which remained stained, appeared
more prominent.

TPP-ase, as the most commonly employed substrate in
the determination of nucleoside diphosphatase activity,
is thought to provide specific marking of the Golgi
apparatus in most cell types (22). According to our light
histochemical investigation, the structural equivalent of
disappeared TPP-ase activity in neurons undergoing "reactive change" was due to disappeared Golgi complex confirmed by ultrastructural observations (3). On the other
hand, increased activity of TPP-ase might be related to
the unaffected neurons of H3 sector which showed unchanged
or even hypertrophic Golgi networks. Only in a few instances a reduced TPP-ase staining was reported in light
microscopic (2, 10, 29) or in combined light and electron
microscopic studies (16).

In conclusion, a study of the histochemical disturbances observed in the pyramidal band of the hippocampus
during regional cerebral ischemia indicates that the
development of the enzymatic alterations is the function
of two factors: ischemia per se and post-ischemic periods.

REFERENCES

1. Barron, K.D., Sklar, S. (1961): Response of lysosomes
   of bulbospinal motoneurons to axon section. Neurology
   (Minneap.) 11: 866-875.

2. Barron, K.D., Tunkboy, T.O. (1962): The histochemistry
   of acid phosphatase and thiamine pyrophosphatase during axon regeneration. Amer. J. Path. 40: 637-652.

3. Bubis, J.J., Fujimoto, T., Ito, U., Mršulja, B.J.,
   Spatz, M. and Klatzo, I. (1976): Experimental cerebral
   ischemia in Mongolian gerbils V. Ultrastructural changes in H3 sector of the hippocampus. Acta neuropathol.
   (Berl.) 36: 285-294.

4. Burstone, M.S. (1958): Histochemical demonstration of
   acid phosphatase. J. Nat. Cancer Inst. 21: 523-539.

5. Burstone, M.S. (1960): Histochemical demonstration of
   cytochrone oxidase with new amine reagent. J. Histochem. Cytochem. 8: 63-70.

6. Burstone, M.S. (1961): Histochemical demonstration of
   phosphatases in frozen sections with naphthol AS-phos-
   phates. J. Histochem. Cytochem. 9: 146-153.

7. Clendenon, N.R., Allen, N., Komatsu, T., Liss, L. and
   Gordon, W.A. (1971): Biochemical alterations in the
   anoxic-ischemic lesion of rat brain. Arch. Neurol.
   (Chic.) 25: 432-448.

8. Friede, R.L. (1962): The cytochemistry of normal and
   reactive astrocytes. J. Neuropath. exp. Neurol. 21:
   471-478.

9. Friede, R.L. (1968): Mappings of oxydative enzymes in
   the brain. In: Pathology of the Nervous System, ed.
   J. Minckler, McGraw-Hill, New York, 306-320.

10. Gadamski, R., Eustachiewicz, R. (1974): Histochemical
    changes in medulla oblongata in rabbit caused by
    circulatory hypoxia (ischemia). Neuropath. Pol. 12:
    603-615.

11. Hess, R., Scarpelli, D.G., Pearse, A.G.E. (1958):
    The cytochemical localisation of oxydative enzymes.
    Piridine nucleotide linked dehydrogenases. J. Biochem.
    Cytol. 4: 101-110.

12. Ito, U., Spatz, M., Walker, J.T. Jr., Klatzo, I.(1975):
    Experimental cerebral ischemia in Mongolian gerbils.
    I. Light microscopic observations. Acta Neuropath.
    (Berl.) 32: 209-223.

13. Kahn, K. (1972): The natural course of experimental
    cerebral infarction in the gerbil. Neurology 22: 510-
    515.

14. Klatzo, I. (1975): Pathophysiological aspects of ce-
    rebral ischemia. The Nervous System, et. Tower, D.B.,
    Raven Press, New York, Vol. I. The Basic Neurosciences
    313-322.

15. Kozik, M. (1972): Doswiadczalny obrzek neuronu w badan
    histienzymatycznych. Neuropat. Pol. 10: 1-15.

16. László, I., Knyihar, E. (1975): Electron histochemis-
    try of thiamine pyrophosphatase activity in the neuro-
    nal Golgi apparatus observed after axotomy and trans-
    neuronal deprivation. J. Neural Transmiss. 36: 123-141.

17. Levine, S., Sohn.D. (1969): Cerebral ischemia in infant
    and adult gerbils. Relation to incomplete circle of
    Willis. Arch. Pathol. 87: 315-317.

18. Little, J.R., Kerr, F.W.L., Sundt, T.M.Jr. (1974):
    The role of lysosomes in production of ischemic nerve
    cell changes. Arch. Neurol. 30: 448-455.

19. Maslinska, D. Thomas, E. (1975): Enzyme histochemical
    studies of "retrograde" reaction in motor neurons of
    immature rats. Acta Neuropath. (Berl.) 33: 317-323.

20. Mossakowski, M.J. (1963): The activity of succinic de-
    hydrogenase in the reactive glia. Acta Neuropath.
    (Berl.) 2: 282-290.

21. Mršulja, B.J., Spatz, M., Walker, J.T. Jr., Klatzo,I.
    (1977): Histochemical investigation on the Mongolian
    gerbils brain during unilateral ischemia (submitted
    to Acta Neuropath. Berl.).   ·

22. Novikoff, A.B., Essner, E. (1962): Pathological chan-
    ges in cytoplasmic organelles. Fed. Proc. 21: 1130-
    1142.

23. Novikoff, A.B. (1963): Lysosomes in the physiology
    and pathology of cells. Contributions of staining
    methods. Ciba Foundation Symposium on Lysosomes, eds.
    A.V.S. de Reuck, M.P.L. Cameron, Little, Brown Comp.
    Boston, 36.

24. Novikoff, P.M., Novikoff, A.B., Quintana, N. and Hauw,
    T.T. (1971): Golgi apparatus, GERL and lysosomes of
    neurons in rat dorsal root ganglia, studied by thick
    section and thin section cytochemistry. J. Cell Biol.
    50: 859-886.

25. Ogawa, K., Barrnett, R.J. (1965): Electron cytochemi-
    cal studies of succinic dehydrogenase and dehydro-
    nicotineamideadenine dinucleotide diaphorase activi-
    ties. J. Ultrastruct. Res. 12: 488-508.

26. Rose, M. (1931): Journal f. Psychol. u. Neurol. 43:
    493-440. Tafeln 9-14.

27. Rubinstein, L.J., Klatzo, I., Wiquel, J. (1962): His-
    tochemical observation on oxidative enzyme ectivity
    of glial cells in a local brain injury. J. Neuropath.
    exp. Neurol. 21: 116-136.

28. Scarpelli, D.G., Pearse, A.G.E. (1958): Cytochemical
    localisation of succinic dehydrogenase in mitochon-
    dria. Anat. Rec. 132: 133-152.

29. Sederholm, U. (1965): Histochemical localisation of
    esterases, phosphatases and tetrasolium reductases in
    the motor neurons of the spinal cord of the rat and
    the effect of nerve division. Acta physiol. scand.
    65, Suppl. 256.

30. Spataro, J. (1966): Anoxic-ischemic encephalopathy
    of the rat brain. Exptl. Neurol. 16: 16-27.

31. Spielmeyer, W. (1922): Histopathologie des Nerven-
    systems. Springer, Berlin 74-79.

32. Van Harreveld, A., Trubatch, J. (1974): Reflex figu-
    res during asphyxial rigidity. Expl. Neurol. 45: 161-
    173.

CEREBRAL WATER AND ELECTROLYTE CONTENT FOLLOWING

ISCHEMIA AND BLOOD-BRAIN BARRIER DISTURBANCES

Hanna M. Pappius, T. Fujimoto, K. Nishimoto,
I. Klatzo, and Maria Spatz

Donner Laboratory of Experimental Neurochemistry
Montreal Neurological Institute
McGill University
Montreal, Canada
                    and
Laboratory of Neuropathology and
  Neuroanatomical Sciences
National Institute of Neurological and
  Communicative Disorders and Stroke
National Institutes of Health
Bethesda, Maryland, U.S.A.

INTRODUCTION

As part of a series of studies on the effects of carotid occlusion in the gerbil (1-3, 5), water and electrolyte contents of the affected hemisphere were determined following 1 h of unilateral occlusion and different periods of recirculation. The results were compared with effects of air embolism (6) on the same parameters to determine whether some of the observed changes were associated generally with opening of the blood-brain barrier or were a more specific consequence of ischemic insult.

METHODS

In the experiments on the effects of carotid occlusion the gerbils were operated on under light pentobarbital anaesthesia (3mg/100g, i.p.). The left common carotid artery was occluded at the neck with a Heiffetz clip for 1 h. Only animals showing severe neurologic signs of cerebral impairment were used (4). The animals were killed by decapitation at different intervals of time after the clip was released. The brain was removed rapidly and the affected hemisphere was divided into a larger part, corresponding to the tissue supplied by the middle cerebral artery, and a smaller frontal part commonly showing less injury after occlusion of the common carotid artery. The present report deals only with results obtained in the most affected cerebral tissues, compared to those obtained in comparable tissue samples from normal animals. The results in sham-operated animals were the same as in the normal group.

In experiments on the effects of air embolism 0.2 ml of air was injected into the left carotid artery of anaesthetized animals (pentobarbital, 3 mg/100 g, i.p.). Animals were killed by decapitation 30 min and 2, 5 or 24 h after the injection of air. The whole affected hemispheres were analyzed in these experiments.

The tissue samples were weighed, dried to constant weight at $100^{\circ}C$ and reweighed to obtain the dry weight percent. The dried samples were digested in concentrated nitric acid and the digests analyzed for sodium and potassium on a Technicon Autoanalyzed Flamephotometer as described previously (7).

RESULTS AND DISCUSSION

The effects of 1-h unilateral carotid artery occlusion followed by different periods of release of the occlusion on the percentage dry weight and potassium and sodium content of gerbil brain tissue are summarized in Fig.1.

During the occlusion dry weight decreased from 21.93 to 20.10 mg percent indicating that a significant increase in water content of the tissue had occurred. At the same time sodium content increased from 41.2 to 48.7 meq per kg while potassium decreased from 92.3 to 74.3 meq per kg.

During the first 8 h following the release of the occlusion there was no significant change in percentage dry weight and sodium content of cerebral tissue in the affected areas. At 10 h there was a secondary fall in percentage dry weight and a sharp increase in sodium content, the latter reaching a peak of 79.1 meq per kg after 15 h of recirculation.

The changes in potassium after release of the occlusion were somewhat different in nature. Between 1

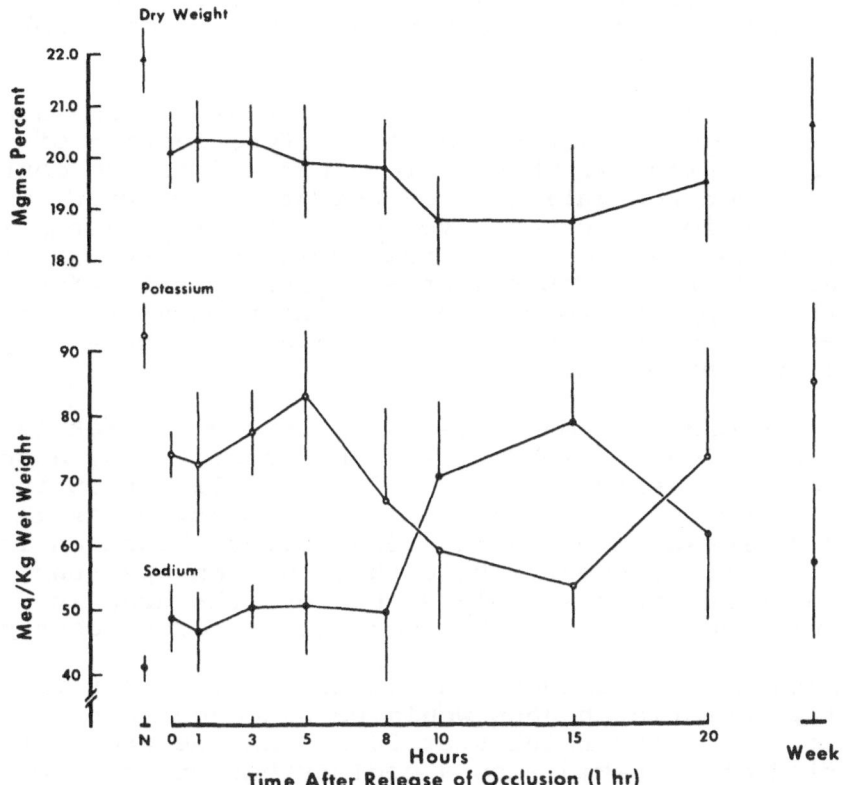

Fig. 1. Percentage dry weight and potassium and sodium
content of gerbil brain after one hour of
unilateral carotid artery occlusion and
different periods of release of the occlusion.

and 5 h following the release of the occlusion there
was a trend toward normal potassium values. When expressed
in terms of dry weight, which corrects for dilutional
effects of the increased water content, ·the potassium
content after 5 h of release was 404 + 34 (av + SD; n=14)
per kg dry weight, no longer significantly different from
the normal value of 421 + 27 (n=14). Between 5 and 15 h
after release of the occlusion there was a sharp loss of
potassium from the tissue (at 15 h potassium content was
53.6 meq per kg wet weight and 282 meq per kg dry weight).

Twenty hours after release of the occlusion a reversal of the earlier changes was seen in percentage dry

weight and especially in sodium and potassium content. However, normal levels were not reached even one week following the occlusion.

These results delineated two distinct phases in post-ischemic changes in water and electrolyte content of gerbil brain tissues. Changes in water and sodium content which had occurred during 1h of unilateral carotid artery occlusion were not reversed during the immediate post-occlusion period, while the potassium level in the affected cerebral tissues increased during this time. Thus, there appears to be a dissociation between changes in water and sodium content on one hand and those in potassium content on the other.

The second delayed phase which reached a peak 15 h after the period of ischemia was characterized by a sharp decrease in potassium content and an equally steep increase in sodium level, with an accompanying definite further decrease in percentage dry weight, hence an increase in water content. While there was an apparent reciprocity in the electrolyte changes, the drop in potassium level preceded the increase in sodium content.

These delayed changes in electrolytes do not reflect necrosis. Necrotic changes would be expected to be irreversible, while the results at 20 h and one week after the period of ischemia show a redistribution of water and electrolytes to immediately post ischemic, although not pre ischemic, levels. The more normal results at 20 h and later cannot be ascribed to longer survival of less affected animals as mortality was not particularly increased between 15 and 20 h.

A very interesting aspect of these results is the fact that the second phase of changes in water and electrolyte content coincides with opening of the blood-brain barrier as demonstrated by increased uptake of protein markers (1, 6). This raises the question whether the delayed second phase of the post ischemic changes is a direct result of the opening of the blood-brain barrier which occurs at the same time or whether the two processes are simultaneous consequences of the ischemic insult.

To shed some light on this problem the effects of air embolism on the same parameters were investigated, since this procedure leads to immediate opening of the blood-brain barrier to protein markers (6).

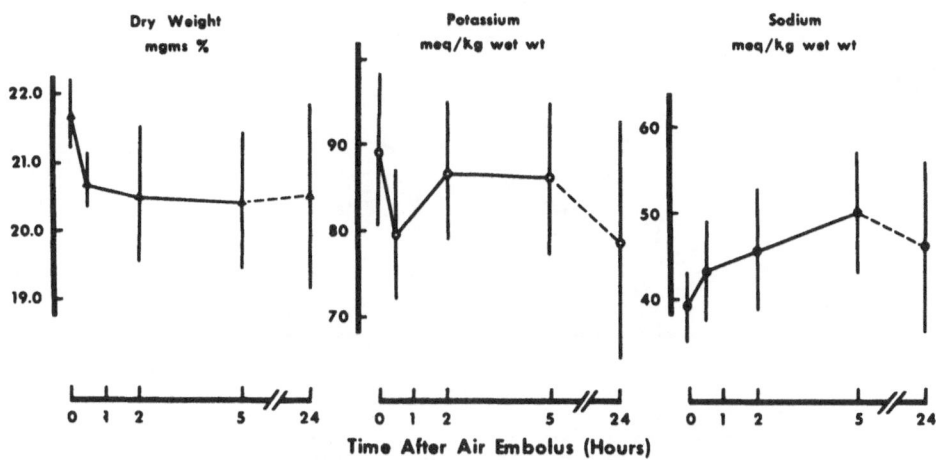

Fig. 2. Effect of air embolism on percentage dry weight
and potassium and sodium content of gerbil brain.

        The results of these experiments, summarized in
Fig.2, show that injection of air into the carotid
artery of the gerbil caused a statistically significant
immediate decrease in dry weight and an increase in sodium. The Potassium content of cerebral tissues showed an
apparent decrease immediately after air embolism; however
the observed change was not statistically significant.
Furthermore, when expressed in terms of dry weight the
values were 403 $\pm$ 37, 424 $\pm$ 7 and 423 $\pm$ 33 meq potassium
per kg at 30 min, 2 h and 5 h after air injection,
respectively. Thus it is quite clear that opening of the
blood-brain barrier to protein markers associated with
air embolism has no effect on potassium content of cerebral tissue and, therefore, that changes observed in
post-ischemic periods are unlikely to reflect generalized
effects of increased blood-brain barrier permeability.

        In Fig. 3, the actual changes in the sodium content of
gerbil brain during the post ischemic period are compared
with changes that may be expected in relation to changes
in the water and potassium content. Two assumptions were
made in calculating the expected changes in sodium. First,
sodium may enter the tissue with the water. If the edema
fluid is derived from the blood and is in equilibrium with
blood as far as its sodium content is concerned, the expected increase in sodium can be calculated from the
increment in water content and the concentration of sodium

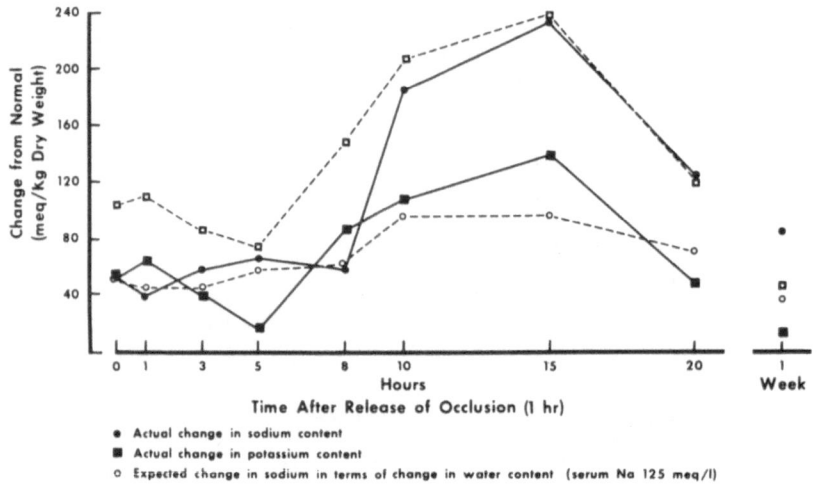

Fig. 3. Actual versus expected changes in sodium content of gerbil brain following ischemia.

in the serum (○). Secondly, in general when potassium is lost from the tissue a replacement by sodium may be expected, if for no other reason than to maintain tonicity of the tissue. If this occurs then the changes in sodium should quantitatively closely approximate those in potassium (■). Finally, it is not unreasonable to expect that sodium may both accompany the influx of water into the tissue and replace the potassium that is lost (□).

It will be seen from Fig. 3, that in the postischemic period initially the changes in sodium agreed very closely with what would be expected in terms of increased water content. During this time there appears to have been no replacement by sodium of the lost potassium. The delayed change in potassium, between 5 and 8 h postischemia, preceeded the secondary increase in sodium (and water) which occurred after 8 h of release of the occlusion. This loss of potassium may represent a sensitive index of delayed but generalized post-ischemic changes in cell membrane permeability. Starting at 8 h postocclusion there appears to have occurred a redistribution of electrolytes in the tissue, with sodium not only increasing with the influx of water but also replacing all of the lost potassium. This would suggest that at the time when the blood-brain barrier permeability to

protein markers is increased the permeability characteristics of cell membranes in general are altered.

SUMMARY

1. Following 1 h of unilateral occlusion of carotid
artery in the gerbil two distinct phases can be distinguished in water and electrolyte changes in the affected
cerebral tissues.

2. Sodium entering the brain during the early postischemic phase appears to be in equilibrium with the
water entering the brain.

3. During the second late phase sodium also appears
to replace potassium lost from the tissue.

4. The second, late, phase coincides with increased
permeability of the blood-brain interface to protein
markers, but does not appear to be a consequence of the
opening of the blood-brain barrier.

5. The secondary loss of potassium from brain tissue
during the post-ischemic period precedes the secondary
increase in both sodium and water, and thus may represent
a very sensitive index of delayed, but rather generalized
changes in cell membrane permeability resulting from
ischemic insult.

ACKNOWLEDGMENTS

The senior author's experimental studies have been
supported by Medical Research Council of Canada Grant
MT-3021 and the Donner Canadian Foundation. Technical
assistance was provided by Mrs. Hanna Szylinger and Mr.
Michael McHugh.

REFERENCES

1. Fujimoto, T., Walker, J.T. Jr., Spatz, M., Klatzo,I.
   (1976): Pathophysiologic aspects of ischemic edema. In:
   Dynamic aspects of Brain Edema. H.M. Pappius and W.
   Feindel (Eds), Springer Verlag, Berlin-Heidelberg,
   171-181.

2. Ito, U., Go, K.G., Walker, J.T. Jr., and Klatzo, I.
   (1976): Experimental cerebral ischemia in Mongolian
   Gerbils. III. Behavior of the Blood-Brain Barrier.
   Acta Neuropathol. (Berlin), 34: 1-6.

3. Ito, U., Spatz, M., Walker, J.T. Jr., Klatzo,I.(1975):
   Experimental cerebral ischemia in Mongolian Gerbils.
   I. Light Microscopic Observations. Acta Neuropath.
   (Berlin) 32: 209-223.

4. Kahn, K. (1972): The natural course of experimental
   cerebral infarction in the Gerbil. Neurology (Minneap.)
   22: 510-515.

5. Mršulja, B.B., Mršulja, B.J., Ito, U., Walker, J.T.R.,
   Spatz, M. and Klatzo, I. (1975): Experimental cerebral
   ischemia in Mongolian Gerbils. II Changes in Carbohydrates. Acta Neuropath. (Berlin) 33: 91-103.

6. Nishimoto, K., Kakari, S., Pappius, H.M., Spatz, M.,
   Walker, J.T. Jr., Klatzo, I. (1978): Behaviour of the
   blood-brain-barrier in cerebral ischemia. This Symposium.

7. Pappius, H.M. and Dayes, L.A. (1965): Hypertonic urea.
   Its effect on the distribution of water and electrolytes
   in normal and edematous brain tissues. Arch. Neurol. 13:
   395-402.

BEHAVIOR OF THE BLOOD-BRAIN BARRIER (BBB) IN CEREBRAL ISCHEMIA

K. Nishimoto, S. Kakari, H. Pappius, M. Spatz,
J. T. Walker, Jr. and I. Klatzo

Laboratory of Neuropathology and
   Neuroanatomical Sciences
National Institute of Neurological and
   Communicative Disorders and Stroke
National Institutes of Health
Bethesda, Maryland 20014, U.S.A.

Although a stationary occlusion of an artery, as well as the introduction of an embolic obstruction into the cerebral circulation can both result in ischemic damage to the brain parenchyma, a number of observations indicate some profound differences regarding the effect of these two conditions on the behavior of the BBB. Whereas after occlusion of the carotid artery in the gerbil the time of the BBB breakdown can be considerably delayed depending on the duration of ischemic insult (1,2), an abnormal permeability of the BBB to various tracers almost immediately follows the passage of emboli (4).

Considering the BBB as a complex phenomenon aimed at assuring an optimal homeostatically controlled environment for brain parenchyma, it can be assumed that dysfunction of various BBB mechanisms can greatly influence the effects of ischemia upon the brain tissue. Such an assumption prompted the present investigation in which the effects of carotid occlusion and air embolism in gerbils were evaluated comparatively with regard to various parameters of BBB dysfunction, disturbances of regional cerebral blood flow (rCBF) and regional changes in glucose utilization by the brain tissue.

MATERIALS AND METHODS

Cerebral ischemia was produced in Mongolian gerbils (Meriones unguiculatus) by clamping the left common carotid artery for different periods of time, the animals being sacrificed at certain time intervals following release of the occlusion.

For introduction of air embolism, 0.03 - 0.05 ml air was injected via the left external carotid artery of gerbils near the bifurcation of the common carotid so that the air was propelled via the internal carotid artery into the brain. In both models various parameters of injury were studied in groups, each consisting of at least ten gerbils.

Water content of the brain tissue was measured either by the specific gravity procedure of Nelson et al. (6) or by wet/dry brain tissue weight ratios.

The BBB changes were assessed with Evans blue (EB), sodium fluorescein (NaF) and C^{14} sucrose (0.2 - 0.5 ml of solution containing 10-50 μCi) tracers administered intravenously 1-30 minutes before the sacrifice of animals at the various time intervals following ischemic insults.

The rCBF was evaluated autoradiographically by the method of Reivich et al. (7) using C^{14} antipyrine (0.2 ml of solution containing 10 μCi) 1 minute before the sacrifice. Regional utilization of glucose was studied by the method of Sokoloff et al. (8) injecting C^{14} 2-deoxy-D-glucose (0.2 ml of solution containing 10 μCi) 45 minutes before the sacrifice.

To compare localization of various tracers simultaneously in the same sections, one of the fluorescent (EB or NaFl) tracers was injected along with one of the C^{14}-labeled substances (sucrose, antipyrine or 2-deoxy-D-glucose). After sacrificing the animals by decapitation the brains were rapidly frozen and the frozen sections (10μm) were mounted on slides and rapidly dried. The photographs were taken under the oblique beam of an HBO 200 mercury vapor lamp provided with blue exciter filters, with a Nikon camera. The same sections were then covered with Kodak film and the autoradiographs were developed approximately 10 days later (9).

RESULTS

Evaluation of <u>brain water content</u> revealed within 5 minutes following carotid occlusion or air embolism a significant increment in water in the affected left hemisphere. The water increment remained at approximately the same level up to 5 hours following release of 1-hour-long occlusion or injection of air. After longer time intervals 3 groups of animals became distinguishable: 1) gerbils in which the increased water content remained unchanged, 2) gerbils in which water content in the left hemisphere returned to the normal level, and 3) animals which showed a drastic further increase in water in the affected hemispheres (Fig. 1). A histopathological study indicated that the affected hemispheres of the animals in the third group contained marked necrotic areas.

The observations on <u>BBB disturbances</u> showed definite differences in duration and intensity with regard to extravasation of individual tracers and this was especially apparent in cerebral air embolism. In this condition, the abnormal leakage of all three BBB tracers was evident in animals sacrificed even 2 minutes after air injection.

Otherwise, the extravasation of EB was mostly rather circumscribed and often in a definite topographical relationship to some blood vessels, whereas the leakage of NaFl and C^{14} sucrose was more diffuse and widespread. In air embolism three hours after air injection an abnormal passage of EB ceased to be observed. On the other hand, the extravasations of the NaFl and C^{14} sucrose were pronounced on the left side even 24 hours after air injection, the sucrose showing definitely more extensive areas of abnormal penetration than the NaFl (Fig. 2).

In experiments with occlusion of the left common carotid artery the permeability of the BBB to EB revealed the previously described (5) "maturation" phenomenon, i.e., some delay in appearance of EB extravasation depending on the intensity of ischemic insult (Fig. 3). As this figure indicates, the opening of the BBB to EB appeared earlier and was more intense in the basal ganglia than in the cortex. Otherwise, the transitory character of the BBB breakdown to EB could not be related either to CBF changes or to necrotic deterioration of the brain parenchyma. Fig. 4 indicates that in the basal ganglia the necrotic changes could be seen at a 5-hour release interval sometimes preceding the EB extravasations. Otherwise, in gerbils sacrificed 20 hours after release of occlusion the necrotic changes persisted, even with greater severity,

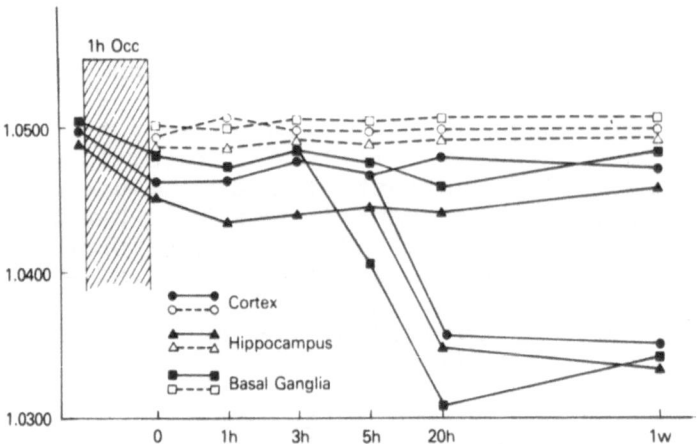

Fig. 1. Specific gravity measurements after one hour occlusion of the left carotid artery and at various periods following clip release. Interrupted lines refer to the right hemisphere.

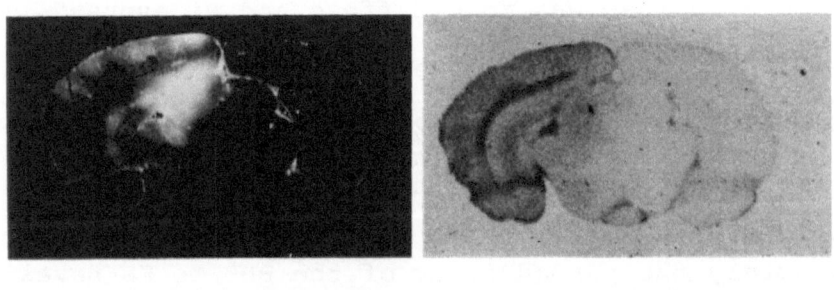

a   b

Fig. 2. Fluorescent picture (a) and autoradiograph (b) from the same brain section of a gerbil sacrificed 24 hrs following air injection into the left internal carotid artery. The distribution of NaFl in the left hemisphere is regionally circumscribed, whereas that of C^{14} sucrose is more widespread and diffuse.

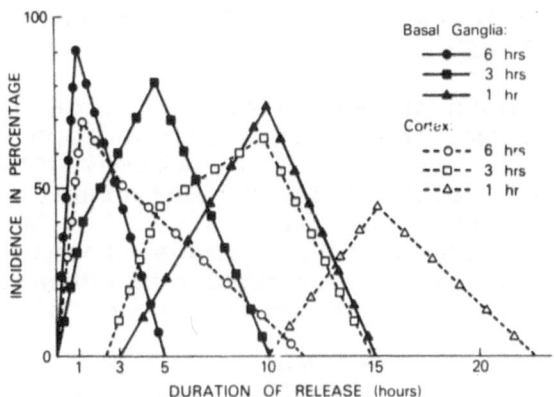

Fig. 3. Incidence of BBB injury after various periods
of occlusion and release. Extravasation of EB
appears stronger and earlier after longer occlusion periods. Also the BBB injury is stronger
and earlier in the basal ganglia than in the
cerebral cortex.

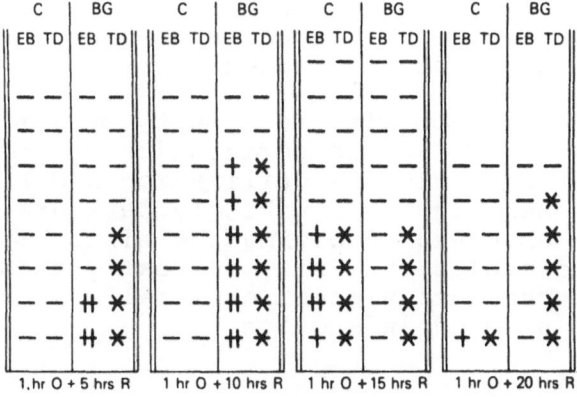

Fig. 4. Appearance of histologically demonstrable necrotic
changes (*) and BBB damage (+ or ++) in individual
gerbils sacrificed after one hour occlusion and
various release periods. (C = cortex; BG = basal
ganglia; EG = Evans blue extravasation; TD = tissue
damage).

although these animals no longer showed an abnormal permeability of the BBB to EB. The C^{14} antipyrine autoradiographs at this release period indicated an increased severe histopathologic changes in the basal ganglia.

Otherwise, both in occlusion of the carotid artery
and in air embolism, there was no clear topographical relationship between areas showing abnormal rCBF and regions indicating extravasation of BBB tracers. Although
occasionally, the regions with reduced C^{14} density corresponded roughly to areas of BBB damage, in many instances
regions with conspiciously reduced or increased C^{14} marker
showed no evidence of an abnormal passage of BBB tracers.

The autoradiographic assay of local cerebral glucose
utilization revealed in both models abnormal patterns
characterized by areas of conspicuously increased or
reduced grain density which showed no spatial relationship to regions marked by abnormal passage of BBB tracers.
In air embolism, the areas of increased grain density frequently involved in a diffuse fashion certain anatomical
regions, especially the hippocampus (Fig. 6) and cerebral
cortex, also on the side contralateral to air injection.
Otherwise, conspicious very dark, small foci could be
seen scattered randomly in various locations in the gray
matter. Areas of increased grain density, located usually
in the cerebral cortex on the side opposite to the air
injection, could be seen occasionally in the animals
sacrificed 3 days after air embolism, when no abnormal
penetration of BBB tracers was demonstrable.

Similarly, in ischemia produced by occlusion of the
carotid artery, there was no correlation between areas with
abnormal patterns of glucose utilization and regions with
disturbed BBB. Here again, strikingly dark grain density
was frequently observed involving the hippocampus on the
side of carotid occlusion. Foci of greatly reduced grain
density were occasionally seen, especially in the basal
ganglia on the side of infarction.

DISCUSSION

The observations presented confirm certain differences in the behaviour of the BBB with regard to ischemic
injury produced by occlusion of a major artery or by
passage of embolic material through the cerebral circulation. The reason for almost instantaneous breakdown of
the BBB in embolism and for a delayed appearance of BBB

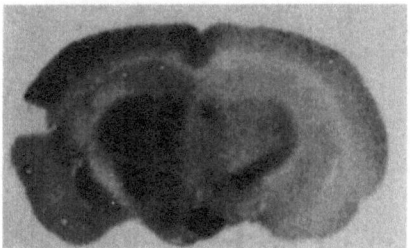

Fig. 5. C^{14} antipyrine autoradiograph of the brain of a gerbil which was sacrificed 20 hours after release of one hour left carotid occlusion. Note the increased radioactivity in the hemisphere on the side of occlusion.

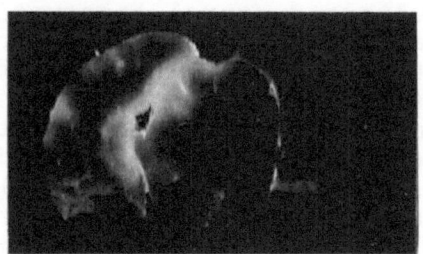 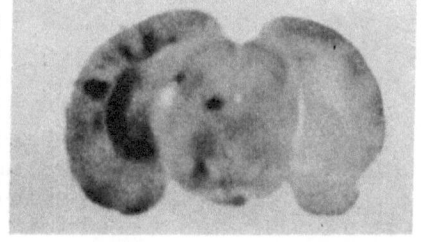

a   b

Fig. 6. Fluorescent picture (a) and autoradiograph (b) from the same brain section of an animal sacrificed 24 hours after air embolism. The NaFl pattern of abnormal passage (left) has no spatial relationship to increased C^{14} 2-deoxy-D-glucose marker visible in the hippocampus and focally in the cerebral cortex and basal ganglia (right).

injury in occlusion of the carotid artery remain at the moment unknown. If, as is suggested by numerous studies, leakage of protein tracers such as EB-albumin complex or horseradish peroxidase depends promarily on abnormally increased vesicular transport across the endothelium, some differences in factors stimulating pinocytotic activity within the endothelium must be assumed. Interestingly, these factors seem to be largely independent of CBF or of damage to surrounding brain parenchyma. Our observations indicated that, after release of carotid occlusion, extravasation of EB stopped at a certain point, whereas the parenchymatous tissue damage was progressing further and there was evidence of rather increased CBF in the ischemic hemispheres. The fact that an increased permeability to micromolecular tracers, such as C^{14} sucrose or NaFl, persists long after abnormal leakage of protein tracers is no longer demonstrable and suggests that an abnormal passage of these micromolecular substances may be related to some mechanism other than a pinocytotic one.

The relationship of BBB dysfunction to changes in CBF remains obscure. It is obvious that in necrotic or edematous regions the CBF may be severely reduced or abolished, thus restricting the entry of BBB tracers. On the other hand, previous ultrastructural observations on ischemia (10) revealed that greatly increased pinocytotic transport of peroxidase could be demonstrated in the endothelium of blood vessels surrounded by brain parenchyma showing severe damage and edematous changes. Correspondingly, in the present study, the regions with reduced CBF showed either no passage or abnormal leakage of BBB tracers. The relationship of BBB changes to an increased CBF was also not clear, since the areas with an increased rCBF were rarely seen in our experimental models. Otherwise, in previous studies on the effect of systemic hypertension on ischemic injury (3) the gerbils in which systemic blood pressure was elevated for one hour following release of occlusion showed an increased rCBF, as well as marked BBB damage, and severe histopathologic changes in the hemisphere on the side of occlusion, whereas in normotensive animals with similar ischemic insult the corresponding hemispheres showed a decrease in rCBF, minimal BBB damage and moderate histopathologic changes.

An ischemic brain tissue injury, whether due to arterial occlusion or due to air embolism, produces a major disturbance of energy metabolism, especially with regard to consumption of glucose. Although in Sokoloff´s procedure (6) autoradiography visualizes essentially the C^{14} label on 2-deoxy-glucose-6-phosphate, which is trapped in the brain tissue following phosphorilation of 2-deoxy-D-glucose by hexokinase, it has been assumed that autoradio-graphic images correspond to relative rates of regional glucose consumption in initial metabolic steps involving phosphorylation. In the present studies autoradiographs in both models revealed areas of greatly increased grain density, as well as some very pale areas. To the degree that the pale areas appeared to be related to regions showing severe tissue damage, the dark areas were observed in histologically well preserved tissue and without any relationship to CBF or BBB changes. They were conspicuous as sharply defined random foci or showed a predilection to certain cytoarchitectonically defined structures, such as hippocampus or cerebral cortex, sometimes on the side opposite to ischemic insult. The latter areas of increased grain density were suggestive rather of some relayed sti-mulation of glucose consumption in these structures than of focal circulatory disturbances. Such an interpretation would then suggest that even some distal areas may react metabolically to a focal ischemic injury.

REFERENCES

1. Fujimoto, T., Walker, J.T.,Jr., Spatz, M., and Klatzo, I. (1976): Pathophysiologic aspects of ischemic edema. In: Dynamics of Brain Edema, H.M. Pappius and W. Feindel (Eds.) Springer-Verlag, Berlin-Heidelberg-New York, 171-180.

2. Ito, U., Go, K.G., Walker, J.T. Jr., Spatz, M. and Klatzo, I.(1976): Experimental cerebral ischemia in Mongolian gerbils. III.Behavior of the blood-brain barrier. Acta Neuropath. 34: 1-6.

3. Ito, U., Mršulja, B.B., Fujimoto, T., Klatzo, I. and Spatz, M.: Effects of hypertension on the post-isch-emic brain following temporary ischemia in gerbils. In: Pathophysiological, Biochemical and Morphological As-pects of Cerebral Ischemia and Arterial Hypertension. Proc. Intern. Symposium Neuropathologia Polska. (in press).

4. Johansson, B.(1974): Damage to the blood-brain barrier in experimental gas embolism. In: L´Embolie Gazeuse du Système Carotidien, G. Arfel and R. Naquet (Eds.) Doin, Paris, 165-170.

5. Klatzo, I.(1975): Pathophysiologic aspects of cerebral ischemia. In: The Nervous System. Vol. I: The Basic Neurosciences, D.B.Tower (Ed.) Raven Press,NY,313-322.

6. Nelson, S.R., Manz, M.L., Maxwell, J.A. (1971): Use of specific gravity in the measurement of cerebral edema. J. Appl. Physiol. 30: 268-271.

7. Reivich, M., Jehle, J.W., Sokoloff, L. and Kety, S.S. (1969): Measurement of regional cerebral blood flow with antipyrine-C^{14} in awake cats. J. Appl. Physiol. 27: 296-300.

8. Sokoloff, L., Reivich, M., Kennedy, C., Des Rosiers, M.H., Patlak, C.S., Pettigrew, K.D., Sakurada, O. and Shinohara, M. (1977): The [^{14}C] deoxyglucose method for the measurement of local cerebral glucose utilization: theory, procedure, and normal values in the conscious and anesthetized albino rat. J. Neurochem. 28: 897-916.

9. Steinwall, O., and Klatzo, I. (1966): Selective vulnerability of the blood-brain barrier in chemically induced lesions. J. Neuropathol. Exp. Neurol. 25: 542-559.

10. Westergaard, E., Go, G.K., Klatzo, I. and Spatz, M. (1976): Increased permeability to horseradish peroxidase induced by ischemia in Mongolian gerbils. Acta Neuropath. 35: 307-325.

REEMPHASIS OF THE ROLE OF 5-HYDROXYTRYPTAMINE

IN CEREBRAL ISCHEMIA

K.M.A. Welch, Eva Chabi, R.J. Gaudet
and T.-P. Wang

Laboratory of Clinical and Experimental
    Cerebral Metabolism
Baylor-Methodist Center for Cerebrovascular
    Research
Department of Neurology, Baylor College of
    Medicine, Houston, Texas 77030, U.S.A.

SUMMARY

Cortical monoamine changes during ischemic episodes
of up to one hour were studied in the gerbil. Norepinephrine (NE) levels decreased after 60 minutes in the
occluded hemisphere of animals with stroke, but dopamine
(DA) levels were unaltered. 5-Hydroxytryptamine (5-HT)
levels became bilaterally reduced in animals with and
without stroke as soon as 5 minutes after occlusion.
Abnormal motor activity suggestive of seizure developed
in some animals that exhibited signs of stroke. Accordingly, the influence of seizure activity on cortical
monoamine changes was studied in a second experiment.
Reduction of DA and NE was only observed when seizures
occurred in association with stroke. 5-HT levels were
reduced bilaterally in animals with and without signs of
stroke and reduced further in animals with stroke plus
seizure. We concluded that seizure activity must be taken
into account when examining the mechanism of disordered
catecholamine metabolism in the gerbil stroke model.
However, 5-HT metabolism appears disordered in ischemic
brain independent of seizure activity.

It was in 1971, after observing 5-hydroxytryptamine
release from brain into cerebral venous blood during
acute ischemia in the primate, that we first confirmed a

suspected disorder of monoamine neurotransmitter metabolism in the pathogenesis of stroke (18) and subsequently performed experiments which indicated that the vaso-
constrictor effect of 5-hydroxytryptamine release onto
intraparenchymatous cerebral vessels in ischemic brain
foci might be responsible for impairment of potential
collateral circulation and for progression of ischemia
(17). Since this time, reports from a number of workers
who have studied prolonged ischemia in the gerbil have
emphasized alterations in catecholamine metabolism, particularly dopamine, as a contributing factor to the
pathogenesis of stroke (5, 9, 20). Although our own studies in the gerbil also revealed catecholamine depletion
in brain cortex after prolonged ischemia (16), two recent
studies reported here and in part elsewhere (1, 19) have
served to emphasize more the role of 5-hydroxytryptamine
changes, particularly during brief and transient cerebral
ischemia.

MATERIALS AND METHODS

Adult male and female Mongolian gerbils (*Meriones
unguiculatus*) weighing 50 to 80 gm were studied. Animals
were caged (three per unit) at constant temperature (22°C)
in simulated day and night conditions and allowed free
access to drinking water and chow.

The first experiment was designed to study periods
of up to one hour of ischemia. The right common carotid
artery (CCA) was dissected free of its accompanying vagus
nerve and jugular vein and occluded by the application
of two miniaturized Heifetz aneurysm clips as gerbils
recovered from brief ether anesthesia. The total time
from induction of anesthesia to operation was never more
than 5 minutes. Animals that exhibited the clinical signs
of ischemia, e.g., ipsilateral circling, splaying of
contralateral limbs, contralateral ptosis, rolling fits
and tonic-clonic convulsions (2, 7, 12, 16) were placed
in the stroke group. If an animal's classification was
in doubt, it was not included in the study. Five minutes
after CCA occlusion was the earliest time that clinical
signs of ischemia could be detected with accuracy. Accordingly, this was the shortest interval of temporary ischemia studied.

Groups of animals (up to 20 in each) were sacrificed
under liquid nitrogen at intervals of 5, 30 or 60 minutes
of occlusion. Sham-operated animals, in which the right
CCA was isolated but not occluded, were sacrificed at

time intervals of 0, 35 and 60 minutes after operation
and used as controls.

In a second experiment, the right CCA was occluded
and all animals were observed for behavioral signs of
ischemic neurological deficit as in the first experiment.
Animals that exhibited ipsilateral deviation of head and
neck plus circling activity, as well as contralateral
ptosis, diminished limb movement, splaying of limbs
and hemisensory neglect, were assigned to the stroke group.
However, on this occasion animals that in addition exhibited abnormal motor behavior that suggested seizure
activity, i.e., wild running, whole body myoclonus, focal
clonic limb movement, standing seizures, rolling fits and
tonic-clonic convulsions (1, 12, 16) were placed in a
stroke-plus-seizure group. If an animal's classification
was in doubt it was not included in the study.

Groups of gerbils with no stroke (N = 46), stroke
(N = 19) or stroke plus seizure (N = 21) were sacrificed
by direct immersion under liquid nitrogen at one and
three hours after operation. Sham-operated animals in
which the right CCA was isolated but not ligated were
sacrificed at identical time intervals.

In both experiments, after the frozen brains were
chiseled out under liquid nitrogen, samples of cerebral
cortex from the right and left hemispheres were analyzed
for 5-hydroxytryptamine (5-HT), dopamine (DA) and norepinephrine (NE) as previously described (16).

For experimental data in the first experiment an
analysis of variance was performed and the values were
compared with sham values from similar post-operation
time intervals. In the second experiment statistical
analysis was performed using Student's t-test.

RESULTS

In experiment one, monoamine levels were unaltered
in sham-operated controls sacrified at similar time intervals to occluded animals and there was no difference
between values from right and left hemispheres.

Cortical monoamine levels during right CCA occlusion
for intervals up to 60 minutes are shown in Fig. 1. DA
was not significantly altered in either the group with
stroke or without stroke. However, the fall in DA levels
five minutes after occlusion to 1.105 $\pm$ 0.459 S.D. (μg/
gm brain tissue) against 1.394 $\pm$ 0.196 S.D. (μg/gm brain

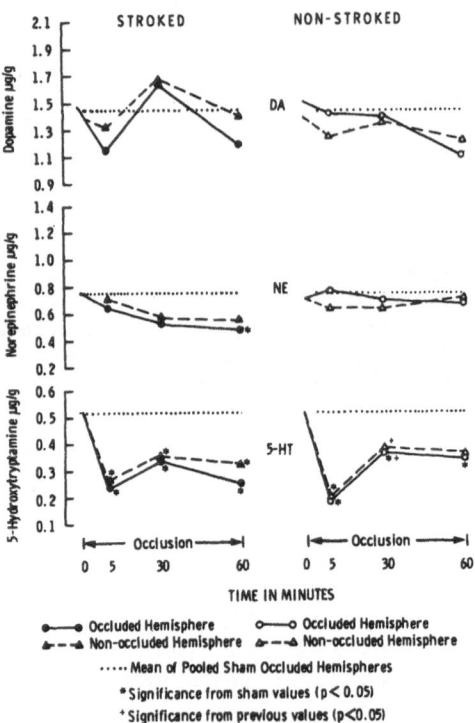

Fig. 1. From experiment one. To show levels of 5-HT, DA
        and NE in the occluded and non-occluded cerebral
        cortex of gerbils with and without stroke during
        right common carotid artery occlusion for periods
        up to 60 minutes.

tissue) for controls closely approached significance, failure to do so probably being related to the large variance
of the ischemic sample. CCA occlusion produced signi-
icant NE decrease in the occluded hemispheres of stroke
animals after 60 minutes, whereas values remained unaltered in animals without stroke. Bilateral hemispheric 5-HT
reduction was measured in both animal groups after 30
and 60 minutes of occlusion, except in the non-occluded
hemisphere of animals without stroke.

     In experiment two, mean values of 5-HT, DA and NE did
not differ in sham-operated control animals sacrificed at
one and three hours after operation, so that pooled means
were calculated. Means and standard deviations in µg/gm
brain tissue of the occluded right hemisphere (RH) and

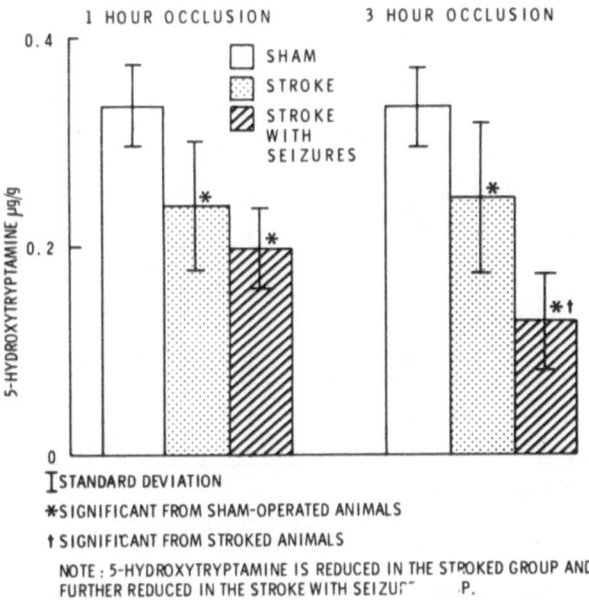

Fig. 2. From experiment two. 5-hydroxytryptamine levels
         in the ipsilateral cortex after unilateral com-
         mon carotid occlusion in the gerbil.

non-occluded left hemisphere (LH) were as follows: Control values for 5-HT were 0.339 + 0.050 (RH) and 0.337 +
0.049 (LH). At one hour after occlusion values in animals
with stroke, 0.240 + 0.067 (RH) and 0.273 + 0.063 (LH),
were significantly reduced bilaterally compared to controls and again at three hours, 0.259 + 0.073 (RH) and
0.264 + 0.053 (LH). There was similar significance of
results at one hour after occlusion in stroke animals
with stroke plus seizures, 0.198 + 0.039 (RH) and 0.281
+ o.111 (LH). After 3 hours levels were further reduced
in the occluded hemisphere to 0.131 + 0.052 (RH). Fig. 2
depicts the 5-HT changes in the occluded hemisphere of
both the stroke and stroke plus-seizure-groups compared
to controls. As in the first experiment without stroke
also showed significant bilateral reduction in 5-HT levels compared to controls.

     DA and NE levels in gerbils with and without signs
of stroke did not differ from control (Figs. 3 and 4). In
animals with stroke plus seizure DA and NE levels were
reduced solely in the occluded hemisphere at one hour.

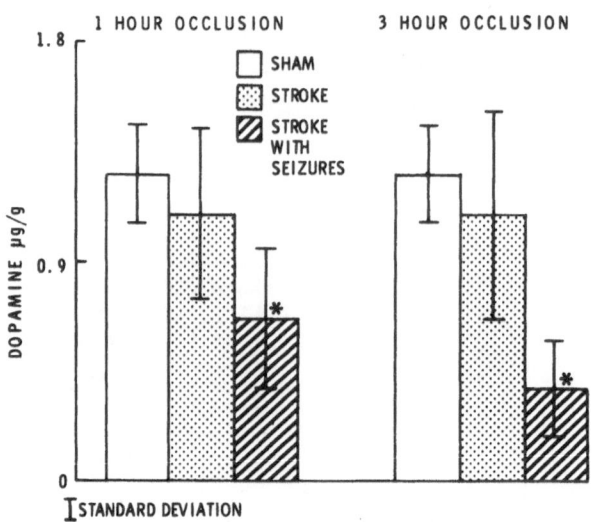

Fig. 3. From experiment two. Dopamine levels in the ipsilateral cortex after unilateral common carotid
occlusion in the gerbil.

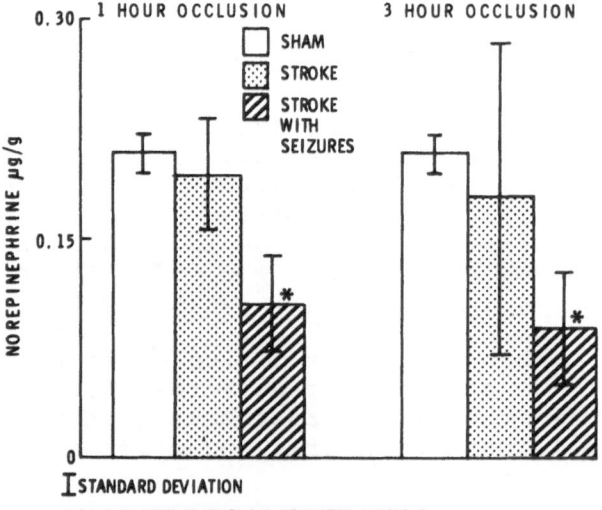

Fig. 4. From experiment two. Norepinephrine levels in the
ipsilateral cortex after unilateral common carotid
occlusion in the gerbil.

After 3 hours of occlusion DA levels were reduced bilaterally (0.372 $\pm$ 0.195, RH; and 0.987 $\pm$ 0.299, LH). NE
levels were similarly affected   (0.085 $\pm$ 0.937, RH; and
0.135 $\pm$ 0.055, LH). However, in both cases levels were
always more reduced in the occluded hemisphere.

DISCUSSION

Catecholamine Changes During Ischemia. No early
changes in cortical DA levels were recorded in the present
study, despite evidence for brain DA depletion during
prolonged ischemia confirmed in a number of laboratories
(5, 9, 16, 20). However, the circling behavior of animals
with stroke, which has been related to decreased DA levels
(6), as well as a strong tendency for DA levels to become significantly reduced in the occluded hemispheres
of those animals 5 min after CCA ligation (0.1, p>0.05),
might nevertheless support some early change in DA metabolism.

Zervas et al. (20) observed no alteration of NE levels in ischemic brain of the gerbil, while later studies
from the same laboratory by Lavyne et al. (5) did record
NE release from ischemic neurons in the same model, although they attributed these contradictory findings to an
additional operative procedure. Mršulja et al. (13), however, demonstrated NE depletion in gerbil cortex during
ischemia and explained this on the basis of NE release
from neurons with failure of reuptake. As in the present
study, Lust et al. (9) have also reported decreased NE
in gerbil cortex after one hour of ischemia.

In the gerbil stroke model, although early monoamine
changes can be attributed to ischemia alone (1), the interpretation of longer-term changes is complicated by
abnormal motor behavior suggestive of seizure activity,
which usually develops for the first time after 30 minutes of arterial occlusion and occurrs up to 30 hours of
occlusion. The second experiment reported here was performed because the contribution of this seizure activity
to cortical monoamine changes therefore needed to be as-
sed in view of the variability in reports of catecholamine changes discussed in the previous paragraph and to
establish which component of the ischemic process is responsible for alteration in tissue monoamines during the
developing infarct. The results showed catecholamine depletion only in animals with seizure. Seizure activity
must therefore be taken into account in the interpretation
of disordered catecholamine metabolism during more pro-

longed ischemia in the gerbil. Pharmacological reduction
of brain monoamines lowers the threshold for seizure activity in brain (10). Though less well documented, seizure
activity itself reduces brain monoamine levels (15) in
models that, however have not employed ischemia. This raises the question of whether the animals which exhibited
seizure activity in the present experiments were more
severely ischemic, with more marked monoamine depletion
which in turn gave rise to seizures, or whether the catecholamine changes were secondary to seizure activity.
However, the first experiment showed that catecholamine
levels were not significantly affected in the time period
preceding the development of seizures.

    5-Hydroxytryptamine Changes During Ischemia: Release
of 5-HT into the cerebral venous blood has been demonstrated to occur within one minute of induction of cerebral
ischemia in the baboon (18). The fall in 5-HT levels in
ischemic cortex in the first series of gerbil experiments
reported here seems in keeping with rapid 5-HT release
into the extracellular compartment, with degradation or
efflux away from the site of release into cerebrospinal
fluid and cerebral venous blood.

    The bilaterality of the more immediate 5-HT changes
after unilateral CCA occlusion might be evidence of diaschisis, i.e., the remote effects of focal ischemia on
cerebral metabolism and function (3, 11). No change of
regional cerebral blood flow was recorded in the non-
occluded hemisphere to account for this (14), and a stress
effect seems unlikely since no monoamine changes were
recorded in sham-operated gerbils. Seizures could contribute to the bilaterality of 5-HT changes recorded at
later time intervals in the first series of experiments.
Kogure et al. (4) have also recorded bilateral changes
in NE levels after unilateral focal embolic ischemia in
the rat and attributed this to synaptic release following
generalized neuronal depolarization. Levy and Duffy (8)
also observed changes in cerebral energy metabolism in
the contralateral cortex of gerbils with unilateral
hemispheric ischemia which they postulated might be due
to neurotransmitter dysfunction.

    Results of the second series of experiments confirm-
ed that 5-HT is depleted in both the occluded ischemic
hemisphere and in the non-occluded hemisphere, independent of seizure activity. The bilateral hemispheric
5-HT decrease in ligated animals with no signs of stroke
was observed in both series of experiments. In these ani-

mals unilateral CCA occlusion caused reduction of ipsilateral regional hemispheric cerebral blood flow, although
not below the critical threshold for production of neurological deficit (14). 5-HT metabolism therefore appears
particularly sensitive to the effects of even minimal ischemia. We believe that the results reported here reem-
phasize the importance of disordered 5-HT metabolism in
ischemic brain.

ACKNOWLEDGMENTS

Reprint requests to: K.M.A. Welch, M.B.Ch.B., M.R.
C.P. (UK), Associate Professor, Department of Neurology,
Baylor College of Medicine, Neurosensory Center, 6501
Fannin, Houston, Texas 77030, U.S.A.

This work was supported by Grant NS 09287 from
National Institute of Neurological and Communicative
Disorders and Stroke, National Institutes of Health,
Bethesda, Maryland 20014.

REFERENCES

1. Gaudet, R., Welch, K.M.A., Chabi, E., Wang, T.-P. (1977):
   Effect of transient ischemia on monoamine levels in
   the cerebral cortex of gerbils. J. Neurochem., in press.

2. Kahn, K. (1972): The natural course of experimental
   cerebral infarction in the gerbil. Neurology, 22:
   510-515.

3. Kogure, K., Busto, R., Scheinberg, P., Reinmuth, O.M.
   (1974): Energy metabolites and water content in rat
   brain during the early stage of development of cerebral infarction. Brain, 97: 103-114.

4. Kogure, K., Scheinberg, P., Matsumoto, A., Busto, R.,
   Reinmuth, O.M. (1975): Catecholamines in experimental
   brain ischemia. Arch. Neurol. 32: 21-24.

5. Lavyne, M.H., Moskowitz, M.A., Larin, F., Zervas,N.T.
   Wurtman, R.J. (1975): Brain H^3-catecholamine metabolism
   in experimental cerebral ischemia. Neurology 25:
   483-485.

6. Lavyne, M. , Moskowitz, M., Zervas, N., Wurtman,R.J.
   (1975): Rotational behavior in gerbils following unilateral common carotid artery ligation. J. Neural
   Transm. 36: 83-89.

7. Levy, D.E., Brierley, J.B., Plum, F. (1975): Ischae-
   mic brain damage in the gerbil in the absence of "no-
   reflow" J. Neurol. Neurosurg. Psychiat. 38: 1197-1205.

8. Levy, D.E., Duffy, T.E. (1977): Cerebral energy metabolism during transient ischemia and recovery in the
   gerbil. J. Neurochem. 28: 63-70.

9. Lust, W.D., Mršulja, B.B., Mršulja, B.J., Passonneau,
   J.V., Klatzo, I. (1975): Putative neurotransmitters
   and cyclic nucleotides in prolonged ischemia of the
   cerebral cortex. Brain Res. 98: 394-399.

10. Maynert, E.W., Marczynski, T.J., Browning, R.A.(1976):
    The role of the neurotransmitters in the epilepsies.
    In Advances in Neurology, Vol. 13, W.J. Friedlander,
    Ed., Raven Press, New York, 153-166.

11. Meyer, J.S., Shinohara, Y., Kanda, T., Fukuuchi,Y.,
    Ericsson, A.D., Kok, N.K. (1970): Diaschisis resulting
    from acute unilateral cerebral infarction. Arch. Neu-
    rol. 23: 241-247.

12. Moskowitz, M.A., Wurtman, R.J. (1976): Acute stroke
    and brain monoamines. In Cerebrovascular Diseases, P.
    Scheinberg, Ed. Raven Press, New York, 153-166.

13. Mršulja, B.B., Mršulja, B.J., Spatz, M., Klatzo, I.
    (1976): Catecholamines in brain ischemia--effects
    of alpha-methyl-p-tyrosine and pargyline. Brain Res.
    104: 373-378.

14. Nakai, K., Welch, K.M.A., Meyer, J.S. (1977): Criti-
    cal cerebral blood flow for production of hemiparesis
    after unilateral carotid occlusion in the gerbil. J.
    Neurol. Neurosurg. Psychiat. 40: 595-599.

15. Scheinberg, P. (1976): Correlation of brain monoamines
    and energy metabolism changes. In Cerebrovascular Di-
    seases, P. Scheinberg, Ed. Raven Press, New York,
    167-171.

16. Welch, K.M.A., Chabi, E., Buckingham,J., Bergin, B.
    Achar, V.S., Meyer, J.S. (1977): Catecholamine and
    5-hydroxytryptamine levels in ischemic brain. Influ-
    ence of p-chlorophenylalanine. Stroke 8: 341-346.

17. Welch, K.M.A., Hashi, K., Meyer, J.S. (1973): Cerebrovascular response to intracarotid injection of serotonin before and after middle cerebral artery occlusion. J. Neurol. Neurosurg. Psychiat. 36: 724-735.

18. Welch, K.M.A., Meyer, J.S., Teraura, T., Hashi, K., Shinmaru, K. (1972): Ischemic anoxia and cerebral serotonin levels. J. Neurol. Sci. 16: 85-92.

19. Welch, K.M.A., Wang, T-P.F., Chabi, E. (1977): Ischemia-induced seizures and cortical monoamine levels. Ann. Neurol. in press.

20. Zervas, N.T., Hori, H., Negora, M., Wurtman, R.J., Larin, F., Lavyne, M.H. (1974): Reduction in brain dopamine following experimental cerebral ischemia. Nature, 247: 283-284.

BRAIN MONOAMINES IN CEREBRAL INFARCTION AND COMA

P. Riederer and K. Jellinger

Ludwig-Boltzmann Institute of Clinical
  Neurobiology   and
Department of Neurology, Lainz Hospital
1 Wolkersbergerstr. A-1130, Vienna, Austria

SUMMARY

Dopamine (DA), serotonin (5-HT), 5-hydroxyindole acetic acid (5-HIAA) and tryptophan (Trp) were assayed spectrofluorometrically in various brain regions of eight human patients who died after acute and old cerebral infarction, in 17 cases of acute metabolic (hepatic, uremic and diabetic) coma, and in three patients with liver cirrhosis without coma. The results were as follows: (1) In both recent and older cerebral infarction a total depletion of DA and 5-HT was associated with slight reduction of DA and 5-HT levels in remote non-ischemic areas and various nuclei of both the injured and contralateral hemispheres. 5-HIAA was significantly reduced in acute ischemic necrosis, while the perifocal edema zone showed accumulation of both 5-HT and 5-HIAA. The degradation zone surrounding old infarcts showed a mild decrease of both 5-HT and 5-HIAA, indicating normalization of 5-HT metabolism after decrease of the complicating edema. (2) In metabolic coma brain DA showed a mild general decrease, while brain Trp was significantly increased in hepatic coma. 5-HT and 5-HIAA were generally increased in all types of coma, most significantly in the brainstem tegmentum and in parts of the limbic system. (3) In liver cirrhosis without coma, brain 5-HT was within normal range, while Trp and 5-HIAA were significantly elevated in the brainstem, their increase being less severe than in hepatic coma.

The data presented in human stroke which confirm
previous findings in experimental cerebral ischemia and
infarction indicate that disorders of brain monoamine
metabolism are contributing to the development of postischemic brain damage and the complicating cerebral edema.
The results in endotoxic coma which are in keeping with
the findings in experimental porto-caval and uremic encephalopathies suggest some common disorders of central
monoamine neurotransmitter metabolism in endotoxic coma
of different etiology usually accompanied by cerebral
edema. Increased 5-HT synthesis and turnover in the ascending serotonergic brainstem systems are considered an
important biochemical substrate of clinical disorders of
consciousness.

INTRODUCTION

Studies in different animal models as well as studies
in clinical patients and in human postmortem materials
have established some abnormalities of brain monoamine
neurotransmitter functions in cerebral edema (22, 23, 31-
33, 47), ischemic brain (7, 20, 23, 25, 29, 30-33, 41,
46, 48), and in metabolic encephalopathies (5, 6, 9, 11,
19, 42-44). Experimental brain infarction and chronic
cerebral ischemia have been shown to reduce noradrenaline
(NA), dopamine (DA) and serotonin (5-HT), with accumulation of their main metabolites homovanillic acid (HVA)
and 5-hydroxyindole acetic acid (5-HIAA)(23, 25, 30, 32,
33, 41, 45, 46), and of gamma aminobutyric acid (GABA)and
cyclic adenosine monophosphate (cAMP) (23, 31, 46). In a
vicious circle these biochemical changes are suggested to
cause progression of cerebral ischemia and of the complicating cerebral edema.

In experimental and human hepatic encephalopathy, increased brain and CSF levels of tryptophan (Trp), 5-HT,
and 5-HIAA have been reported (5, 6, 9, 11, 12, 19, 44),
and have been attributed to amino acid imbalance in plasma
and brain (6, 12, 43). Similar changes of brain monoamines
have been found in both uremic and diabetic coma in human
postmortem brain (19), while in chronically uremic rats
increased serum and brain Trp and 5-HIAA without considerable changes in brain 5-HT have been observed (42). Although these studies strongly suggest profound alterations
in monoamine central neurotransmitter synthesis and metabolism during both cerebral ischemia and metabolic catastrophes associated with neurologic symptoms and cerebral
edema, both the responsible mechanisms and the roles they
may play in the pathogenesis of ischemic and metabolic

brain dysfunctions remain to be established.

Here we report the changes of brain monoamine levels in human patients dying after cerebral infarction, and in liver disease without coma.

MATERIALS AND METHODS

Human postmortem brain samples were obtained from three series:

1. Eight cases of brain infarction due to thrombotic occlusion of large cerebral arteries. They included four recent infarctions (6 to 48 hours), three older infarctions (7 days to 4 months), and one case with both recent (about 20 hours) and older infarctions (5 days) in different brain regions. The age range of the patients was 66 to 84 years (average 75.1 years). Clinically, none of the patients showed metabolic catastrophes (diabetic coma, uremia), and only two fell into final coma due to complicating brainstem infarction or hemorrhage. Drug treatment included hemodiluting and dehydrating agents, antibiotics, heart drugs and antihypertonic agents. The clinico-pathologic data and time intervals between death and freezing of the brain samples (2 to 14 hours) are summarized in Table I.

2. 17 patients who died in acute coma: 10 cases of hepatic coma with serum ammonia levels up to 352 µg/ml; 3 cases of uremic coma with BUN up to 300 mg/ml, 4 cases of diabetic coma including one case each of hyperosmolar and ketoacidotic coma. The age range of these patients was 47 to 86 years (average 69.6 years). The essential clinicopathologic data and postmortem intervals (3 to 24 hours) are summarized in Table II.

3. Three patients aged 63 to 71 years with liver cirrhosis (compensated or decompensated) with serum bilirubin levels up to 16.5 mg/ml and serum ammonia levels up to 152 ug/ml but without clinical signs of coma (35) who died from acute gastrointestinal hemorrhage or acute myocardial infarction (Table II).

Brains were obtained at autopsy and immediately processed. In group 1, tissues were taken from various regions of the cerebral infarct (cortex, white matter), the perifocal edema zone, and adjacent areas of degradation and glio-mesenchymal proliferation, as well as from remote intact areas of both cerebral hemispheres. From all brains tissue samples were taken from 15 different brain

Table I.   CLINICO-PATHOLOGICAL FEATURES OF CASES WITH AND WITHOUT METABOLIC COMA

Case No.	Age	Sex	Type of Coma	General autopsy	Neuropathology	Autopsy interval
2102/76	72y	F	hepatic	alcoholic cirrhosis	not examined	24 hr
363/77	62y	M	hepatic	alcoholic cirrhosis	hepat. encephalop.	20 hr
383/77	68y	M	hepatic	alcoholic cirrhosis	hepat. encephalop.	17 hr
980/77	64y	F	hepatic	hepatic carcinoma	hepat. encephalop.	6 hr
1035/77	74y	M	hepatic	portocaval cirrhosis hemorrh.esoph.varices	hepat. encephalop.	7 hr
1477/77	47y	M	hepatic	same	hepat. encephalop.	16 hr
1817/77	56y	F	hepatic	same	hepat. encephalop.	11 hr
1847/77	76y	M	hepatic	postnecrotic cirrhosis	hepat. encephalop.	9 hr
2118/76	64y	M	hepatic	postnecrotic cirrhosis	hepat. encephalop.	18 hr
1702/77	66y	F	hepatic	postnecrotic cirrhosis hemorrh.esoph.varices	hepat. encephalop.	18 hr
2892/77	76y	M	uremic	plasmacytoma nephrosis	brain edema	14 hr
3218/76	82y	F	same	pyelonephritis	brain edema	12 hr
240/77	83y	F	uremic	nephrosclerosis,uremic enteritis,blad.carcin.	brain edema cer. atheroscler.	10 hr
3018/76	79y	M	diabetic	diab. glomeruloscler. diab. polyneuropathy	brain edema	7 hr
1922/76	56y	F	diabetic	ketoacidotic coma	brain edema	4 hr
1269/77	75y	M	diabetic	diab. glomeruloscler. diab. polyneuropathy	brain edema senile brain	7 hr
2347/77	86y	F	diabetic (hyperos)	diabet.glomeruloscler. general atherosclerosis	brain edema toxic encephalop.	16 hr
1928/77	65y	F	no coma	postnecrotic cirrhosis ascites,GItract hemor.	brain atrophy slight hepatic encephalop.	17 hr
2770/77	71y	M	no coma	portocaval cirrhosis hemorrh.esoph.varices	brain atrophy no brain edema	14 hr
2811/77	63y	M	no coma	postnecrotic cirrhosis acute myocard.infarct.	slight brain atrophy no brain edema	3 hr

Table II.   CLINICO-PATHOLOGICAL FEATURES OF CASES WITH CEREBRAL INFARCTION

Case No.	Age	Sex	Clinical features	General autopsy	Neuropathology	Brain weight	Aut. int.
32/77	66y	M	acute rt hemiparesis 6-7 hr b.d. terminal coma,midbrain syndr.	gen.atheroscleros. thrombosis lft ICA and MCA	recent inf.lft. F-T-P-region, br.edema,hemor.	1500g	10 hr
30/77	68y	M	acute rt hemianopia 8 hr b.d.;acute tetra paresis and fin.coma	chron.lymphocytic leukemia;thrombos. basil.art.+lft PCA	recent inf.rt temp.-occ.lobe acute pont.inf.	1200g	3 hr
97/77	72y	M	ac.lft.hemiparesis 24 hr bef. death	gen.atheroscleros. thrombosis rt MCA	rec.infarct.rt par.lob+lentic. nucl.,brain edema	1350g	7 hr
214/77	70y	M	ac.aphasia + rt hemi- paresis 2 ds b.d. ac.myocard.infarct.	ac.myocard. inf., thrombosis lft ICA +MCA	rec.inf. lft F-P-reg. brain edema	1350g	5 hr
45/77	76y	M	ac.aphasia 5 ds b.d. ac. rt hemiplegia 20 hr b.d.	gen.atheroscleros. thrombosis lft MCA	older and recent inf.lft pariet. inc.lent.n.edema	1500g	2 hr
87/77	84y	F	ac.rt hemiparesis 7 ds b.d.,diab.mell. ac.myocard.infarct.	gen.atheroscleros. thromb.lft MCA ac.myoc.infarction	older inf.lft front.region no cerebr.edema	1050g	6 hr
134/77	81y	F	ac. aphasia, lft he- mipar. 17 ds b.d., ac.myocard.infarct.	gen. atheroscleros. thrombosis rt MCA ac.myocard. inf.	old inf. rt pariet.lobe sen. brain,no edema	1150g	3 hr
20/77	84y	F	ac.rt hemiparesis 4 mos b.d., acute cardial infarction	gen.atheroscleros. stenosis lft MCA ac.myocard.infarct.	old inf. lft par.lobe,sen. no edema	1000g	14 hr

b.d. = before death   ICA = internal carotid artery   F = frontal region
rt = right   MCA = middle cerebral artery   P = parietal region
lft = left   PCA = posterior cerebral artery   T = temporal region

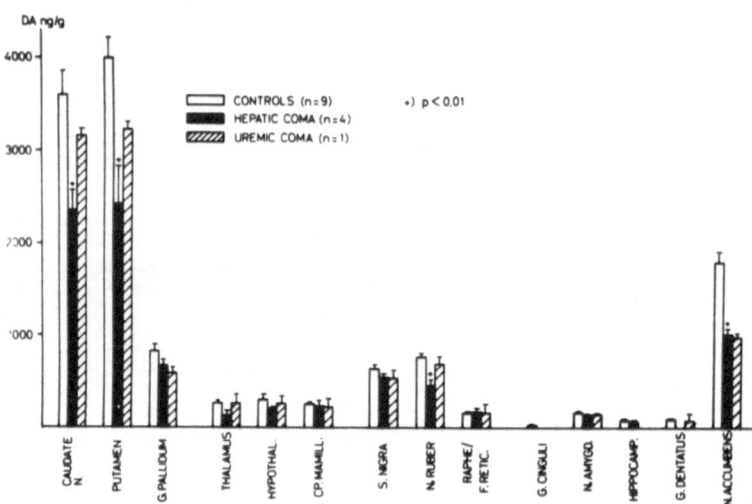

Fig. 1.   Brain dopamine in controls and metabolic coma

regions, and were stored at -30°C until analysis. For details regarding removal of brain samples, processing and
biochemical methods see (40). DA, 5-HT, 5-HIAA, and Trp
were extracted and spectrofluorometrically assayed using
established methods (1, 2, 16). GABA was determined in 3
cases of stroke by double-enzymatic fluorometric assay
(18). The results were compared with a series of 9 age-
matched controls (patients without metabolic, nervous or
cerebrovascular disorders who rapidly died of myocardial
infarction) with similar autopsy intervals and storage
time (3, 40). Brain Trp levels were not essentially altered during the postmortem intervals used (39). From all
cases corresponding brain areas were processed for histological examination using routine neuropathological methods.

RESULTS

(1) Cerebral infarction. The results are summarized
in Table 3 and 4.

a) In recent infarction a complete depletion within
the necrotic area (cortex and white matter) of DA and 5-HT
was associated with considerable reduction of 5-HIAA to
about 40% of control values. The perifocal edema zone
showed almost normal DA values, with accumulation of both
5-HT and 5-HIAA, the average elevation in the edematous

Table III. BRAIN MONOAMINES IN HUMAN CEREBRAL INFARCT

Brain area		DA ng/g f.w. (n) mean ± SEM		5-HT ng/g (n) mean ± SEM		5-HIAA ng/g (n) mean ± SEM	
Recent infarct (6-8 hrs)	cortex	n.d.	(10)	n.d.	(16)	47.5 ± 10.3*	(4)
	white m.	n.d.	(9)	n.d.	(9)	33.7 ± 1.7	(3)
Perifocal edema	cortex	26.1 ± 6.95	(7)	112.6 ± 22.4	(6)	166.0 ± 28.7*	(6)
	white m.	28.7 ± 6.8	(4)	41.5 ± 9.6	(4)	144.5	(2)
Old infarct (5 ds.-4 mos)	cortex	n.d.	(8)	n.d.	(8)	114.8 ± 21.5	(7)
	white m.	n.d.	(8)	n.d.	(8)	n.e.	
Perifocal area	cortex	38.6 ± 1.86	(5)	70.0 ± 7.0	(2)	78.8 ± 4.2	(4)
Intact cortex	ipsilat.	31.0 ± 2.55	(5)	36.7 ± 1.7	(3)	69.5 ± 1.55*	(4)
	contralat.	30.0 ± 0.0	(2)	44.0 ± 9.25	(4)	97.5 ± 15.2	(4)
Controls	cortex	25.0 ± 5.0	(9)	55.0 ± 7.0	(9)	119.0 ± 8.0	(9)
	white m.	n.e.		25.0 ± 4.2	(3)	38.0 ± 4.0	(3)

n = number of assayed samples   n.d. = not detectable   *) = $p < 0.01$
f.w. = fresh weight   n.e. = not examined

Table IV.   REGIONAL BRAIN DOPAMINE AND SEROTONIN IN HUMAN CEREBRAL INFARCTION

Region		Controls (n=9) DA ng/g x ± SEM	Recent/Old Infarcts (n=1/1) DA ng/g		Controls (n=9) 5-HT ng/g x − SEM	Recent/Old Infarcts (n) 5-HT ng/g	Recent/Old Infarcts (n) 5-HT ng/g
Caudate n.	i.l.	3600 ± 265	2753	—	285 ± 16	207 (1)	—
	c.l.		2850	2970		220 (1)	193 (1)
Putamen	i.l.	4005 ± 220	3120	—	266 ± 15	163 (2)	106 (1)
	c.l.		—	—		165 (2)	240 (1)
Gl.pallid.	i.l.	825 ± 42	637	680	364 ± 27	241 (2)	310 (2)
	c.l.		674	438		317 (2)	380 (2)
Thalamus	i.l.	260 ± 31	163	155	340 ± 18	293 (2)	275 (2)
	c.l.		185	195		307 (1)	148 (2)
Cp.mamill.	i.l.	250 ± 18	—	160	250 ± 18	266 (1)	—
	c.l.		—	200		496 (1)	—
Amygdal.n.	i.l.	156 ± 9	95	—	264 ± 15	101 (1)	180 (1)
	c.l.		125	—		130 (1)	212 (2)
S. nigra	c.l.	630 ± 40	—	—	583 ± 22	223 (2)	440 (1)
Raphe nucl.		150 ± 12	—	140	476 ± 25	631 (1)	1145 (2)
Front.cx.	i.l.	25 ± 5	30	30	55 ± 7	41 (2)	30 (2)
	c.l.		30	30		30 (2)	30 (2)

i.l. = ipsilateral, c.l. = contralateral; n = number of cases examined;
− = not examined

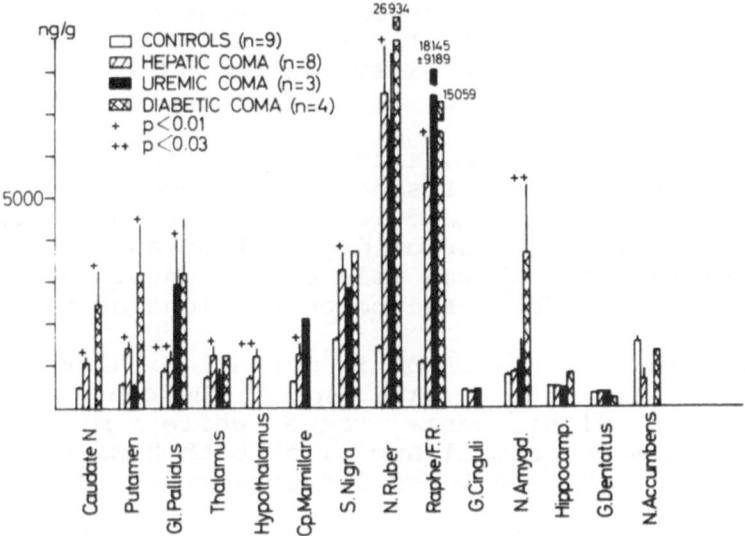

Fig. 2.   Brain 5-hydroxyindole acetic acid in controls
          and metabolic coma

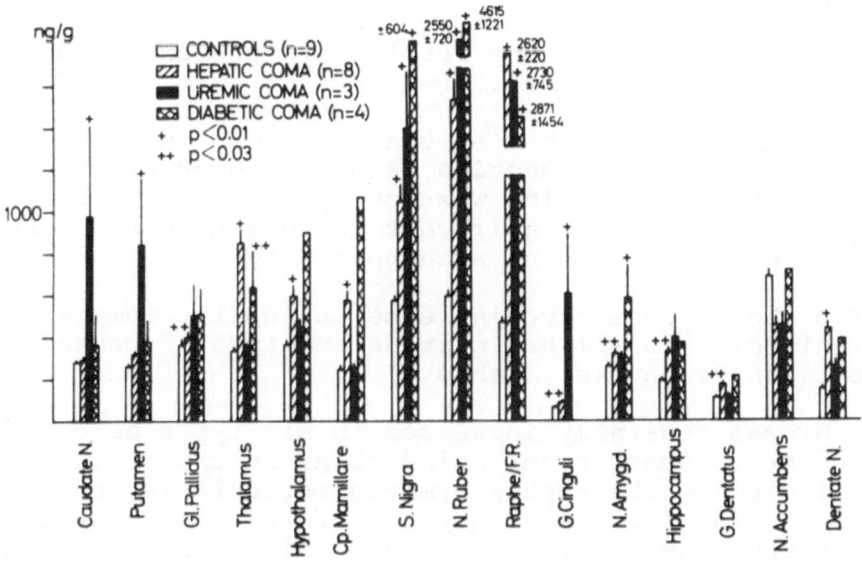

Fig. 3.   Brain serotonin (5-HT) in controls and meta-
          bolic coma

cortex being 200% and 145%, respectively. In the edematous white matter, a less severe increase of 5-HT and
5-HIAA was observed. A similar increase of GABA was present in both the necrotic tissue (about 130%) and the
perifocal edema zone (about 160% of controls) (20).

b) In older infarcts, a total depletion of DA and
5-HT within the necrotic tissue was accompanied by almost
normal or slightly decreased levels of 5-HIAA, while the
surrounding cortical areas of degradation and glio-mesenchymal proliferation showed mild elevation of DA and 5-HT
levels, opposed by decreased concentrations of 5-HIAA.

c) Cortical levels of DA in morphologically intact
cortex remote from the infarction and the contralateral
hemisphere were within normal range, while a mild reduction of DA in most central nuclei of both hemispheres was
seen in brain with recent and old infarcts.

d) Cortical levels of both 5-HT and 5-HIAA were slightly decreased in remote intact areas of both hemispheres,
while regional 5-HT levels in brains with acute and old
infarction usually showed a mild decrease which was often
more pronounced in the injured hemispheres (Table 2). Increased 5-HT was only found in the raphe nuclei of the
brainstem.

(2) Metabolic coma. The results are presented in Figs.
1-3 and in Tables V and VI.

DA was decreased in most brain regions in all types
of coma, the average reduction being 20 to 30 percent of
the controls. DA depletion was most pronounced in hepatic
and uremic coma, with significant reduction in the corpus
striatum, n. ruber, and n. accumbens (Fig. 1).

Trp showed excessive increase in hepatic coma which
was significant in all brain areas examined, especially
in the brainstem nuclei (Table V).

5-HT was generally increased in all types of coma,
the highest concentrations being found in the brainstem
nuclei including the raphe/formatio reticularis, n. ruber
and substantia nigra, and in parts of the limbic system.
Although there were notable regional differences in the
individual brains, similar changes of 5-HT were found in
hepatic, uremic, and diabetic coma (Fig. 2).

5-HIAA showed similar or even higher increase, particularly in the brainstem and corpus striatum, with the

Table V. BRAIN TRYPTOPHAN IN CONTROLS AND HEPATIC FAILURE
WITH AND WITHOUT COMA

Brain area	Controls (n = 11)	Cirrhosis without coma (n of cases)	Hepatic coma (n = 5)
Lenticular n.			
Caudate n.	10.9 ± 0.78	6.0 (1)	39.2 ± 9.7*
Putamen	14.9 ± 0.89	18.0 (1)	40.3 ± 11.5*
Gl.pallidus	15.4 ± 0.24	11.1 (1)	69.8 ± 21.6*
Diencephalon			
Thalamus	16.8 (2)	14.2 (1)	31.1 (2)
Hypothalamus	9.8 (2)	7.0 (1)	13.4 (1)
Cp.mamillare	12.7 (2)	3.3 (1)	n.e.
Brainstem			
S. nigra	19.6 ± 1.42	35.0 ± 16.0 (3)	153.2 ± 48
N. ruber	20.3 ± 1.45	38.0 ± 5.0 (3)	143.6 ± 37*
Raphe/F.ret.	15.8 ± 0.82	31.0 ± 4.0 (3)	80.0 ± 22*
Limbic system			
G. cinguli	14.0 ± 0.96	4.2 (1)	42.2 ± 6.1*
N. amygdalae	15.7 ± 0.8	18.4 (1)	71.6 ± 18.5*
Hippocampus	11.2 (2)	10.2 (1)	6.8 (1)
G. dentatus	11.5 (2)	27.4 (1)	16.0 (1)
N. accumbens	12.6 (2)	7.5 (1)	n.e.
Frontal cortex	10.0 (2)	15.2 (1)	12.0 ± 1.0
White matter	7.8 (2)	n.e.	8.6 (2)

Values are means ± S.E.M.   * = p < 0.01
Values in µg/g fresh weight

highest concentrations in both uremic and diabetic coma
(Fig. 3).

(3) <u>Liver cirrhosis without coma</u>. The results are presented in Tables V and VI.

<u>Trp</u> was significantly elevated in the brainstem nu-
clei as compared to controls, but this regional increase
was significantly lower than in cases of hepatic coma
(Table V).

<u>5-HT</u> was not significantly altered as compared to
controls and, in particular, no definite changes were seen
in the brainstem (Table VI).

<u>5-HIAA</u> showed an increase, particularly in the brain-
stem nucluei, but this elevation of 5-HIAA in non-coma-
tose liver cirrhosis was considerably lower than in hepatic coma (Table VI).

DISCUSSION

Recent experimental studies indicate that continued
<u>cerebral ischemia</u> leads to profound depletion of central
monoamine neurotransmitters coupled with progressive brain
edema (30-33, 41, 45, 46, 48). Increased concentrations
of NE and 5-HT in the lumbar CSF have been recorded in
clinical patients early after the onset of cerebral infarction; these changes may have correlated with reduction
in the blood flow to infarcted and non-infarcted hemispheres (29). Our results in human postmortem brains confirm a total depletion of DA and 5-HT in both acute and old
cerebral infarctions, as well as a slight DA reduction in
most non-ischemic areas of <u>both</u> hemispheres, which was
associated in some instances with edema. While in experimental cerebral ischemia HVA and 5-HIAA, the main metabolites of DA and 5-HT in the brain, were accumulated in
ischemic tissues (32, 33), in human brains 5-HIAA was considerably reduced in the area of acute infarction, and
after about 6 days increased again to almost normal or
only slightly reduced levels, indicating some normalization of this substance after decrease of brain edema. HVA
was not examined in the present series.

There is increasing experimental evidence that focal
cerebral ischemia causes synaptosomal release of neurotransmitters early after the onset of ischemia (25). Neuronal membrane depolarization with ionic shifts as well
as synaptosomal mitochondrial dysfunction (26), conditions
which may cause abnormal release of neurotransmitters,

Table VI.   BRAIN SEROTONIN (5-HT) AND 5-HYDROXYINDOLE ACETIC ACID (5-HIAA) IN LIVER CIRRHOSIS WITHOUT COMA

Brain area	5 - HT (ng/g) (means ± S.E.M.)		5 - HIAA (ng/g) (means ± S.E.M.)	
	Controls (n = 9)	Liver cirrhosis without coma	Controls (n = 9)	Liver cirrhosis without coma
Caudate n.	285 ± 16	267 (1)	442 ± 26	492 (1)
Putamen	266 ± 15	218 (2)	535 ± 28	672 (2)
Gl. Pallidus	364 ± 27	169 (1)	875 ± 61	782 (1)
Thalamus	340 ± 18	350 (1)	725 ± 39	1033 (1)
S. nigra	583 ± 22	663 ± 96 (3)	1595 ± 47	2524 ± 843 (3)
N. ruber	603 ± 29	666 ± 111 (3)	1379 ± 59	3095 ± 571 (3)
Raphe/F.ret.	476 ± 25	1467 ± 287 (3)	1035 ± 42	4587 ± 1569 (3)
N. amygdalae	264 ± 15	303 (1)	760 ± 27	963 (1)
G. dentatus	110 ± 9	114 (1)	288 ± 11	870 (1)
Hippocampus	200 ± 11	131 (1)	480 ± 21	389 (1)

Number of cases in parentheses

are well documented in cerebral ischemia. Other factors
which could cause intraneuronal monoamine depletion include failure of energy-dependant synaptosomal re-uptake as
well as synthesis impairment (8, 25). Synaptosomal release
of neurotransmitters occurs not only from neuronal terminals involved in the damaged tissue but also from those
in remote non-ischemic areas of the brain, as typified by
reduction of cortical DA and 5-HT levels in the injured
hemisphere (30, 32, 33, 45, 46) and by bilateral decrease
of cortical DA in animals subjected to unilateral carotid
artery occlusion (45) or unilateral cerebral embolism (23).
These data seem evidence of a remote effect of focal ischemia which may be due to cerebral hemodynamic shifts,
i.e. reduced regional cerebral blood flow (rCBF), even in
the absence of clinical and morphological signs of ischemia (45). Monoamine changes in the injured hemisphere
may be related to transient increase of water content and
ultrastructural signs of edema which were demonstrated in
the contralateral hemispheres of symptomatic animals in
the gerbil stroke model (45).

Experimental studies have further shown that monoamine synthesis and metabolic rate are inhibited by lack
of oxygen (31-33). A reduced activity of tyrosine hydroxylase and of tryptophan hydroxylase is likely to occur in
the ischemic brain (30, 33). Inhibition of these enzymes
responsible for the formation of HVA and 5-HIAA by ischemia would explain the decreased levels of these indole
metabolites found in acute infarcts in human brain. In experimental ischemia, 5-HIAA was not changed initially,
while in the later phase of ischemia HVA and 5-HIAA were
accumulated in the damaged tissue (32, 33), and in the adjacent edema zone, even when inhibition of synthesis of
their precursors was produced by AMPT (alpha-methyl-p-
tyrosine) and PCPA (p-chloro-phenylalanine) (33, 45, 46).
It is suggested that these results indicate initial release together with reduced synthesis of monoamines in the
ischemic brain, and that accumulation of monoamine metabolites in ischemic brain is, at least in part, due to
inhibition of transport mechanisms from the brain to blood
(33). Both increased release and extracellular accumulation of indoleamines and their metabolites may explain
the significant increase of 5-HT and 5-HIAA in the edematous tissue surrounding acute infarction in human brain
while degradation zones adjacent to old infarcts show
normal or slightly reduced contents of these substances,
which may correlate with a decrease of acute perifocal
edema.

The increased release and turnover of 5-HT and extracellular accumulation of 5-HT and its metabolites in and around tissues damaged by ischemia, and in the peri-focal edema zone of acute cerebral infarction my induce microcirculation disorders due to vasoconstriction, and cerebral edema with increased permeability of the blood-brain barrier, thus contributing to the progression of ischemia (22, 32, 33, 46, 47). Among other chemical brain constituents, GABA shows marked elevation in both experimental ischemia (31), and acute cerebral infarction in man, with later return to normal values (20).

In experimental cerebral ischemia, cAMP is significantly increased (23, 31, 46) and in addition to depletion of monoamines has been closely related to edema in later stages of ischemia (23, 46). In a vicious circle, these events may cause progression of ischemia and edema, thereby contributing to the development of postischemic brain injury.

Our studies in human brains after hepatic coma confirm previous findings in experimental porto-caval encephalopathy, indicating a significant general elevation of brain Trp, a uniform decrease in brain DA, and regionally increased 5-HT content in the brainstem and striatum with rather uniform increase of 5-HT turnover, as measured by the level of brain 5-HIAA (5, 6). Disorders of brain neurotransmitter synthesis in hepatic failure are attributed mainly to the imbalance of plasma amino acids which compete for transport and uptake mechanisms across the blood-brain barrier (5, 10-12). In both chronic hepatic and renal insufficiency, there is an excessive accumulation in serum and CSF of free phenoles and phenolic acids (27, 38), and of free Trp (13) which may interfere with brain DA and 5-HT metabolism (17) suggesting common disorders of indoleamine metabolism in the brain.

The relations between plasma and brain Trp, 5-HT and 5-HIAA appear to be very complex, and factors other than the competing plasma amino acids may control brain DA and 5-HT metabolism (14, 15, 42). Some basic disorders in insulin and glucagon metabolism have been suggested for hepatic, uremic and diabetic encephalopathies which may interfere with brain monoamine metabolism in all three types of metabolic coma (43).

General elevation of brain 5-HT with significant increase in the brainstem seen in all types of metabolic coma of our human postmortem series was associated with

generalized increase of 5-HIAA level showing excessive
local increase in the brainstem nuclei, thus indicating
a uniform and regional increase of 5-HT turnover in the
brain with prominent involvement of the brainstem. Similar increase of 5-HIAA levels in the brainstem was recently reported in human postmortem cases of hepatic, uremic
and diabetic coma (9).

Interestingly enough, three cases of liver cirrhosis
without coma, in spite of significant elevation of Trp in
the brainstem did not show significant changes of brain 5-
HT as compared with controls. On the other hand, there was
significant elevation of 5-HIAA in the brainstem of these
cases, indicating a rather localized increase of 5-HT turnover in the brainstem. As the brain levels of Trp and 5-
HIAA in patients with liver cirrhosis elevated without
clinical signs of coma, the clinical syndrome of coma
associated with hepatic failure is suggested to represent a dynamic process related at least in part to alterations of brain indoleamine metabolism. As the changes
of biogenic amines in the human brain do not considerably differ in cases of hepatic, uremic and diabetic
coma, one can speculate that these changes may reflect
some common disturbances of central neurotransmitter metabolism in endotoxic coma of different origin.

Both portosystemic and uremic encephalopathies are
characterized by watery swelling of the astroglia and
increased permeability of the blood-brain barrier (24,
37) which have been related to increased brain 5-HT content, the pathophysiologic effects of 5-HT on the BBB
permeability being well established (47). The rather
uniform accumulation of brain 5-HT in different types of
metabolic coma suggests its basic importance for the development of brain edema, a common pathologic feature of
toxic coma (28).

In both experimental hepatic encephalopathy and human endotoxic coma, the most pronounced changes of the
indoleamine transmitter 5-HT are found in the nuclei of
the brainstem tegmentum (raphe/formatio reticularis) and
in parts of the limbic system, i.e. regions involved in
the regulation of consciousness. Recent studies suggest
that vigilance is regulated by the ascending monoamin-
ergic (dopaminergic and serotonergic) brainstem and meso-
limbic systems, indicating some relationship between
brain 5-HT and sleep (21, 34, 36). The marked elevation
in human endotoxic coma of both 5-HT and 5-HIAA levels
in the brainstem tegmentum and in functionally connected

parts of the hippocampus (34), suggest some common regional metabolic disorders in the ascending serotonergic
brainstem-mesolimbic system which may be related to clinical disorders of consciousness. This hypothesis is supported by unpublished personal studies in a case of hypo-
glycemic coma showing a general decrease of brain 5-HT
and 5-HIAA, with slight, but definite elevation of both
indoleamines in the raphe/formatio reticularis of the
brainstem. In two cases of cerebral infarction who died
in a final comatose state, slight uniform decrease of
brain 5-HT and 5-HIAA was opposed by increased levels of
5-HT and 5-HIAA in the brainstem tegmentum. In Parkin-
sonian patients with L-Dopa psychosis without coma increased levels of 5-HT and 5-HIAA in the brainstem were
noted and disorders of consciousness were related to regional imbalance of DA and 5-HT in the extrastriatal regions (4). From these data it can be speculated that increased 5-HT synthesis or turnover in the ascending serotonergic brainstem system may represent an important regional biochemical substrate of disorders of consciousness in metabolic catastrophes and, possibly, in other
disorders associated with coma. The pathogenesis and basic metabolic dysfunctions underlying the clinical syndrome of coma remain to be further elucidated.

ACKNOWLEDGMENTS

The authors are indebted to Dr.St. Wuketich, chief
of the Dept. of Pathology, Lainz-Hospital, for providing
the brain material and autopsy data, to Prof. Dr. K.
Irsigler and Dr. H. Schuster, chiefs of the Depts. of
Medicine I and III of Lainz-Hospital, for providing the
clinical data, to Mrs. E. Müller for excellent execution
of the assays, and to Mrs. M. Fuchs for secretarial work.

REFERENCES

1. Anton, A.H., Sayre, D.F. (1964): The distribution of
   dopamine and dopa in various animals and a method for
   their determination in diverse biological material.J.
   Pharm. exp. Ther. 145: 326-336.

2. Ashcroft, G.W., Sharman, F.D. (1962): Drug induced
   changes in the concentration of 5-OR indolyl compounds
   in the cerebrospinal fluid and caudate nucleus. Brit.J.
   Pharmacol. 19: 153-160.

3. Birkmayer, W., Danielczyk, W., Neumayer,E. (1974):
   Nucleus ruber and L-Dopa psychosis: biochemical postmortem findings. J. Neural Transm. 35: 93-116.

4. Birkmayer, W., Jellinger, K., Riederer, P. (1977):
   Striatal and extrastriatal dopaminergic functions.
   In: Psychobiology of the Striatum (Cools, A.R.,
   Lohman, A.H.M., Van den Bercken, J.H.L. eds.) Amster-
   dam-New York-Oxford: North Holland Publ. Comp. 141-
   153.

5. Cummings, M.G., Soeters, P.B., James, J.H., Klane,J.
   M., Fischer, J.E. (1976): Regional brain indoleamine
   metabolism following chronic portocaval anastomosis
   in the rat. J. Neurochem. 27: 501-509.

6. Curzon, G., Kantamaneni, B.D., Fernando, J.C., Woods,
   M.S., Cavanagh, J.B. ((1975): Effects of chronic portocaval anastomosis on brain tryptophan, tyrosine and
   5-hydroxytryptamine. J. Neurochem. 24: 1065-1070.

7. Davis, J.N.(1975): The adaptation of brain monoamine
   synthesis to hypoxia in the rat. J. Appl. Physiol. 39:
   215-220.

8. Davis, J.N., Carlsson, A. (1973): Effect of hypoxia
   on tyrosine and tryptophan hydroxylation in unanesthetized rat brain. J. Neurochem. 20: 913-915.

9. DiReda, N., Livrea,P.,De Blas, A. (1977): Effects of
   premortem conditions on HVA and 5-HIAA levels in human brain areas at autopsy. Abstr. 11th World Congr.
   Neurol. W.A.den Hartog-Jager, G.W. Bruyn, A.P.J.
   Heijstee (eds): Excerpta medica, Amsterdam, ICS Nr.
   427: 215.

10. Fernstrom, J.D., Wurtman, R.J.(1972): Brain serotonin
    content: physiological regulation by plasma neutral
    amino acids. Science 178: 414-416.

11. Fischer, J.E., Baldessarini, R.J. (1976): Pathogene-
    sis and therapy of hepatic coma. In: Progress in
    Liver Disease, Popper, H., Schaffner, F. (eds) New
    York: Grune & Stratton.

12. Fisher, J.E., Funovics, J.M., Falcao, H.A., Wesdorp,
    R.I.C. (1976): L-Dopa in hepatic coma. Ann. Surg.
    183: 386-391.

13. Gulyassy, P.F., Aviram, A., Peters, J.H. (1970): Eval-
    uation of amino acid and protein requirements in
    chronic uremia. Arch. Int. Med. 126: 855-859.

14. Hamon, M., Bourgoin, S., Morot-Gaudry, Y., Hery, F., Flowinski, J. (1974): Role of active transport of tryptophan in the control of 5-hydroxytryptamine synthesis. Adv. Biochem. Psychopharmac. 11: 153-162.

15. Heller, A. (1972): Neuronal control of brain serotonin. Fed. Proc. 31: 81-90.

16. Hess, S., Udenfried, S. (1959): A fluorometric procedure for the measurement of tryptophan in tissues. J. Pharm. exp. Ther. 127: 175-181.

17. Holm, E. (1975): Ammoniak und hepatische Enzephalopathie. Biochemie, Elektrophysiologie, Toxikologie. Stuttgart: G. Fischer.

18. Jakoby, W.B., Scott, E.M. (1959): Aldehyde oxidation. III. Succinic semialdehyde dehydrogenase. J. Biol. Chem. 234: 936-940.

19. Jellinger, K. and Riederer, P. (1977): Brain monoamines in metabolic (endotoxic) coma. J. neurol. Transm. 41: 275-286.

20. Jellinger, K., Riederer, P., Kothbauer, P.: Brain monoamines in human cerebral infarction. A preliminary study. Acta neuropath. (Berl.) 41: (in press).

21. Jouvet, M. (1969): Biogenic amines and the states of sleep. Science, 163: 301-306.

22. Klatzo, I. (1075): Pathophysiologic aspects of cerebral ischemia. In: The nervous system, D.B. Tower, ed. New York: Raven Press, 313-321.

23. Kogure, K., Scheinberg, P., Kishikawa, H. and Busto, R. (1976): The role of monoamines and cyclic AMP in ischemic brain edema. In: Dynamics of Brain Edema. H.M. Pappius and W. Feindell (eds.) Berlin-Heidelberg-New York: Springer-Verlag, 203-214.

24. Laursen, H., Westergaard, E.(1977): The blood-brain barrier to horseradish peroxidase in rats with porto-systemic encephalopathy. Neuropathol. appl. Neurobiol. 3: 20-43.

25. Lavyne, M., Moskowitz, M.A., Larin, F., Zervas, N. and Wurtman, R.J. (1975): Brain ^{3}H-catecholamine metabolism in experimental cerebral ischemia. Neurology (Min.) 25: 483-485.

26. Lee, L.W., Yatsu, F.M. (1974): ATP synthesis by mito-
    chondria of brain synaptosomes. J. Neurochem. 23:
    1081-1082.

27. Lesch, P. (1973): Toxische Metaboliten im Gehirn von
    Patienten mit Leberzirrhose und porto-cavalem Shunt.
    Dtsch.med.Wschr. 98: 1929-1931.

28. Manz, H. (1976): The pathology of cerebral edema.
    Human Path. 5: 291-313.

29. Meyer, J.S., Okamoto, S. and Shimaza, K. (1974): Dis-
    ordered neurotransmitters function demonstrated by
    measurement of norepinephrine and 5-hydroxytryptamine
    in CSF of patients with recent cerebral infarction.
    Brain 97: 655-664.

30. Moskowitz, M.A. and Wurtman, R.J. (1976): Acute stroke
    and brain monoamines. In: Cerebrovascular diseases, P.
    Scheinberg (ed.) New York: Raven Press: 153-166.

31. Mršulja, B.B., Lust, W.D., Mršulja, B.J. and Passonn-
    eau, J.V, (1977): Effect of repeated cerebral ische-
    mia on metabolites and metabolic rate in gerbil cortex.
    Brain Res. 119: 480-486.

32. Mršulja, B.B., Mršulja, B.J., Spatz, M., Ito, U.,
    Walker, J.T. Jr. and Klatzo, I. (1976): Experimental
    cerebral ischemia in mongolian gerbils. IV Behavior of
    biogenic amines. Acta neuropath. (Berlin) 36: 1-8.

33. Mršulja, B.B., Mršulja, B.J., Spatz, M. and Klatzo, I.
    (1976): Monoamines in cerebral ischemia in relation
    to brain edema. In: Dynamics of Brain Edema. H.M.
    Pappius and W. Feindell (eds.) Berlin-Heidelberg-
    New York: Springer-Verlag, 187-192.

34. Pasquier, D.A. and Reinoso-Suarez, F. (1977): Differ-
    ential efferent connections of the brainstem to the
    hippocampus in the cat. Brain Res. 120: 540-548.

35. Plum, F., Hindfelt, B. (1976): The neurological com-
    plications of liver disease. In: Handbook of Clinical
    Neurology (Binken, P.J., Gruyn, G.W. eds,) vol. 25
    Amsterdam-New York: North Holland, 349-377.

36. Radulovački, M., Buckingham, R.L., Chen, E.H. and
    Kovačević, R.(1977): Similar effects of tryptophan
    and sleep on costernal cerebrospinal fluid 5-hydroxy-

indolacetic acid and homovanillic acid in cats. Brain
Res. 129: 371-374.

37. Raskin, N.H. and Fishman, R.A. (1976): Neurologic disorders in renal failure. New Engl. J. Med. 206: 204-209.

38. Record, N.B., Prichard, J.W., Gallagher, B.B., Selig-
son, D. (1969): Phenolic acids in experimental uremia.
Arch. Neurol. 21: 387-400.

39. Riederer, P., Jellinger, K., Wuketich, St. (1978):
Postmortem changes of tryptophan in human brain. In
preparation.

40. Riederer, P., Wuketich, St. (1976): Time course of
nigrostriatal degeneration in Parkinson´s disease.
J. Neural Transm. 38: 277-301.

41. Robinson, R., Shoemaker, W., Schlupf, M., Valk, T. and
Bloom, F.E. (1975): Effect of experimental cerebral
infarction in rat brain on catecholamines and behavior.
Nature, 255: 332-334.

42. Siassi, F., Wang,M., Kopple, J.D. and Swendseid, M.E.
(1977): Plasma tryptophan levels and brain serotonin
metabolism in chronically uremic rats. J. Nutr. 107:
840-845.

43. Soeters, P.B., Fischer, J.E. (1976): Insulin, glucagon,
aminoacid imbalance and hepatic encephalopathy. Lancet
2: 880-882.

44. Sourkes, T.L., Young, S.N., Garelis, E., Lal, S.(1975):
Gradients of concentrations of tryptophan and 5-hy-
droxyindoleacetic acid (5-HIAA) in cerebrospinal fluid
(CSF). Acta vitam. enzym.(Milano) 29: 97-99.

45. Welch, K.M.A., Chabi,E., Buckingham,J., Bergin, B.,
Achar, V.S., Meyer, J.S. (1977): Catecholamine and
5-hydroxytryptamine levels in ischemic brain. Influence of p-chlorophenylalanine. Stroke 8: 341-346.

46. Welch, K.M.A., Chabi,E., Dodson, R.F., Wang, T.P.F.,
Nell,J. and Bergin, B. (1976): The role of biogenic
amines in the progression of cerebral ischemia and
edema: modification by p-chlorophenylalanine, methy-
sergide and pentoxyfilline. In:Dynamics of Brain Edema,
H.H. Pappius and W. Feindell (eds). Berlin-Heidelberg-
New York: Springer-Verlag, 193-202.

47. Westergaard, E. (1977): The blood-brain barrier to horseradish peroxidase under normal and experimental conditions. Acta neuropath. (Berlin) 39: 181-186.

48. Wurtman, R.J., Lavyne, M.H. and Zervas, N.T. (1975): Brain catecholamines in relation to cerebral blood vessels. In: Cerebral Vascular Disease, J.P. Whisnant and B.A. Sandok (eds). New York: Grune and Stratton: 13-27.

TRANSPORT PHENOMENA IN CEREBRAL ISCHEMIA

M. Spatz, D. Mićić, T. Fujimoto,
B.B. Mršulja and I. Klatzo

Laboratory of Neuropathology and
   Neuroanatomical Sciences
National Institute of Neurological and
   Communicative Disorders and Stroke
National Institutes of Health
Bethesda, Maryland 20014, U.S.A.

The primary sequelae of cerebral ischemia are due to the lack of adequate blood circulation, which is the cause of insufficient supply of oxygen and nutrients to the brain. This circulatory disturbance is also one of the greatest obstacles to direct in vivo studies of processes involved in the transport of substrates into and out of the brain. Therefore, our previous investigations concerned with glucose transport in cerebral ischemia were performed at various periods following the reestablishment of cerebral blood circulation (11, 12, 14). Moreover, these limitations on the in vivo studies and the fact that the observed transport phenomena could have occurred in one or more sites such as capillary endothelium, glia and neurons have led so far to separate investigations of the cerebral microvessels and synaptosomes function.

In this report, we will review briefly: (1) the transport of glucose analogues across the blood-brain barrier (BBB) after long- and short-term ischemia as well as (2) the uptake of 2 deoxy-D-[^{3}H]glucose ([^{3}H]2-DG) and of (3) some neutral amino acids in synaptosomes and in cerebral microvessels separated from ischemic and postischemic brains. In addition, we will (4) describe the effect of oxygen deprivation on the capillary uptake of [^{3}H]2-DG and compare it with the

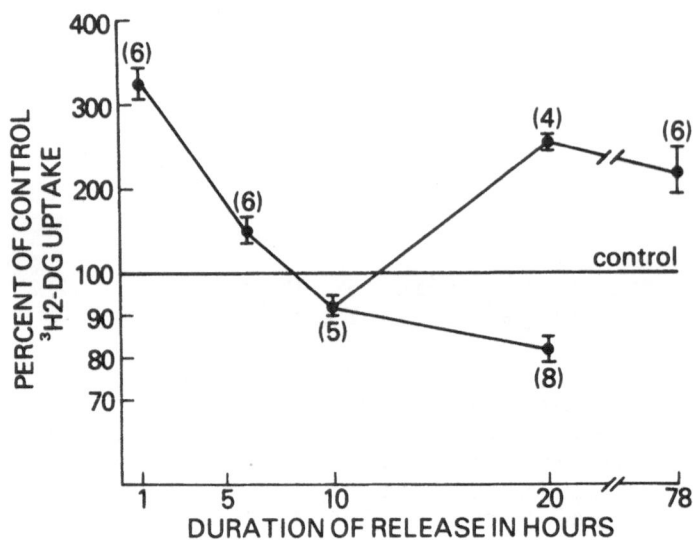

Fig. 1. Effect of postischemia on 2-deoxy-D-[^{3}H]glucose
brain uptake. The gerbils were subjected to unilateral common carotid artery clipping for 1
hour and various periods of clip release. The
brain uptake was determined by Oldendorf's
double-isotope technique using [^{14}C]butanol as
a reference substance. The values are means ±
S.E.M. of 4-8 experiments (as indicated by the
numbers in parentheses).

effect on the amino acids tested in order to shed some
light on capillary function as the mediator of brain
transport phenomena.

1. POSTISCHEMIC EFFECT ON 2-DEOXY-D-[^{3}H] GLUCOSE
BRAIN UPTAKE

A. Long-term ischemia. An increased [^{3}H]2-DG brain
uptake index (BUI) was found in the gerbil's cerebral
hemisphere ipsilateral to left common carotid artery
clipping for 6 and 18 hours and 1-hour clip release, the
increase amounting to 155 and 172% of control brain uptake, respectively. The augmented [^{3}H]2-DG uptake was
inhibited with unlabeled 3-0-methyl-glucose (3-MG) to
the same degree in experimental as in control animals,

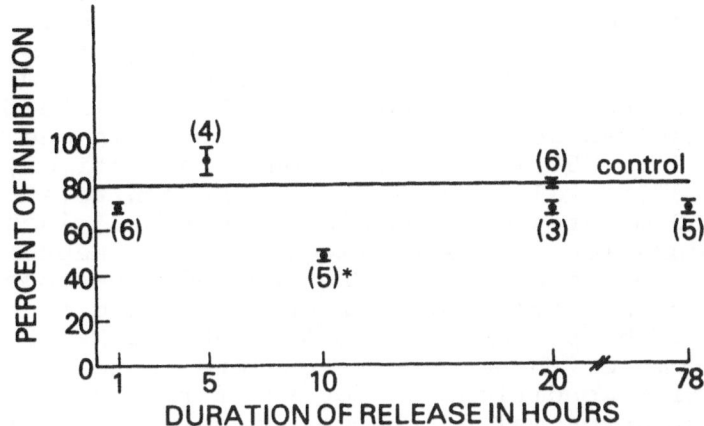

Fig. 2. Effect of unlabeled 3-0-methyl-D-glucose on the
2-deoxy-D-[^{3}H]glucose brain uptake. The experimental procedures were the same as described in
Fig. 1 except that 60 mM of 3-0-methyl-D-glucose
was added to the labeled injectate.

with the exception of 18-hour-clipped gerbils, suggesting a saturable uptake in the former but not in the
latter group (12).

    B. <u>Short-term ischemia</u>. As can be seen from Fig. 1,
an increased BUI for [^{3}H]2-DG was seen in the affected
cerebral hemisphere of gerbils with reestablished cerebral blood circulation for 1-72 hours, except for 10
hours after common carotid artery occlusion of 1-hour
duration. This uptake was inhibited by cold 3-MG, except
for 10 hours of postischemia (Fig. 2). At the same time
(10 hours after clip release), the peak of the altered
BBB permeability to substances of molecular size similar
to that of albumin was evident by the percentage increase
in relative uptake of dextran and by the extravasated
Evans blue dye in the affected cerebral hemisphere. However, an increased BBB permeability to small molecules
such as sucrose occurred earlier and persisted for a
longer period of time than for albumin, suggesting a
selective vulnerability of the BBB during the postischemic period (14).

2. SYNAPTOSOMES

A. Effect of ischemia and postischemia on [^{3}H]2-DG uptake. The specific uptake of [^{3}H]2-DG was progressively decreased in affected synaptosomes obtained from the brains of gerbils subjected to unilateral ischemia for 30-180 min. The specific entry of [^{3}H]2-DG into ischemic synaptosomes was decreased from 23 to 53% in one group and from 45 to 75% in another group, respectively, when compared to the relative uptake of the controls (100%) after 30-60 minutes of ischemia. In cerebral ischemia of 3 hours' duration, the specific [^{3}H]2-DG synaptosomal uptake was completely lost. The reduced synaptosomal [^{3}H]2-DG uptake could not be restored by addition of high-energy phosphates such as ATP, ADP, cAMP and PEP, although some of these substrates were found to decrease under similar conditions. Recovery of the specific [^{3}H]2-DG synaptosomal uptake was seen in 4 out of 6 and in 2 out of 4 gerbils exposed to 5 minutes of cerebral blood flow recirculation after 60 and 180 minutes of carotid artery occlusion. Complete restoration of the synaptosomal transport function for [^{3}H]2-DG was observed 60 minutes following the release of carotid artery occlusion of 60-180 minutes' duration (13).

A decrease in the specific [^{3}H]2-DG synaptosomal entry was also observed in animals with bilateral occlusion of the common carotid artery for 15 minutes, which ischemic effect is usually equivalent to 30 minutes of unilateral arterial clipping.

B. Effect of ischemia on neutral amino acid uptake. No significant alteration in the uptake of [^{3}H]isoleucine, [^{14}C]cycloleucine and [^{3}H] phenylalanine was observed in synaptosomes isolated from the brains of animals subjected to bilateral carotid artery clipping for 1-30 minutes, except for 3 minutes of arterial occlusion. At this time the synaptosomal uptake of [^{3}H]isoleucine, [^{14}H]-cycloleucine and [^{3}H]phenylalanine was increased 25-46% above the control values, depending on the buffer used for the incubation of the synaptosomes (a higher level of uptake was found in Ringer than in K phosphate sucrose buffer, unpublished observation).

3. CEREBRAL MICROVESSELS

A. Effect of ischemia and postischemia on [^{3}H]2-DG uptake. The capillary uptake of [^{3}H]2-DG was found to be decreased (40% lower than control values) in cerebral

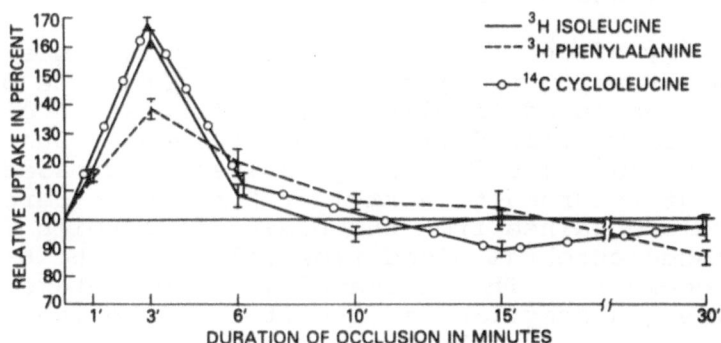

Fig. 3. Uptake of neutral amino acids in cerebral
capillaries isolated from brains of gerbils with
bilateral occlusion of the common carotid artery.
Each point represents the mean ± S.E.M. of 4-16
experiments.

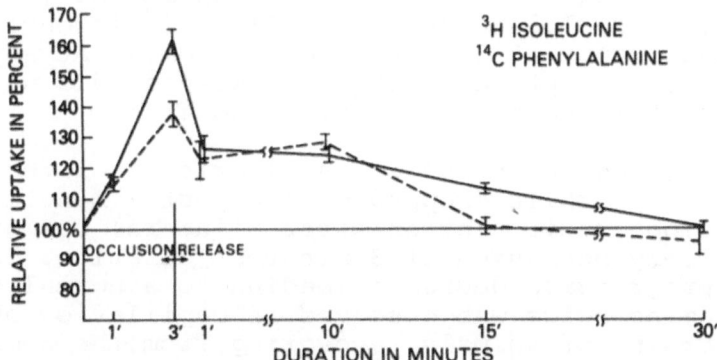

Fig. 4. Uptake of isoleucine and phenylalanine in cerebral capillaries isolated from brains of gerbils
subjected to bilateral common carotid artery
occlusion for 3 minutes and 1-30 minutes of reestablished cerebral blood circulation. Each point
represented the mean ± S.E.M. of 3-8 experiments.

microvessels isolated from the brains of gerbils exposed
to bilateral common carotid artery occlusion for 6 minutes
but not for 1 and 3 minutes.

Recovery of the specific $[^3H]$2-DG uptake was found
at 5-15 minutes in capillaries isolated from the brains
of animals subjected to bilateral recirculation of cerebral blood flow after 6 minutes of occlusion. Subsequently, the capillary $[^3H]$2-DG progressively increased up to
3 hours, but returned to normal at 5 hours in cerebral
microvessels obtained from the brains of gerbils with
reestablished cerebral blood flow following the 6-minute
arterial occlusion. The increased $[^3H]$2-DG uptake in the
capillaries was specific because it could be inhibited
to the same degree in the experimental as in the control
animals (15).

B. Effect of ischemia and postischemia on neutral
amino acid uptake. Since it has been generally accepted
that many of the amino acids enter and/or leave the brain
by energy-dependent process (6, 8, 10), we investigated
the effect of ischemia and postischemia on their uptake
into the cerebral microvessels. In order to assess the
effect on transport rather than on metabolism, we had
studied so far some representatives of neutral amino
acids because of the availability of the nonmetabolizable
analogue cycloleucine. The capillary uptake of $[^3H]$iso-
leucine, $[^{14}C]$cycloleucine and $[^{14}C]$ or $[^3H]$phenylalanine
was increased only at 3 minutes of ischemia. At any other
time tested, the uptake of these substances did not differ from that in controls (Figs. 3 and 4).

This increased uptake in capillaries isolated from
the brains of gerbils dropped within just 1 minute of
cerebral blood recirculation after bilateral common
carotid artery occlusion of 3 minutes' duration. There-
after, a progressive decrease leading to a normal uptake
of these amino acids was observed in capillaries obtained
from the brains of animals recovering from the 3-minute
occlusion for 10-30 minutes (Fig. 4). In contrast, $[^{14}C]$-
glutamine capillary uptake was not altered during ischemia.

Considering the postischemic increase of $[^3H]$2-DG
capillary uptake after 6 minutes of ischemia, we also
determined the uptake of these amino acids into cerebral
capillaries isolated from the brains of gerbils subjected
to 6-minute release of bilateral carotid clipping that
had continued for 1-120 minutes.

As can be seen from Fig. 3, the capillary uptake of [^{3}H]isoleucine, [^{14}H]cycloleucine and phenylalanine was unchanged at 6 minutes of ischemia as compared to controls. However, the entry of these amino acids progressively increased up to 30 minutes and returned to normal at 1 hour of postischemia. Fo example, the capillary [^{3}H]isoleucine uptake was 97.7 + 5.2, 124 + 2.4, 135 + 4.7 and 159.7 + 8.2% of control values (100%) at 1, 10, 15 and 30 min.

Both the ischemic and postischemic increased amino acid entry into the cerebral capillaries was specific, since they could be inhibited to the same degree as in the controls by various concentrations of the unlabeled substances added to the isotope incubation solution.

In summary, the entry of [^{3}H]2-DG into the cerebral microvessels and synaptosomes was affected differently than the entry of the amino acids tested since the [^{3}H]2-DG uptake was decreased while that of the amino acids was transiently increased in ischemia. These studies suggested an oxygen-dependent uptake of the former and an oxygen-independent entry of the latter substances under pathological conditions. In order to evaluate this possibility, cerebral microvessels isolated from normal brains were exposed to pure nitrogen and to various oxygen-containing gases, such as breathing and exhaling air, N_2 95%/O_2 5% and 100% O_2.

4. ANOXIA

A. Effect on [^{3}H]2-DG capillary uptake. Cerebral microvessels incubated in pure nitrogen atmosphere showed a 71% reduction in [^{3}H]2-DG uptake, which could be completely recovered by substituting the nitrogen with oxygen-containing gas mixture. Addition of free fatty acid albumin (FFA Alb) to the incubation solution prevented such a decreased uptake of [^{3}H]2-DG into the cerebral capillaries under anaerobic (anoxic) conditions. The protective effect could be easily disturbed by addition to the medium of one or more saturated or unsaturated fatty acids and FFA Alb. Moreover, some of the fatty acids such as palmitate, stearate and oleate interfered with the [^{3}H]2-DG capillary uptake under aerobic conditions. A reduction of the aerobic and anaerobic [^{3}H]2-DG capillary uptake was also observed with the addition of ATP and EDTA. However, partial recovery of the [^{3}H]2-DG uptake was seen with cAMP or $MgCl_2$ in the aerobic phase. The improved anoxic capillary hexose uptake with cAMP

suggests that the entry of $[^3H]$2-DG is not only an oxygen-
dependent but also to some degree an energy-dependent
process, especially since the addition of metabolic inhibitors such as dinitrophenol and KCN inhibited, in 93
and 95%, respectively, the capillary $[^3H]$2-DG uptake
under aerobic and anaerobic conditions. Furthermore, the
interference of free fatty acids, ATP, and EDTA with the
normal aerobic and/or anaerobic $[^3H]$2-DG capillary uptake
suggests that glucose uptake may rely on the structural-
chemical integrity of the plasma membrane and may be related to fatty acid metabolism (irrespective of whether
the fatty acids are of cellular or membranous origin),
because partial recovery of $[^3H]$2-DG capillary uptake
took place not only with cAMP but also with either $MgCl_2$
or FFA Alb. Details of this study will be published elsewhere (9), and we shall not dwell on it here. However,
the FFA impairment of the FFA Alb protective effect is
worth reiterating, since the release of free fatty acids
in ischemia could be one of the main factors responsible
for the decreased glucose uptake across the blood-brain
barrier. This possibility is substantiated by the described release of free fatty acids such as stearic and
arachidonic in so-called cerebral ischemia due to decapitation or microwave irradiation (1, 2, 5). The mechanism
responsible for the postischemic increase of hexose uptake remains unknown, but most probably is also related
to a metabolic event which may be associated with the
carrier-mediated process.

B. Effect on capillary uptake of neutral amino acids.
The exposure of cerebral microvessels to a nitrogen atmosphere for 15-30 minutes did not interfere with the
normal uptake of the tested amino acids in a Na^+-free
medium. However, an increase of $[^3H]$isoleucine and $[^3H]$-
phenylalanine uptake (20-30% above the control levels)
was seen in the presence of Na^+ in the incubating solution under anoxic conditions only. The addition of metabolic inhibitors such as dinitrophenol, KCN and NaFl had
no effect on the uptake of these amino acids in cerebral
capillaries exposed to pure oxygen or nitrogen.

Therefore, it appears that the uptake of the tested
neutral amino acids is Na^+-dependent in anoxia and possibly in cerebral ischemia. Moreover, these studies suggest that the passage of these amino acids across the
BBB may occur by oxygen- and most probably energy-inde-
pendent mechanism, which also was surmised from the
study of cerebral anoxia in the dog by Betz et al. (3).
However, our previous studies of hypoxic rabbits in vivo

showed a decreased brain uptake index for L-leucine when
compared to the controls. Furthermore, Lajtha and co-
workers described leucine brain efflux against a concentration gradient of elevated plasma levels in rats and
mice (6, 8) and a partially Na-dependent high-affinity
uptake of neutral amino acids in brain slices (7). These
contrary observations could be the result of species
differences and/or blood flow in the former, or, in the
latter model, the expression of the overall function,
which might be lost in the fractions of the brain such
as microvessels and synaptosomes, especially since
Chaplin et al. (4) also described a Na-independent leucine transport but not metabolism in the synaptosomes.
Furthermore, the observations made in the brain slices
may reflect the function of glia, which has not yet been
studied, or the result of anoxia.

Nevertheless, on the basis of our observations it
can be assumed that the specific entry of neutral amino
acids into the cerebral capillaries is saturable (unpublished observation) and oxygen- and Na-independent, while
the capillary uptake of glucose is an oxygen-dependent
but Na-independent carrier-mediated process as determined under similar conditions. Therefore, the oxygen
deprivation most likely dominates the mechanism by which
the carrier-mediated transfer of these substances from
blood to brain may or may not be affected during cerebral ischemia.

SUMMARY

The purpose of this paper has been to review the investigations on the effect of cerebral ischemia on 2-DG
uptake in the brain as well as on the uptake of this
glucose analogue and some neutral amino acids in isolated
synaptosomes and cerebral microvessels. The most important observations were as follows: (1) the noted postischemic increase of 2-DG uptake in the brain was also
found to occur in isolated cerebral microvessels; (2)
cerebral ischemia led to a decreased 2-DG uptake and a
transiently increased uptake of some neutral amino acids
in synaptosomes and cerebral capillaries; (3) a similar
effect on the capillary uptake of these substances was
observed when isolated cerebral microvessels were exposed
to a nitrogen atmosphere in vitro; (4) the effect of
cerebral ischemia on the capillary uptake of 2-DG and the
tested amino acids is most likely the result of an altered
plasma membrane function due to the deprivation of oxygen,
which appears to be essential for the normal capillary
uptake of 2-DG but not for that of the neutral amino
acids.

ACKNOWLEDGMENTS

The authors wish to thank Miss Madora Swink for technical assistance and Mrs. J. Phelps and Ms. J. Yutzey for secretarial help.

REFERENCES

1. Aveldano, M. I. and Bazan, N. G., Jr. (1975): Rapid production of diacylglycerols enriched in arachnidonate and stearate during early brain ischemia. J. Neurochem. 25: 919-920.

2. Bazan, N. G., Jr. (1971): Changes in free fatty acids of brain by drug-induced convulsions, electroshock and anaesthesia. J. Neurochem. 18: 1379-1385.

3. Betz, A. L., Gilboe, D. D. and Drewes, L. R. (1975): Kinetics of unidirectional leucine transport into brain: effects of isoleucine, valine, and anoxia. Am. J. Physiol. 228: 895-900.

4. Chaplin, E. R., Goldberg, A. L. and Diamond, I. (1976): Leucine oxidation in brain slices and nerve endings. J. Neurochem. 26: 701-707.

5. Galli, C. and Spaguolo, C. (1976): The release of brain free fatty acids during ischemia in essential fatty acid deficient rats. J. Neurochem. 26: 401-404.

6. Lajtha, A. (1968): Transport as control mechanism of cerebrl metabolite levels. Prog. Brain Res. 29: 201-214.

7. Lajtha, A. and Sershen, H. (1975): Inhibition of amino acid uptake by the absence of Na^+ in slices of brain. J. Neurochem. 24: 667-672.

8. Lajtha, A. and Toth, J. (1961): The brain barrier system - II Uptake and transport of amino acids by the brain. J. Neurochem. 8: 216-225.

9. Mićić, D., Mićić, J., Swink, M. E. and Spatz, M. (1978): The anoxic effect on 2-deoxy-D-[^{3}H]glucose uptake in the isolated cerebral capillaries. Proc. Soc. Exp. Biol. Med. 158: 318-322.

10. Quastel, J. H. (1965): Molecular transport at cell membranes. Proc. R. Soc. London (Biol.) 163: 169-196.

11. Spatz, M., Berson, F., Fujimoto, T. and Klatzo, I.
    (1976): Transport of nutrients and non-nutrients
    across the blood-brain barrier in pathological con-
    ditions. In: The Cerebral Vessel Wall, Cervós-Navarro,
    J., ed. Raven Press, New York, 225-232.

12. Spatz, M., Go, G. K. and Klatzo, I. (1974): The ef-
    fect of ischemia on the brain uptake of [^{14}C]glucose
    analogues and [^{14}C]sucrose. In: Pathology of Cerebral
    Microcirculation, Cervós-Navarro, J., ed. Walter de
    Gruyter & Co., Berlin, 361-336.

13. Spatz, M., Mršulja, B. B., Mršulja, B. J. and Klatzo,
    I. (1976): Recovery of decreased synaptosomal 2-deoxy-D-
    [^{3}H]glucose uptake after cerebral ischemia in Mon-
    golian gerbils. Brain Res. 103: 193-198.

14. Spatz, M., Fujimoto, T. and Go, G. K. (1976): Trans-
    port studies in ischemic cerebral edema. In: Dynamics
    of Brain Edema, Pappius, H. M. and Feindel, W., eds.
    Springer-Verlag, Berlin-Heidelberg-New York, 181-186.

15. Spatz, M., Mršulja, B. B., Mićić, D., Mršulja, B. J.
    and Klatzo, I. (1977): Ischemic and post-ischemic
    effects on 2-deoxy-D-[^{3}H] glucose uptake in cerebral
    capillaries. Brain Res. 120: 141-145.

ENERGETICS, THERMAL SENSITIVITY AND TOLERANCE MECHANISMS

OF BRAIN SYNAPTOSOMES: A COMPARATIVE STUDY

R.K. Andjus and T. Ćirković

Department of Cryophysiology and Bioenergetics
Institute for Biological Research
29 Novembra 142, 11 000 Belgrade

Department of Physiology, University of Belgrade
School of Sciences, Belgrade, Yugoslavia

ABSTRACT

In an attempt to correlate neuronal energy metabolism with thermal tolerance by means of a comparative approach, the rate of D-[^{14}C]glucose conversion to $^{14}CO_2$ has been studied as a function of temperature (38-0°C) in brain synaptosomes from animal species characterized by different tolerances. The rat served as a reference homeotherm species; the heat-sensitive but cold-tolerant dogfish shark (Scyliorhinus canicula) represented a poikilotherm model of temperature in lower vertebrates, and was also chosen because of its known refractoriness to edema-inducing agents peculiar to the shark brain; the ground squirrel (Citellus citellus) was chosen for its coupled tolerance to hypothermia and anoxia, peculiar to hibernating mammals. In order to contribute by thermodynamic considerations to the understanding of molecular mechanisms underlying tolerance, apparent energies of activation (E_a), temperature coefficients (Q_{10}), transition or critical temperatures (T_c), and apparent dissociation constants have been derived from comparative experiments with synaptosomes in standard incubation media, and in media altered by the elimination of Na^+ or the addition of ouabain. The consequences of uncoupling oxidations by means of 2,4-dinitrophenol (2,4-DNP) were also investigated. In addition, the effects of reduced concentrations of the substrate (D-glucose) and of the competitive

inhibition of its utilization by 2-deoxy-D-glucose (2-DG)
were explored.

Sharp discontinuities in Arrhenius plots of glucose
oxidation rates occurred at critical temperatures (T_c)
coinciding with thermal limits of survival in hypothermia
(rat) or with upper tolerance (dogfish). Such discontinuities were absent in the case of synaptosomes from hibernating ground squirrels known to tolerate the entire
range of temperatures explored. These results suggest a
limiting role of neuronal energy metabolism in determining tolerance to extreme body temperatures. The comparison of E_a and Q_{10} values obtained with synaptosomes of
different animal origin and in particular thermal ranges
also revealed a close correlation with the general thermophysiological features of the respective species.

Particularly pertinent to the understanding of underlying mechanisms were the results concerning the
Na^+-stimulated fraction of oxidations linked to active
cation transport. Resistance to the inhibitory effects
of cooling, indicated by the maintenance of relatively
low E_a values, appeared in all our animal models to be
due to an increasing contribution of Na^+-stimulated oxidations with falling temperature. The transition to
greatly elevated Q_{10} values below T_c appeared, on the
other hand, to be a corollary of the failure at subcritical temperatures of this sodium-stimulated and
transport-linked fraction of oxidations. Increased E_a
values were also recorded after the addition of 2,4-DNP,
which is known to interfere with active cation transport.
However, the dose and temperature dependence of the effect of uncoupling appeared in addition to be a correlate
of the glycolytic capacity of brain tissue and of tolerance to hypothermia and anoxia within the comparative
animal series. The known effect of ouabain in preventing
Na^+ stimulation in rat synaptosomes, which emphasizes
the responsibility of the transport ATPase for the so-
dium-stimulation phenomenon, has been confirmed in experiments with shark synaptosomes. In rat brain synaptosomes, thermal sensitivity parameters and thermodynamic
functions showed a close relationship to the temperature-
dependence characteristics of an isolated rat-brain transport ATPase preparation described by other authors. Reduced concentrations of the substrate, as well as inhibition by 2-DG markedly depressed D-glucose oxidation
rates in rat synaptosomes irrespective of temperature,
but had comparatively little effect on thermal sensitivity. However, apparent dissociation constants derived
from Dixon plots, as well as Q_{10} and E_a values obtained

with reduced substrate concentrations, indicated an increase with falling temperature, down to the T_c level, of the underlying enzyme-substrate affinities. In contrast, decreasing affinities were indicated below T_c. Coupled to the reduction of available thermal energy, they suggest a rapidly increasing incompatibility with the maintenance of efficient energy-yielding metabolic processes at subcritical temperatures.

INTRODUCTION

The results reported herein stem from a broader long-term study undertaken to elucidate, from the standpoint of survival and revival mechanisms, the role of brain energetics in tolerance to extreme conditions, particularly those created by anoxia and extreme hypothermia (2, 4). The basic approach is a comparative one, involving different animal species, developmental stages and physiological states as comparative models of tolerance. It combines kinetic studies of energy metabolism of brain in situ, studies of physiological tolerance parameters and investigations of energy metabolism at the neuronal level. Isolated synaptosomes have been used within the framework of these studies primarily as a model system capable of reflecting basic metabolic features of the neuron involving its energy-yielding and energy-utilizing capacity.

It was previously shown with synaptosomes of the rat that glucose enters the synaptosome by a high-affinity carrier-mediated system, becomes oxidized by intact synaptosomes and supports intrasynaptosomal synthesis of ATP and creatine-phosphate (7). The rate of D-[^{14}C]glucose conversion to $^{14}CO_2$ in synaptosomes has been chosen as particularly suitable for our thermophysiological analysis of tolerance phenomena, based on thermodynamic considerations.

Thermodynamic functions, the apparent energy of activation in particular, have been used in our study primarily as useful comparative indicators of thermal sensitivity (5). The sodium-stimulated fraction of oxidations has been chosen as a possible indicator of the involvement in tolerance mechanisms of the active transport linked to membrane activities. This analysis was completed by the use of an uncoupling agent (2,4-dinitrophenol) and of ouabain. Other procedures of experimental analysis, analogous to those frequently used with isolated enzyme preparations and also providing useful information when applied to the system, have been included in our analysis of tolerance mechanisms. Such was the procedure of deriving apparent dissociation constants from Dixon plots of synaptosomal glucose oxidation rates at reduced substrate concentrations in the medium (7). Low substrate concentrations were also used in view of their frequently underlined role in revealing physiologically meaningful patterns of thermal sensitivity (11).

The European ground squirrel (Citellus citellus) was used as an experimental model of the coupled tolerance to extremely low body temperatures and to anoxia peculiar to hibernating animals. The European small-spotted dogfish (Scyliorhinus canicula), characterized by its primitive shark brain, was used, after acclimation to cold (10°C) in thermostatistically controlled aquaria, as a cold-tolerant and heat-sensitive poikilotherm model of temperature tolerance. In addition, a remarkable refractoriness to agents inducing experimental brain edema, demonstrated by Klatzo and Steinwall (18), represents another tolerance characteristic of the shark brain which determined the inclusion of a shark species in our comparative animal series. Hooded rats served as the reference homeotherm species, with well-established survival and revival limits of tolerance to hypothermia. Although still incomplete, comparative data concerning characteristics and underlying mechanisms have been obtained with synaptosomes of these three animals.

MATERIALS AND METHODS

Adult rats (Mill-Hill hooded) weighing 170-203 g were acclimated to a temperature of 23°C prior to being sacrificed by decapitation (71 rats; average brain tissue yield per animal: 1.013 g). Ground squirrels (Citellus citellus) were used after prolonged periods of hibernation (5 to 6 months) in artificial hibernacula at 6 to 7°C. They were decapitated while in the state of deep hibernation (rectal temperature 6-7°C; 8 males and 4 females; 151.7 ± 9.9 g mean body weight; average weight loss during hibernation: 90.4 g; mean brain weight: 1.861 g). In other ground squirrels, hibernation was prevented by constantly elevated environmental temperature (30°C). Their body temperature was maintained at the euthermic level during the season of hibernation and they were sacrificed at the same time as the hibernating animals (6 males; 241.3 ± 9.3 g mean body weight; average brain tissue yield: 1.967 g per animal). Dogfish sharks (Scyliorhinus canicula), caught near the South Adriatic coast at a depth of approximately 100 m, were used after 1 month of adaptation to thermostatically controlled cold-water

aquaria (10°C) with recirculating, biologically filtered
sea water. They were decapitated after being rapidly cooled in sea water to about 3°C (25 females and 42 males;
168.0 $\pm$ 4.3 g mean body weight; average individual brain
tissue yield: 0.534 g).

In experiments with rats and ground squirrels, synaptosomes were prepared and tested for D-[^{14}C]glucose oxidation to $^{14}CO_2$. The procedures used are described in
detail by Diamond and Milfay (8) and Diamond and Fishman
(7), so that only the main modifications need be mentioned. Instead of sucrose gradients, discontinuous Ficoll-
sucrose gradients were used, consisting of 2 ml 13% (wt/
vol) Ficoll in 0.32 M sucrose and 2 ml 7.5% (wt/vol)
Ficoll 0.32 M sucrose. The final concentration of NaCl in
the incubation media was slightly increased and amounted
to 80 mM. In experiments with Na$^+$-free media, NaCl was
replaced by an osmotically equivalent concentration of
sucrose. The concentration of unlabeled glucose was 8mM
in standard media, and it was reduced to 0.14 and 0.035mM
in experiments with low substrate concentrations. In experiments with high and low glucose concentrations, the
amount of the added D-[U-^{14}C]glucose was adjusted to obtain specific activities of the order of 10^8 and 10^{10}
cpm/mM, respectively. Synaptosomal suspensions, prepared
from homogenized pooled brains, were used in aliquots of
0.25 or 0.3 ml, combined with the incubation medium in a
total volume of 2.5 ml per test vial. Prior to use, the
suspension was adjusted by dilution with the homogenizing
medium so as to provide amounts of synaptosomes per aliquot equivalent to 200-250 mg of the initial material
("brain weight equivalents"). Incubation was carried out
in closed vials immersed in thermostatically controlled
shaking baths. A circular piece of Whatman No. 1 filter
paper was attached to the inner surface of the vial lid
and saturated with 10% KOH to trap evolved $^{14}CO_2$. It was
subsequently transferred to counting vials containing
the scintillation solution. Since low incubation temperatures were included in the study, longer incubation
times were required (up to 2 hours at 0°C).

For the preparation of synaptosomes from dogfish
brains, a different procedure had to be worked out, and
will be described in detail in a separate publication,
together with other data concerned specifically with
dogfish synaptosomes and their functional and ultrastructural characteristics (Andjus, Diamond, Ćirković and
Rogač). The main differences, in comparison to the already described procedures applied to mammalian synaptosomes, were as follows: In order to match the internal

environment, characterized as in other marine elasmo-
branchs by an elevated osmotic pressure and urea content,
we used urea-sucrose homogenizing media, similar to those
applied earlier to Octopus preparations (15) and consisting of 0.7 M sucrose and 0.3 M urea. The supernatant obtained after centrifugation at 1000g was subjected to a
60-min centrifugation at 17300g, and synaptosomes were
recovered directly from the floating pellicle. The pellicle material was transferred to another tube and allowed to form again, after standing 15 min, a floating synaptosome-rich layer. After the clear portion underneath
was discarded, the synaptosome-rich material was diluted
with the homogenizing medium. Aliquots of the synaptosomal suspension amounted to 150-200 mg of the initial
brain weight. They were added to the incubation medium
to make a total volume of 2 ml per test vial. The medium
differed from the one used with rat and ground squirrel
synaptosomes by the following concentrations: 300 mM NaCl,
330 mM urea, 170 mM sucrose and 10 mM glucose.

RESULTS

Apparent energy of activation and thermophysiological tolerance. Figure 1 shows that sharp discontinuities
in Arrhenius plots of glucose oxidation rates occur at
different "critical" temperature levels in homeotherm
(rat) and poikilotherm (shark) synaptosomes. In the case
of rat synaptosomes (uppermost curve) the break is situated in the vicinity of 15°C (see also Fig. 2). It marks
the transition from a relatively warm region of continuous Arrhenius linearity (heavy uninterrupted straight
lines in the figures), characterized by physiological E_a
and Q_{10} values, to a cold region of grossly elevated
activation energies and more than doubled temperature
coefficients (see also Table I). The temperature at which
this transition occurs coincides strikingly with the body
temperature known as survival-limiting in the hypothermic
rat. According to Adolph (1), the colonic temperature of
$15.1 \pm 0.5°C$ represents the lethal limit of cold immersion
in the adult rat, and our own whole-animal studies (3)
showed that below 15°C the rat loses the capacity for
spontaneous recovery unassisted by external heat: even
in cold-acclimated rats, the critical body temperature
of rewarming is situated between 14 and 15°C.

In dogfish synaptosomes (lowest curve in Fig. 1),
the break in the Arrhenius plot occurs at a higher level
(above 22°C), and it is of a different nature. It marks
the transition from a cold region of Arrhenius linearity
and physiological E_a and Q_{10} values to a warm region of

Table I. Thermodynamic Indicators of Thermal Sensitivity and of Its Experimentally Induced Alterations

Medium	μM	Range (°C)	Energy of activation Cal/mole^{-1}	Temperature coefficient 25-15°	15-0°
RAT					
Control		38-15	16,499+1109	2.63	5.50
(8 mM gluc., 80 mM NaCl)		38-15	16,064∓ 375	2.56	-
Low glucose	140	38-15	17,831+ 157	2.85	6.31
	35	38-15	15,418∓ 514	2.47	9.02
Low glucose + 2-DG	140-100	38-15	15,922+1298	2.64	6.07
	35-100	38-15	15,418∓1575	2.47	6.17
2,4-dinitrophenol	80	38-15	24,465+ 915	4.19	8.71
Na$^+$-free (sucrose)		38-11	25,110+ 647	4.36	-
GROUND SQUIRREL (hibernating)					
Control		38-0	18,578+ 915	2.97	3.21
(8 mM gluc., 80 nM NaCl)		38-0	18,340∓ 361	2.92	3.16
		38-0	16,269∓1756	2.60	2.78
2,4-dinitrophenol	80	38-11	19,563+1269	3.14	9.87
	80	15-0	36,430∓2060	3.14	9.87
Na$^+$-free (sucrose)		38-0	21,984+ 761	3.63	3.98
SHARK (adapted to 10°)					
Control		21-0			
(10 mM gluc., 300 mM NaCl		21-0	16,607+ 549	(1.92)	2.84
2,4-dinitrophenol	80	13-0	27,851+2711	(1.58)	5.76
	40	13-0	27,630∓2086	(1.36)	5.69
Ouabain	100	21-0	17,976+ 523	(2.50)	3.10
Na$^+$-free (sucrose)		21-0	22,726+ 310	(3.63)	4.17

Data derived from temperature-dependence studies of D-[^{14}C]glucose oxidation rates in rat, ground squirrel and dogfish synaptosomes.

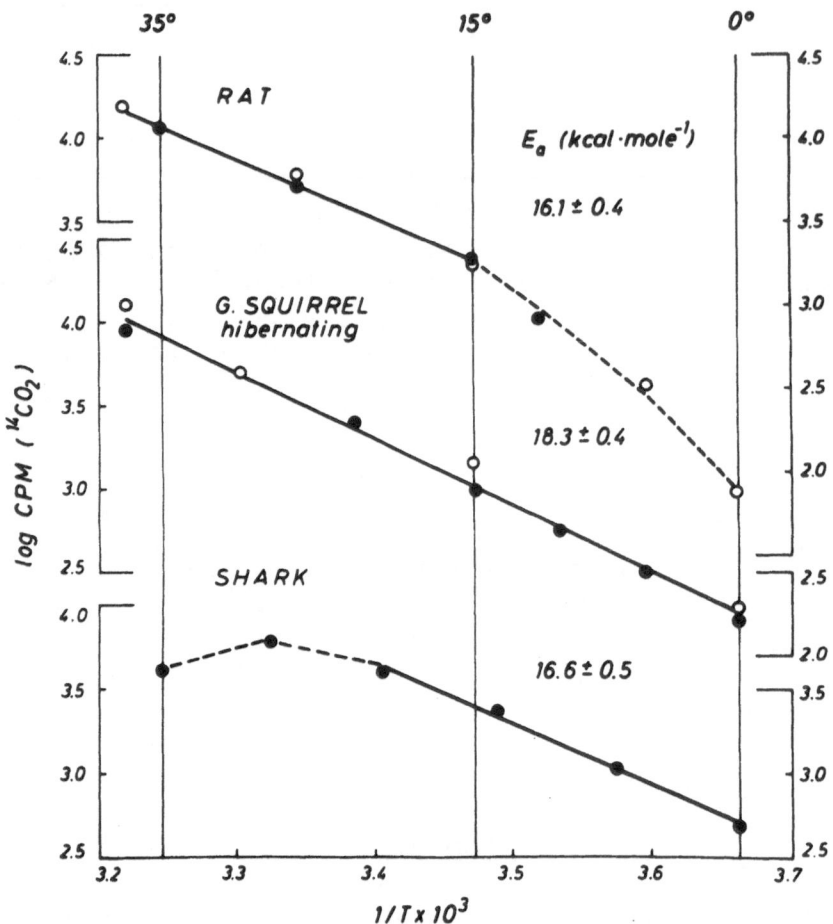

Fig. 1. Interspecific differences in thermal sensitivity
of brain synaptosomes as revealed by Arrhenius
plots of D-[^{14}C]glucose oxidation rates. Apparent
energies of activation indicated in the graph
(E_a) concern only the ranges of temperature covered by uninterrupted straight Arrhenius lines.
(•) Same pool of synaptosomes; (o) separate pools
at each temperature.

abnormally low or even inverted temperature coefficients
and "inactivation energies" (Table I). The temperature
at which the break occurs coincides with the upper limit
of heat tolerance of this cold-resistant but heat-sensitive bottom-dwelling shark species. According to our own
experience in adapting this species to different aquar-
ium temperatures, the European small-spotted dogfish can
hardly tolerate tempeatures above 25°C for longer periods.
It should be emphasized that the shark data in Fig. 1
were obtained with synaptosomes originating from fishes
adapted to 10°C in the course of 1 month prior to the
experiment.

In contrast to rat and shark data, brain synaptosomes
obtained from hibernating ground squirrels (middle line in
Fig. 1) provided data which could be fitted to a continuous straight Arrhenius line covering the whole range of
temperatures explored, from 38°C down to 0°C, and are
therefore characterized by a single activation energy
figure ($18,340 \pm 361$ cal/mole^{-1}). A comparison restricted
to the range below 15°C (see also Table I) reveals much
lower Q_{10} values in the hibernator (3.21 and 3.16 in
Table I) than in the rat (5.50) and close in the ground
squirrel to the value recorded in fish synaptosomes
(2.84)). Results obtained with ground squirrel synaptosomes therefore reflect the well-known tolerance of
hibernators to low temperatures, particularly during
actual hibernation. In our laboratory, European ground
squirrels hibernate even at subzero surrounding temperatures (-1 to -20°C), allowing their rectal temperature
to fall to -1°C, without, however, losing the capacity
of spontaneous periodic arousals.

Competitive inhibition and reduced substrate concentration. The competitive inhibition by 2-deoxy-D-glucose
is a potent means of altering the rate of D-glucose oxidation. It was previously shown (7) that 2-DG markedly
reduced the rate of D-[^{14}C]glucose oxidation in rat synaptosomes studied at a temperature of 37°C, which corresponds to the euthermic level of body temperature in this
animal. Our results in Fig. 2 show that the competitive
inhibition by 2-DG persists in rat synaptosomes at reduced temperatures. Dixon plots in Fig. 2 testify to the
competitive nature of the 2-DG inhibition of glucose oxidation even at the level of 15°C, shown in the preceding
section to be critical for rat synaptosomes, and indicate
that the apparent dissociation constants are even lower
at the reduced temperature. This means of altering the
rate of glucose oxidation could therefore be used in our

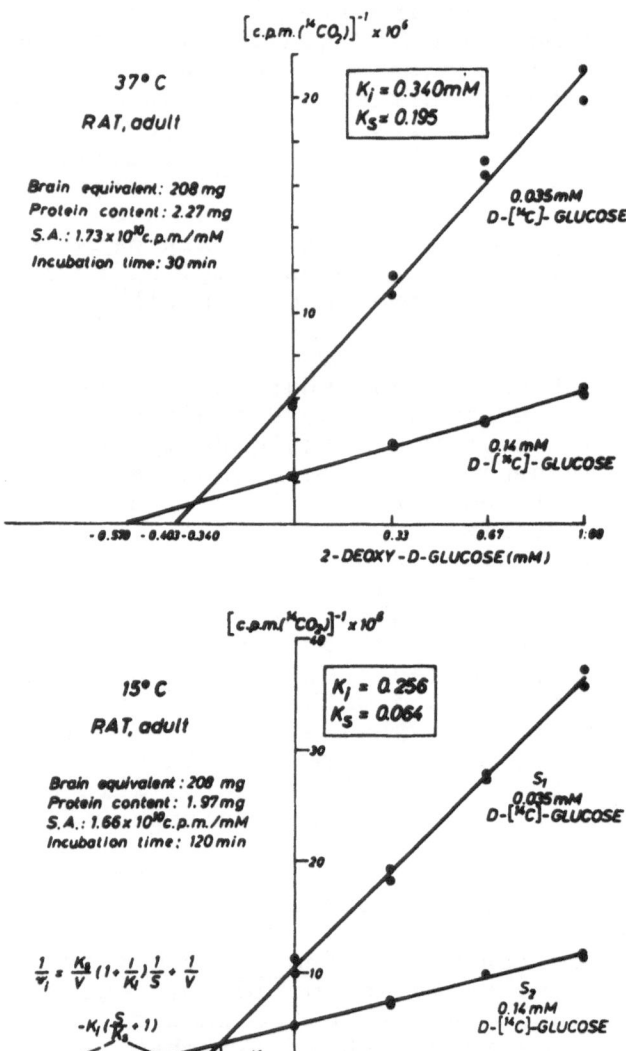

Fig. 2. Dixon plots of 2-deoxy-D-glucose inhibition of D-[¹⁴C]glucose oxidation by rat synaptosomes at 38°C (upper graph) and at 15°C (lower graph). Indicated in the lower graph are procedures used for calculating the apparent dissociation constants, K_i and K_s. Two separate pools of synaptosomes were used at the two incubation temperatures.

analysis of the thermophysiological behavior of synaptosome metabolism. At the same time, we checked the influence of reduced substrate concentrations by using exceedingly low concentrations of D-glucose in the medium,
concentrations that were capable in themselves of substantially reducing the rate of glucose oxidation.

Figure 3 shows that marked changes in the rate of
glucose oxidation by rat synaptosomes, elicited either
by the competitive inhibition of its utilization in the
presence of 2-deoxy-D-glucose or by changes in the concentration of D-glucose itself, have comparatively little
influence on the basic thermodynamic characteristics of
synaptosomal metabolism. The inhibitory action of 1 mM
2-DG, even though it substantially reduces the rate of
glucose oxidation (by 74% in the case of the lowest, 35
mM, concentration of D-glucose), does not affect appreciably the thermal sensitivity parameters. The range of
linear Arrhenius plots remains practically unchanged,
and apparent energies of activation do not differ considerably from those obtained in the absence of inhibition. Similarly, a fourfold reduction of substrate concentration (from 140 to 35 mM D-glucose), which resulted
in a 48-64% reduction of glucose oxidation rates in the
absence of the inhibitor, exhibited comparatively little
effect on the Arrhenius slope and linearity range. However, a closer inspection of Arrhenius slopes and Q_{10}
figures (Table I) concerning comparative experiments
with 140 and 35 uM glucose indicates that the lower of
the two concentrations provided lower E_a and Q_{10} values
above the critical temperature of 15°C. At lower temperatures, in contrast, Q_{10} values found with the lower concentration of substrate (35 µM) were considerably higher
than those provided by the higher concentration (140 µM).

Effects of uncoupling. In contrast to the effect of
competitive inhibition and of reduced substrate concentrations, the uncoupling action of 2,4-dinitrophenol
(2,4-DNP) exhibited a profound effect on the synaptosome
metabolism not only by displaying its well-known stimulating effect on oxidation rates, but also by altering,
at the same time, their temperature dependence (Fig. 4
and Table I). Thus, in the rat, and within the range of
temperatures covered in control experiments by a straight
Arrhenius line (38-15°C), uncoupled oxidations provided
a 50% higher E_a value (24465 + 915 cal/mole^{-1} versus
16499 + 1109 in control experiments). Even in the range
below the critical level of 15°C, substantially higher
temperature coefficients have been obtained after the
addition of 2,4-DNP (8.71 versus 5.50 in the controls).

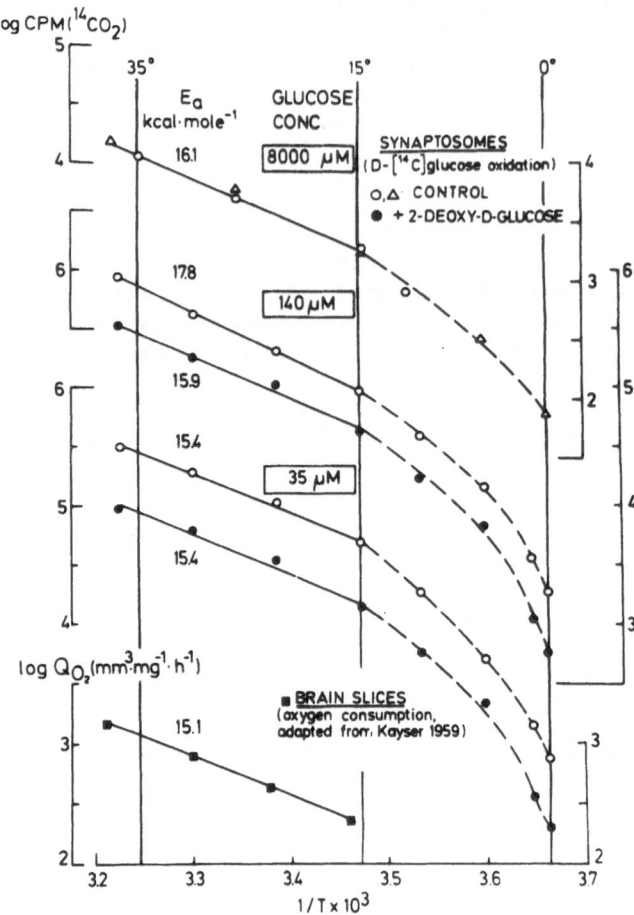

Fig. 3. Arrhenius plots of D-[^{14}C]glucose oxidation rates
in brain synaptosomes of the rat. Effect of 1 mM
2-deoxy-D-glucose at different D-glucose concentrations in the medium. Uppermost curve: Control
data obtained in the presence of the standard
high concentration of D-glucose in the physiological medium [8 mM, final specific radioactivity: 2.9 x 10^8 cpm/mmole; (o) same synaptosome
preparation; (△) separate preparations at each
temperature]. Two pairs of curves below: Effect
of 2-DG (●) and control data (o), recorded at two
levels of reduced substrate concentration (140
and 35 μM, but in the presence of one and the
same final specific preparations at each temperature. Bottom curve: Brain slice data from Kayser
(16).

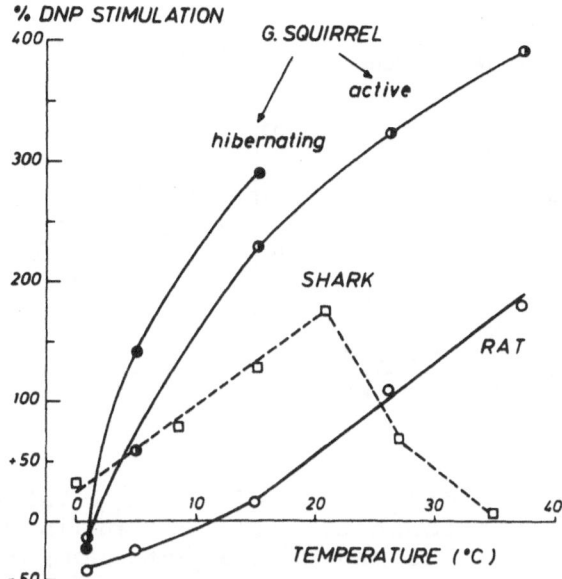

Fig. 4. Temperature dependence of the effect of 2,4-di-
nitrophenol. D-[^{14}C]glucose oxidation rates obtained in the presence of 0.08 mM 2,4-DNP are
expressed as percent changes of the rates recorded in control experiments without the uncoupler.
At each temperature, a different pool of synaptosomes was used, provided by a separate group of
animals.

This greatly increased thermal sensitivity elicited by
the uncoupler resulted, as shown in Fig. 4, in a continuous decrease with falling temperature of the stimulatory
effect of 2,4-DNP on the rate of glucose oxidation by
synaptosomes. In the rat, the stimulatory effect becomes
virtually abolished in the vicinity of the critical temperature of 15°C, and appears even as inhibitory at still
lower temperatures. This may be taken as an additional
expression of the already demonstrated correlation with
survival limits of reduced body temperature as defined
by whole-animal studies. The same type of correlation
was also revealed by our experiments with 2,4-DNP-stimu-
lated synaptosomes of other species within our comparative

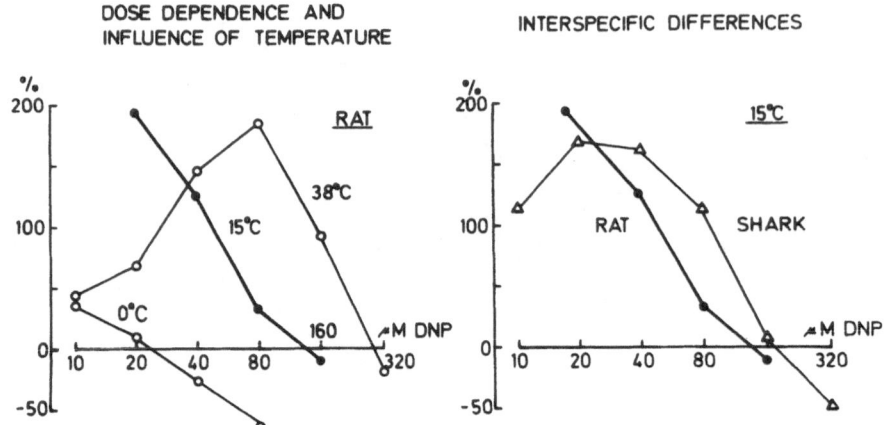

Fig. 5. Temperature dependence of the 2,4-DNP dose-and-
         effect relationship. Left: Rat brain synapto-
         somes at 38, 15 and 0°C. Right: Comparison of
         rat and shark synaptosomes at 15°C. Each curve
         was obtained with a separate pool of synaptosomes.
         2,4-DNP effects expressed as percent changes of
         glucose oxidation rates recorded in control,
         DNP-free media.

animal series. As shown in the same figure (Fig. 4,
dashed curve), in the cold-tolerant but heat-sensitive
dogfish, the stimulatory 2,4-DNP effect on glucose oxidation by brain synaptosomes persists down to 0°C, but
it also decreases above 22°C and is abolished by still
higher temperatures. In synaptosomes originating from
the cold-tolerant hibernator, the stimulatory effect of
2,4-DNP is considerably higher at all temperatures than
in the nonhibernator (rat). It also decreases, however,
with falling temperature, particularly steeply below
15°C, but does not disappear before temperatures in the
vicinity of 0°C are reached. The difference between synaptosomes originating from actually hibernating ground
squirrels (●) and those provided by active animals (◑)
reflects the difference in cold tolerance known to exist
between these two physiological states of hibernators.
In synaptosomes from hibernating animals the stimulatory

effect of the uncoupler remained higher at reduced temperatures than in synaptosomes originating from active
ground squirrels, and it is known that hibernators tolerate lower body temperatures during actual hibernation
than after artificial cooling from their active state
(3).

The reactivity of synaptosomes to 2,4-DNP proved to
be highly and characteristically dose-dependent (Fig. 5).
Typical bell-shaped dose-and-effect curves were obtained
with rat synaptosomes at 37°C (left graph), as well as
with shark nerve endings at 15°C (right graph). Note that
higher doses lead not only to a reduction of stimulation,
but also to the inversion of the effect: depressed rather
than enhanced oxidation rates are observed with the highest doses. Figure 5 illustrates in addition a highly pronounced temperature dependence of the dose-and-effect relationship: reduced temperatures provoke a shift of the
dose-dependence curves to the left. At lower temperatures
lower doses become more effective in enhancing glucose
oxidation rates, at least above the critical temperature
of 15°C in rat synaptosomes. At still lower temperatures
(0°C in the left-hand graph) the stimulatory effect is
reduced altogether, and the inhibitory one prevails,
while appreciable enhancements are recorded with the
smallest doses. The right-hand graph in Fig. 5 illustrates interspecific differences. The curve obtained with
rat synaptosomes at 15°C appears again as shifted to the
left, but this time in comparison to the shark curve,
obtained at the same temperature of 15°C which is now
acting as physiologically lower for rat synaptosomes.
This can be taken as an additional expression, at the
level of synaptosome metabolism, of the correlation with
cold tolerance in the respective species.

Sodium-stimulated and ouabain-blocked oxidations.
Rat, ground squirrel and shark synaptosomes exhibited
substantially reduced glucose oxidation rates when deprived of sodium chloride in their incubation medium. In
contrast, however, to oxidations depressed by competitive
inhibition or by low substrate concentration, reduced
oxidation rates in Na^+-free media appeared to be linked
to a markedly altered thermal sensitivity. Energies of
activation and Q_{10} figures (Table I) were substantially
increased, testifying to a greatly increased susceptibility to thermal influences. As shown in the preceding
section, the same was true for the effect of 2,4-DNP,
although in the case of uncoupling the increased thermal
sensitivity was linked to enhanced rather than depressed
oxidation rates, at least within the thermal range of

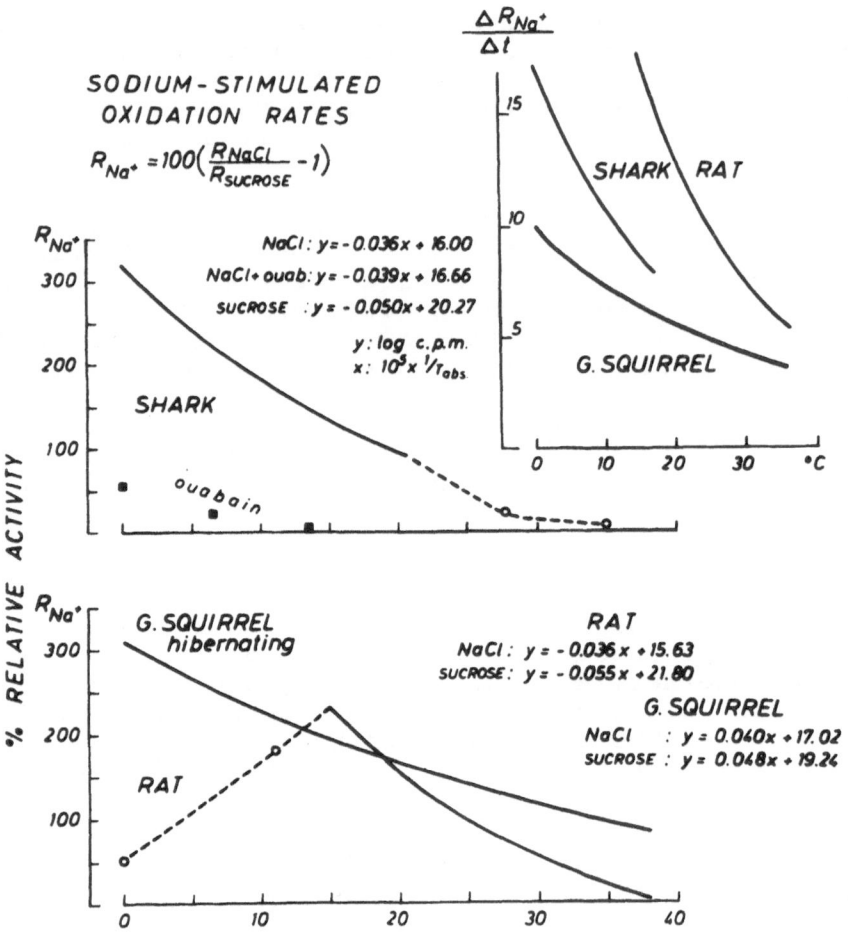

Fig. 6. Sodium-stimulated oxidations in brain synaptosomes as affected by temperature. Data expressed
as percent changes of D-[^{14}C]glucose oxidation
rates in NaCl-containing media compared to rates
recorded in media in which NaCl was replaced by
equivalent concentrations of sucrose. The continuous parts of the curves unaccompanied by dots,
were derived from the corresponding pairs of
straight Arrhenius lines (equations in the graphs)
to which experimental cpm values have been fitted.
(■) Data derived from measurements of oxidation
rates in NaCl-containing media, in the presence
of 0.1 mM ouabain and in its absence. Upper inset:
Change of the Na$^+$-stimulated fraction of oxidations per 1°C within the range of physiological
tolerance (Arrhenius linearity).

physiological tolerance. In addition, the reduced oxidation rates obtained in Na^+-free media showed in general
an extension of the range of linear Arrhenius plots into
the cold region, far beyond the limits shown to be critical in the presence of Na^+, or after the addition of
2,4-DNP to sodium-containing media.

If reduced oxidation rates in Na^+-free media are
considered as reflecting primarily the absence of the
transport-linked fraction of oxidations, the difference
between oxidation rates recorded in the presence and in
the absence of sodium ions in the medium becomes a measure of transport-linked or sodium-stimulated oxidations
(7). Figure 6 shows, by the discrepancy between Arrhenius
slopes provided by oxidation rates recorded in the presence and in the absence of Na^+ in the medium, that there
is a substantial increase in the sodium-activated fraction
of oxidation with falling temperature. In shark synaptosomes (upper graph), within the range of linear Arrhenius
plots (22-0°C), the sodium-activated fraction of oxidations increases continuously with falling temperature
and reaches three times higher values at 0°C than at the
upper limit of tolerance. As shown by the solid rectan-
gles in the same graph, the addition of 0.1 mM ouabain
to Na^+-containing media largely prevents the stimulation
of glucose oxidations by sodium ions. This effect of
ouabain was previously demonstrated in experiments with
synaptosomes of the rat studied at 37°C (7), and it was
considered as indicating that the Na^+-stimulated glucose
oxidation in rat synaptosomes is linked to the membrane
Na^+-K^+-ATPase activity and its role in active cation
transport. Our results, then, show that the same reason-
ing is applicable to brain synaptosomes of the shark.

The lower graph in Fig. 6 stresses the difference
between the rat and the cold-tolerant hibernator. In rat
synaptosomes the continuous, cold-induced, increase of
sodium-activated oxidations is interrupted in the vicinity of the critical level of 15°C. In synaptosomes originating from actually hibernating ground squirrels, however, sodium-activated oxidations start at 38°C with a
considerably higher value in comparison to rat synaptosomes, and rise continuously with falling temperature
over the entire range of reduced temperatures tolerated
by shark synaptosomes. Moreover, a threefold increase
of Na^+-stimulated oxidations was recorded with synaptosomes from hibernating ground squirrels, although this
time the same increase was spread over a considerably
broader range of temperatures (38-0°C). Hibernator synaptosomes therefore appear to be characterized by a

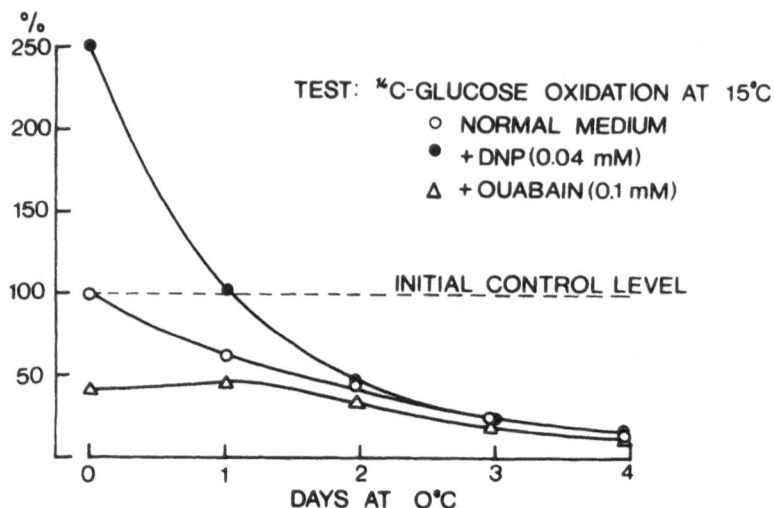

Fig. 7. Decrease of viability during cold storage as revealed by measurements of D-[^{14}C]glucose oxidation rates. Shark synaptosomes were stored at
0°C in the sucrose-urea medium free of glucose
(see Methods). They were tested at daily intervals, at 15°C, and in standard NaCl-containing
media, as well as after the addition of 0.04 mM
2,4-DNP or 0.1 mM ouabain.

considerably restricted increase of transport-linked
oxidations per degree of reduced temperature (Fig. 6,
inset). This is yet another expression of cold-tolerance
mechanisms peculiar to hibernation.

Figure 7 shows that both drug-induced effects used
in our thermophysiological analysis, the effect of 2,4-DNP
and that of ouabain, are sensitive indicators of the functional status of brain synaptosomes. Before glucose oxidation rates of control shark synaptosomes are reduced by
one half as a consequence of cold storage (0°C), decreased
viability is indicated not only by the loss of reactivity
to uncoupling, but also by the abolishmnet of the sodium-
dependent, transport-linked fraction of oxidations represented in the graph by the difference between control and
ouabain-treated synaptosomes, tested at 15°C. After the
first 24 hours of storage, oxidation rates deprived by
ouabain of their transport-linked fraction remain practically unchanged, while control oxidation rates appear to

be considerably reduced. The loss of the transport-linked
fraction coincides more or less with the loss of reactivity to 2,4-DNP. Note also that in the course of the
deterioration process, the 2,4-DNP effect does become
inverted (inhibitory).

DISCUSSION

Critical temperatures

The correlation of the so-called "transitional" or
"critical temperatures" (T_c) for glucose oxidation in
brain synaptosomes with thermal limits of survival revealed by whole-animal studies favors a limiting role of
neuronal energy metabolism in determining tolerance to
extreme body temperatures. The existence of critical
temperatures, as revealed by sharp discontinuities in
Arrhenius plots, has been reported previously for a variety of isolated enzymes and mitochondrial and tissue
preparations, although T_c values did not coincide to
the same extent with body temperatures limiting survival
in the respective species as in our experiments with synaptosomes. Since intrasynaptosomal mitochondria are responsible for the respiratory activity of the synaptosome
(28), data on mitochondrial oxidation rates appear to be
particularly relevant for comparison with our results.
Data reported by South (27), for instance, clearly show
discontinuity in the Arrhenius plots for heart mitochondria from the rat and from active hamsters between
24 and 26°C, i.e., at temperatures which are rather distant from hypothermic levels limiting survival of these
species during cooling. Despite considerably higher T_c
values, however, a similar type of correlation with
thermal limits of survival has been demonstrated in still
other experiments with isolated mitochondria (19, 23).
Thus, in the case of succinate oxidation by liver and
heart mitochondria from rats and ground squirrels (Citellus lateralis), transition temperatures were also situated
at a relatively high level (around 23°C), but it was found
that in association with hibernation a considerable shift
toward lower temperatures takes place: Arrhenius plots
are virtually deprived of discontinuities down to 5°C
(23). A similar phenomenon was observed with some enzyme
preparations. In the case of muscle pyruvate kinase from
bats, T_c values appeared to be shifted from 17° to 5°C
when nonhibernating bats were compared to torpid animals
(5). It should be emphasized, however, that other enzyme
studies did not reveal discontinuities in Arrhenius plots
obtained with preparations from either hibernators or non-
hibernators (21).

Also at variance with our results concerning synaptosomes from hibernating ground squirrels, as well as
with those of other studies using mitochondrial preparations from hibernating animals (23), are the results
published by South (27), in which a discontinuity in the
Arrhenius plot was apparent even in the case of mitochondria from actually hibernating hamsters. It was already pointed out (23) that the difference in results
obtained with mitochondrial preparations from hibernating
animals concerning the presence or absence of discontinuities in the Arrhenius plot may be due to differences
in experimental procedures (polygraphic versus manometric
determination of oxidation rates, use of the same or of
different preparations of mitochondria for each experimental temperature). It has also been emphasized that
not all tissues of a given hibernator show the same relationship to temperature and that, in addition, variations should be expected in the temperature response of
homologous tissues among hibernating species showing different patterns of hibernation (bats, hamsters, ground
squirrels) and having a different phylogenetic origin
(13, 14). Accordingly, different findings, concerning
the pattern of Arrhenius plots obtained with preparations
from hibernators may also reside in species differences,
the time spent in hibernation prior to tissue sampling
and the level of body temperature during hibernation.

Also in contrast to our results concerning dogfish
synaptosomes and indicating breaks in Arrhenius plots at
temperatures close to those representing the upper limits
of thermal resistance in this particular heat-sensitive
dogfish species, experiments with liver mitochondria from
two teleost fish species (trout and catfish) demonstrated
uniform Arrhenius plots from 4 to 30°C and led to the
far-reaching conclusion that the discontinuity at a critical temperature is a unique property of succinate oxidation in homeotherms which is not observed in poikilotherms (19). With glucose oxidation by synaptosomes, this
is certainly not the case. In experiments with isolated
enzymes from other poikilotherms, including a fish species
(the cod), breaks in Arrhenius plots have been described
in association with elevated temperatures (20). However,
the point of intersection of E_a and E_i lines coincided
with temperatures considerably above the upper limit of
thermal resistance of the respective species.

Although the existence of "critical temperatures,"
revealed by sharp breaks in Arrhenius plots, has been
extensively reported, the responsible underlying mechanisms are still poorly understood (5). The transition

temperatures revealed by studies on isolated mitochondria
were regarded as related to configurational changes of
membrane-associated enzyme proteins induced by a phase
change in the membrane lipids (23). Such a phase change
would not occur at reduced temperature in poikilotherms
(trout and catfish) because of a higher proportion of
unsaturated fatty acids in their membrane lipids. It was
suggested that a small increase in the proportion of unsaturated fatty acids in the mitochondrial membrane lipids
could abolish the temperature-induced phase change indicated by the discontinuity in the Arrhenius plot. An al-
ternative explanation for shifts of the T_c values toward
lower temperatures, such as those associated with hibernation, has been found in the supposed incorporation of
a "cryoprotective agent" into the mitochondrial membranes.
Cholesterol was specifically mentioned on the basis of
experiments with synthetic phospholipids in which the
incorporation of a small amount of cholesterol eliminates
the temperature-dependent phase change. Since, however,
the same types of transition temperatures and of their
shifts have been demonstrated in enzymes that are not
membrane-bound, their relation to a lipid phase change
has been questioned (5).

The significance, from the standpoint of survival,
of the abrupt metabolic changes at critical temperatures
is also uncertain. In the case of mitochondria, sudden
changes of oxidation rates at transition temperatures,
indicating an abruptly suppressed mitochondrial respiration, have been considered as causing a detrimental imbalance between mitochondrial and extramitochondrial
metabolism, and the time-temperature relationship limiting survival of the homeotherm in hypothermia has been
regarded as a function of the ability of cells to cope
with this imbalance (23). Even, however, in the absence
of discontinuities in the respective Arrhenius plots,
differences in Arrhenius slopes between intra- and extramitochondrial enzymes (malic dehydrogenase) have been
taken as indicating marked alterations in the relative
activity levels at low temperatures which might be responsible for disturbed cellular function in hypothermia (21).

It must be taken into consideration, when comparing
thermal sensitivity figures for isolated systems from
different organ sources, that not all tissues from the
same animal should be expected to display the same relationship to temperature. Pyruvate kinases from flight
muscles and liver of the bat appeared to be quite different and demonstrated specific molecular strategies in
adapting to prevailing thermal conditions associated with

the active or hibernating state (5). In addition, under
in vivo conditions, some tissues may alter the internal
environment in such a way as to confer upon other tissues
protection against deleterious temperature effects. The
temperature coefficients recorded in vitro have been
shown to be critically dependent on a number of factors
in the artificial medium, and even rigorous attempts to
approximate in vivo conditions may fail to prevent alterations of kinetic properties and temperature sensitivity (29).

Thermodynamic functions

Apparent energies of activation (E_a), derived from
the Arrhenius equation, are often used as useful indicators of thermal sensitivity (5). Reduced E_a values are
taken as testifying to a lower thermal sensitivity, since
they indicate a smaller change for each degree of decrease
or increase in temperature. Lower apparent energies of
activation in the cold range are considered as reflecting
lesser structural modifications at the molecular level
in response to low temperature, and as indicating increased catalytic efficiency of enzymes. It is well recognized that the temperature dependence of complex biological systems, in which the total effect arises from
several elementary steps characterized by their own thermal sensitivity, refers to a mesh of interrelated reactions, and that the magnitude of the E_a value alone can
hardly be used to identify a "master" reaction (22). In
the present study, apparent energies of activation were
derived only from the straight-line regions of the Arrhenius plot, and they served the purpose of an empirical
thermophysiological characterization of the synaptosome
system. Temperature coefficients were derived from Arrhenius equations when referring to the straight-line region
of the plot. When dealing with temperatures outside this
region, Q_{10} values were calculated from a pair of individual experimental results obtained at temperatures 10
degrees apart, and therefore represent only a rough estimate of the temperature dependence. Since Q_{10} values
increase with decreasing temperature even within the
range of temperatures characterized by a single temperature-independent E_a value, the magnitude of Q_{10} alterations observed when passing from a straight-line Arrhenius slope to the adjacent region below the transition
temperature can be best appreciated by comparing the
altered Q_{10} value with the one which would be expected
in the same temperature zone by extrapolating into it
the straight Arrhenius line.

Energy activation values obtained in our experiments with synaptosomes studied in standard physiological media were of the same order as those frequently reported for isolated tissues, mitochondria or enzyme preparations. Particularly impressive is the great similarity between our E_a values for $D[^{14}C]$glucose oxidation in synaptosomes of the rat within the 38 to 15°C range (16.1, 16.4 kcal), and the E_a value for oxygen consumption by rat brain slices within almost the same temperature range (16-38°C), which we derived from a graph of Arrhenius plots published by Kayser (16). This would indicate, therefore, that syn-aptosomal E_a values can be taken as fairly representative of the oxidative metabolism of the brain tissue as a whole. However, for a broader range of temperature (10-43°C), South (26) reported a lower value for brain slices (10.81 kcal/mole^{-1}), and similarly low E_a values for brain slices from the torpid bat, the nonhibernating hamster and the hibernating hamster (9.90, 10.66 and 11.11 re-spectively, the last figure referring, however, to a narrower temperature range, 17-43°C. It was concluded that the similarity of Arrhenius slopes obtained with all these brain preparations indicates that there is probably no significant alteration in the overall kinetics of energy production in brain tissue. Our comparative experiments with synaptosomes revealed, however, marked differences between the nonhibernator (rat), and the hibernating hibernator (ground squirrel) as to the respec-tive energies of activation in particular thermal ranges. The remarkable extension of the straight-line continuity of the Arrhenius plot into the cold region was linked in synaptosomes from hibernating ground squirrels to an appreciably increased E_a value in comparison to the one obtained with rat synaptosomes under identical conditions and within temperatures providing a straight-line Arrhe-nius relationship. Within the thermal range above the critical temperature for the rat (38-15°C), therefore, hibernator synaptosomes were characterized by higher E_a values, indicating a slightly increased thermal sensi-tivity. In the cold region, however, below the transition temperature for rat synaptosomes, the opposite was true, as demonstrated by considerably lower Q_{10} values in the hibernating hibernators: in the cold region its synapto-somal glucose oxidation rates appeared to be less sensi-tive to temperature change. At the enzyme level, this would mean that in the hibernating hibernator as compared to the nonhibernator, a lower catalytic efficiency is present at higher temperature (38-15°C), but that a con-siderably greater efficiency of the enzymes appears in the cold region (15-0°C). It is highly indicative in this respect that in experiments dealing directly with tissue enzymes, isolated from another hibernator, the bat (Myotis lucifugus), a strikingly comparable situation was encountered, although this time the hibernating animals were compared to nonhibernating individuals of the same species and the enzyme was prepared from muscle tissue (5). The straight-line Arrhenius relationship concerning the activity of muscle pyruvate kinase appeared, in torpid as compared to nonhibernating bats, considerably extended into the cold region (from 17 down to 5°C), but it was also linked to a considerably increased E_a value (13.7 instead of 8.62 kcal/mole^{-1}). Again, a comparison restricted to the cold region (5-17°C) reveals lower E_a values and consequently a decreased thermal sensitivity of the enzyme from actually hibernating bats (13.7 versus 19.5 kcal), but the same enzyme from the torpid animals appears to be more temperature-sensitive at higher temperatures (13.7 versus 8.6 kcal/mole^{-1} within the range above 17°C). It should be pointed out, as well, that the same type of thermal sensitivity differences between hibernating and active animals (Citellus lateralis) has been revealed by earlier studies with liver mitochondria (23). Below the transition temperature for succinate oxidation by mitochondria from active animals, situated at a relatively high temperature (near 23°C), the E_a value for mitochondria from hibernating individuals was lower (13.8 versus 16.7 kcal/mole^{-1}), while above 23°C the same E_a value of 13.8 cal, still valid for the hibernating animals, appeared to be significantly higher than the one recorded with mitochondria from active ground squirrels (9.1 cal). Thus, energies of activation recorded in our experiments with synaptosomes from hibernating ground squirrels, as compared to those from the rat, seem to reflect a general thermodynamic peculiarity of the hibernating state encountered at the subcellular (mitochondrial) and enzyme level. At the same time, however, data obtained with synaptosomes show, as already pointed out, a closer correlation with the respective survival limits of tolerance to reduced body temperature.

Reduced substrate concentration and competitive inhibition

Our experiments with rat synaptosomes revealed that reduced substrate (D-glucose) concentrations and the competitive inhibition by 2-deoxy-D-glucose, although they suppress D-glucose oxidation rates markedly, do not appreciably alter their temperature dependence within the range of physiological tolerance characterized by the straight-line Arrhenius relationship. At lower temperatures, however, a considerably greater increase of Q_{10}

values has been recorded at reduced D-glucose concentrations, and after 2-DG inhibition. In this respect, experiments with isolated enzyme preparations, studied at different substrate concentrations, appear to be particularly revealing. It was shown with the muscle pyruvate kinase preparation from normothermic bats that the reduction of substrate concentrations toward K_m or physiological levels is not followed by appreciable changes in the slope of the Arrhenius plot (5). Parallel lines obtained for the same enzyme in the presence of different concentrations of a substrate (PEP), indicating a constant proportional change in velocity, were taken as suggesting that K_m is temperature-independent. However, with another substrate (ADP), discrepancies appeared between Arrhenius slopes obtained with different substrate concentrations, but only in the cold region, below the critical temperature. They were taken as indicating an increase of K_m at low temperatures (below 17°C). By analogy, therefore, and to the extent that changes in glucose oxidation rates are representative in rat synaptosomes of the basic properties of the underlying enzyme-catalyzed processes, the considerably greater increase of Q_{10} values at reduced than at high glucose concentrations in the 15-0°C range, demonstrated in our experiments, would suggest substantially decreasing enzyme-substrate (E-S) affinities below the critical temperature. Cold tolerance would then appear to be limited by incompatible kinetics of enzyme-catalyzed processes: the marked increase of the underlying E-S affinities, coupled to the decrease of available thermal energy, leads to exceedingly large reaction-rate decreases and to biological inactivation. On the other hand, almost parallel Arrhenius lines, obtained with different substrate concentrations and after 2-DG inhibition, also demonstrated by our experiments, but within the range of physiological tolerance, suggest the persistence of relatively high, temperature-independent E-S affinities (low K_m values) above the critical temperature. Alternatively, this latter range might even be characterized by increasing affinities with falling temperature, a phenomenon known to occur with enzymes of poikilotherms at low substrate concentrations (11). This phenomenon is an important component of the cold-tolerance mechanisms, since it counteracts the decrease of available thermal energy. Strongly suggesting that this may be the case with the glucose oxidation system of rat synaptosomes are our data in Fig. 2, which show decreased values of the apparent dissociation constants at the lower temperature extreme of the range of physiological tolerance. Further extensive studies are needed in order to definitely clarify this point.

It has also been suggested that the fall of Q_{10} values observed at elevated temperatures with enzymes from poikilotherms is a consequence of increased K_s values which, in the presence of low concentrations of substrate, become exceedingly high, indicating very low enzyme-substrate affinity (11). The marked fall of Q_{10} values for glucose oxidation rates in shark synaptosomes at temperatures above those tolerated by the dogfish might be interpreted in an analogous way, by supposingly increased underlying K_m values.

It is also worth noting that Kayser's early experiments with rat and hamster brain slices (16) revealed no change in the Arrhenius slope after malonate inhibition of oxygen consumption rates, although a shift of the slopes in vertical placement was well pronounced, since the lowest temperature in the plots was 16°C.

Effects of uncoupling

Uncoupling agents have already been shown to markedly stimulate glucose oxidation by rat synaptosomes at 37°C (7). On the other hand, iodoacetate, a powerful inhibitor of glucolysis, reduced glucose oxidation rates to a limited extent only. These and other findings (28) indicated that rat nerve endings exhibit coupled respiratory activity. Our study showed that the effect of uncoupling was markedly temperature-dependent in rat synaptosomes, and linked to a marked increase of the energy of activation value. It also revealed that higher doses may lead not only to reduced stimulatory effects, but also to the inversion of the effect. This high-dose phenomenon was enhanced by low temperatures.

The dose-dependence curves in Fig. 5 can be interpreted formally by assuming that the effect of 2,4-DNP, as observed at the level of overall oxidation rates, is composed of two main, mutually opposed components. The suppressive (negative) component prevails at high doses and low temperatures, while the enhancing, positive one appears as a low-threshold component, less opposed by its antagonizing counterpart at lower doses. Such an interpretation is consistent with the known fact that uncoupling not only releases mitochondrial oxidations from their feedback control, but also concomitantly reduces the energy charge and the efficiency of active transport mechanisms.

The markedly stimulated glucose conversion to carbon dioxide observed in cells under the effect of uncoupling

agents is considered to be due to the inhibition of ATP
production in mitochondria, the stimulation of mitochondrial ATPase activity and the release of respiration from
acceptor control, all these processes being potentially
capable of stimulating glycolysis (24). It was found that
the rate of uncoupling oxidations is limited by the availability of activated substrate, and the role of glycolysis
in providing ATP for priming was also underlined. In the
presence of the uncoupler there may be no extra ATP for
substrate activation, to the extent that the effect of
uncoupling may appear as inhibiting respiration in the
presence of a potential substrate if ATP is not provided
by glycolysis (24). Differences in the glycolytic capacity may therefore underline a number of differences in
various 2,4-DNP effects observed in the present study.
The effect of the same dose of 2,4-DNP in stimulating
glucose conversion to CO_2 in synaptosomes of ground squirrels, hibernating or active, was considerably greater, at
all temperatures, than in the rat. It was shown previously
that brain slices from hibernators (hamsters) in comparison to the same tissue from the rat exhibit a more intense glycolysis and higher glucose comsumption rates
(17). Our earlier experiments also indicated a greater
glycolytic efficiency in the ground squirrel brain in
situ as compared to the brain of the rat, and demonstrated
its correlation with tolerance to both anoxia and hypothermia (2). It appears, then, that the extent to which
uncoupling promotes the glucose conversion rate in brain
synaptosomes does indeed reflect the glycolytic capacity
of neurons. It is worth noting in this respect that a
greater stimulation by 2,4-DNP of the respiration rates
of brain slices from hamsters than of those from the rat
has been recorded in early experiments by Kayser (16),
although in a later study the same difference between the
rat and hibernator appeared to be less pronounced and not
statistically significant (17).

The difference, however, observed in our experiments
between 2,4-DNP-treated synaptosomes from active and hibernating ground squirrels is also strikingly reminiscent
of the difference found between uncoupled mitochondria
isolated from the livers of hibernating and nonhibernating
ground squirrels (25). In the presence of succinate the
uncoupler stimulated oxygen consumption to an appreciably
greater extent in the case of mitochondria originating
from actually hibernating squirrels, while the effect of
ADP was, in contrast, much smaller. This was found difficult to interpret, but it was stressed that the respiration appeared not to be limited by the amount of cytochrome, although the latter was reduced by one half in

comparison to the amount found in nonhibernating animals.
Alterations of the coupling mechanism itself were suggested, as well as a possible inhibition of the translocation of adenine nucleotides through mitochondrial
membranes by fatty acids.

It is well known that in nerve fibers a sufficient
extra ATP concentration needed for active sodium extrusion cannot be maintained without the support of oxidative phosphorylation, and that uncoupling leads to the
suppression of sodium efflux (12). Accordingly, the
sodium-stimulated fraction of glucose conversion to carbon dioxide should be expected to be particularly affected by 2,4-DNP. Our experiments show indeed that within the range of physiological tolerance the addition of
the uncoupler to the medium results in the same marked
increase of thermal sensitivity, indicated by the substantially elevated E_a value, as after the elimination
of Na^+ from the medium (Table I), despite great differences between the underlying Arrhenius slopes in vertical
placement. In experiments with rat synaptosomes, at temperatures above the critical level of 15°C, the decrease
of the stimulatory effect of 2,4-DNP with falling temperature (Fig. 4) was proportional to the concomitant
increase of the sodium-activated fraction of oxidations
(Fig. 6). A similar correlation appeared, within the
range of physiological tolerance, in experiments with
shark synaptosomes. It appears, then, as though the uncoupler were, at the same time, depriving overall oxidation rates of their sodium-dependent component and
adding to them its own, seemingly temperature-insensitive,
contribution. Our results showed also that high DNP concentrations in the medium, in the presence of which the
enhancement of uncoupled glucose oxidation rates cannot
manifest itself, may depress the conversion of glucose
to carbon dioxide even below the level recorded before
uncoupling in the presence of extrasynaptosomal Na^+. Such
doses tend to eliminate the difference between oxidation
rates recorded in the presence and in the absence of Na^+
in the medium. It should be noted in this respect that
small (10^{-6} M) concentrations of 2,4-DNP appeared to be
incapable of significantly altering the Arrhenius slopes
of the respiration rates of hamster and rat brain slices,
although a shift in vertical placement was well pronounced
(16, 17). On the other hand, it was found with rat liver
slices at 38°C that the dose dependence of the effect of
2,4-DNP on respiratory rates is linked to its inhibitory
effect on cation transport. The respiration increased
with increasing concentrations of 2,4-DNP up to a concentration of 50 µM, which did not affect the net movement

of sodium. Higher concentrations, however, brought about
a decrease of the stimulating effect on respiration and
a concomitant reduction of the net cation movements. The
published curves show that concentrations approaching
10^{-3} M resulted in the reduction of respiratory rates
below control levels and a complete abolishment of the
net movement of sodium (9).

Na^+-stimulation and thermal resistance

Glucose enters synaptosomes by a high-affinity Na^+-
independent carrier-mediated transport system, but after
it is transported into the synaptosome its oxidation is
markedly stimulated by sodium ions (7). Since ouabain
largely prevents Na^+ stimulation of glucose oxidation,
without, however, inhibiting the conversion of glucose
to carbon dioxide in the absence of Na^+ in the incubation
medium, Na^+ stimulation appears to be linked to the activity of the membrane-bound cation-dependent ATPase. The
stimulation was shown to be dependent on the resulting
extramitochondrial production of ADP and its translocation
across the inner mitochondrial membrane. Thus, the Na^+-
stimulation phenomenon appears to be not an intrinsic
feature of the mitochondria, but a property of the synaptosomal system capable of regulating by stimulatory
feedback the energy-yielding processes linked to the
cation-transport activities of the synaptosomal membrane
(28).

Sodium stimulation appeared in our experiments with
rat synaptosomes to be contributing substantially to the
resistance of synaptosomal glucose oxidation rates to
falling temperature, but only down to a critical level.
Above this level, E_a values obtained with control, Na^+-
containing, media appeared to be markedly reduced in comparison to those observed in Na^+-free media or after the
addition of 2,4-DNP. These relatively low E_a values, testifying to a reduced thermal sensitivity, appeared to be
the consequence of an increase with falling temperature
of the sodium-stimulated fraction of oxidations. However,
after reaching a maximum at a critically low temperature,
this increase was interrupted and followed by a steep
decline of the sodium-dependent fraction with further
cooling. As a consequence, the transition to markedly
higher E_a and Q_{10} values took place. A similar relationship between the maintenance of a relatively low E_a value
at reduced temperatures and the cold-induced increase
of the sodium-stimulated fraction of glucose oxidations
has been found with shark and ground squirrel synaptosomes. In both, however, this relationship persisted

without interruption down to the vicinity of 0°C. Synaptosomes of the poikilotherm were in addition characterized by the abolishment of the sodium fraction in
the range of temperatures above 25°C, associated also
with abnormally low Q_{10} values and the appearance of
"inactivation energies."

It is well known that low temperatures interfere
with cation-transport mechanisms and may lead to considerable alterations of the tissue electrolyte content.
A considerable gain of intracellular sodium and loss of
potassium ions takes place in isolated tissues of the
rat in the course of incubation at temperatures approaching zero (+1°C). These changes have been shown to be
promptly and largely reversible at higher temperatures
(38°C) (9).

Cation pumping may be slowed down by low temperature
not because of a shortage of ATP, but because of a depressed transport ATPase activity, as shown by experiments with rat brain microsomal preparations (6). Moreover, it has been shown that the cation-activated ATPase
isolated from the brain of the rat exhibits a break in
the Arrhenius plot at about 20°C and yet another in the
vicinity of 6°C. These data are in close correlation
with our results illustrated in Fig. 3 concerning the
temperature dependence of glucose oxidation rates in
rat brain synaptosomes incubated in Na^+-rich media. Since
in our experiments the breaks in the Arrhenius plots appeared as a consequence of a cold-induced failure of the
sodium-stimulated fraction of glucose oxidations, the
responsibility of the underlying temperature dependence
of the Na^+-K^+-ATPase is strongly suggested. A sharp cold-
induced decrease of the transport ATPase activity below
the critical temperature would be expected to be followed
not only by a less efficient extrusion of sodium, but
also by a depressed sodium stimulation of glucose oxidations, since it has been shown (28) that the sodium-
stimulation phenomenon depends on extramitochondrial
production of ADP and its stimulatory feedback action.
In the range above the critical temperature, the stimulatory influence on glucose oxidation rates of the transport-linked activity of ATPase would be expected to ef-
ficiently oppose the inhibitory effect of falling temperature. This effect of ATPase is itself depressed by
reduced temperatures, but published figures testify to
a relatively low thermal sensitivity of the enzyme in
the range above 20°C (E_a = 7800 cal/mole^{-1}). The considerably higher thermal sensitivity of glucose oxidation
rates found in our experiments with synaptosomes in

sodium-rich media (E_a = 16,470 cal/mole^{-1}) appears, however, to be intermediate between the high thermal sensitivity of synaptosomal glucose oxidations in the absence of Na$^+$ stimulation (25,164 cal/mole^{-1}) and the already mentioned low thermal sensitivity exhibited by the isolated, cation-stimulated, membrane ATPase preparation (7800 cal/mole^{-1}).

The absence of breaks in Arrhenius plots obtained with ground squirrel synaptosomes, and the persistence of an increasing sodium-dependent fraction of glucose oxidation down to the lowest temperatures explored, were similar to the situation encountered in experiments with dogfish synaptosomes in the low-temperature range. This is consistent with earlier findings concerning an increased resistance of excised tissues from hibernators to cold-induced swelling and uptake of solutes. It has been shown that in hibernators cation transport still influences respiration in tissue slices at low temperatures, and the fraction of respiration associated with transport was shown to be the same at 5 and 38°C (29). In these experiments, however, only two extreme temperatures were compared, and tissue slices from a nonhibernating hibernator (hamster) were used. Thus, the chosen low-temperature extreme (5°C) may have been already below the limit compatible with an increase of the transport-associated fraction with falling temperature.

ACKNOWLEDGMENTS

These studies were supported by NIH Research Agreement No. 02-042-1, and by a joint grant of the Serbian Academy of Science and the Serbian Science Association. The technical assistance of A. Karakašević and D. Mitić was invaluable and it is highly appreciated. The studies were partly accomplished at the Marine Biological Station of the Institute for Experimental Biology and Medicine in Kotor. The use of its working facilities and the technical assistance of its members are gratefully acknowledged.

REFERENCES

1. Adolph, E.F. (1948): Lethal limits of cold immersion in adult rats. Am. J. Physiol. 155: 378-388.

2. Andjus, R.K. (1969): Some mechanisms of mammalian tolerance to low body temperatures. Symp. Soc. Exp. Biol. 23: 351-394.

3. Andjus, R.K., Matić, O., Petrović, V. and Rajevski V. (1964): Influence of hibernation and of intermittent hypothermia on the formation of immune hemagglutinins in the ground squirrel. Ann. Acad. Sci. Fenn. Helsinki Ser. A: IV, 71/1: 27-35.

4. Andjus, R.K., Ristanović, D. and Ćirković, T. (1974): Brain metabolism and survival in hypothermia and anoxia. Contribution IV. A.3. in Problems of Hypothermia (6th Dortmund Workshop) J.A. Miller et al. eds. Drug Res. 24: 961-971.

5. Borgman, A.I. and Moon, T.W. (1976): Enzymes of the normothermic and hibernating bat, Myotis lucifugus: temperature as a modulator of pyruvate kinase. J. Comp. Physiol. 107: 185-199.

6. Browler, K. and Duncan, C.J. (1968): The effect of temperature on the Mg^{2+}-dependent and Na^+-K^+-ATPase of a rat brain microsomal preparation. Comp. Biochem. Physiol. 24: 1043-1054.

7. Diamond, I. and Fishman, R.A. (1973): Development of Na^+-stimulated glucose oxidation in synaptosomes. J. Neurochem. 21: 1043-1050.

8. Diamond, I. and Milfay, D. (1972): Uptake of ^{3}H-methylchlorine by microsomal, synaptosomal, mitochondrial and synaptic vesicle fractions of rat brain. J. Neurochem. 19: 1899-1909.

9. Elshove, A. and Van Rossum, G.D.V. (1963): Net movements of sodium and potassium, and their relation to respiration, in slices of rat liver incubated in vitro. J. Physiol. 168: 531-553.

10. Gruener, N. and Avi-Dor, Y. (1966): Temperature-dependence of activation and inhibition of rat-brain adenosine triphosphate activated by sodium and potassium ions. Biochem. J. 100: 762-767.

11. Hochachka, P.W. and Somero, G.N. (1968): The adaptation of enzymes to temperature. Comp. Biochem. Phys. 27: 659-668.

12. Hodgkin, A.L. and Keynes, R.D. (1955): Active transport of cations in giant axons from Sepia and Loligo. J. Physiol. (London) 128: 28-60.

13. Horowitz, B.A. (1964): Temperature effects on oxygen uptake of liver and kidney tissues of a hibernating and nonhibernating mammal. Physiol. Zool. 37: 231-239.

14. Horowitz, B.A. and Nelson, L. (1968): Effect of temperature on mitochondrial respiration in a hibernator (Myotis austroriparius) and a non-hibernator (Rattus rattus). Comp. Biochem. Physiol. 24: 385-394.

15. Jones, D.G. (1967): An electron-microscope study of subcellular fraction of Octopus brain. J. Cell. Sci. 2: 573-586.

16. Kayser, C. (1959): Effect du malonate et du dinitrophénol sur la respiration de coupe d'encéphale du rat adulte, de rat en croissance et du hamster adulte. C.R. Acad. Sci. 248: 1219-1222.

17. Kayser, Ch. and Malan, A. (1963): Central nervous system and hibernation. Experientia 19: 441-451.

18. Klatzo, I. and Steinwall, O. (1965): Observation on cerebrospinal fluid pathways and behaviour of the blood-brain barrier in sharks. Acta Neuropathol. 5: 161-175.

19. Lyons, J.M. and Raison, J.K. (1970): A temperature-induced transition in mitochondrial oxidation: contrasts between cold and warm-blooded animals. Comp. Biochem. Physiol. 37: 405-411.

20. Olsson, S.-O.R. (1975): Comparative studies on the temperature dependence of lactate and malate dehydrogenases from a homeotherm, guinea pig (Cavia porcellus); two hibernators, hedgehog (Erinaceus europeus) and bat (Nyctallus noctula); and two poikilotherms, frog (Rana temporaria) and cod (Gadus callaria). Comp. Biochem. Physiol. 51B: 5-18.

21. Paulsrund, J.R., Mann, K.G. and Dryer, R.L. (1970):
    A comparison of rat and bat malic dehydrogenase iso-
    enzymes. In: Brown Adipose Tissue. O. Lindberg (ed.).
    American Elsevier, New York, 197-206.

22. Precht, H. (1973): Constant systems. In: Temperature
    and Life, by H. Precht, J. Christophersen, H. Hensel
    and W. Larcher. Springer-Verlag, New York, 302-310.

23. Raison, J.K. and Lyons, J.M. (1971): Hibernation:
    Alteration of mitochondrial membranes as a requisite
    for metabolism at low temperatures. Proc. Natl. Acad.
    Sci. U.S.A. 68: 2092-2094.

24. Reed, N. and Fain, J.N. (1970): Hormonal regulation
    of the metabolism of free brown fat cells. In: Brown
    Adipose Tissue. O. Lindberg (ed.). American Elsevier,
    New York, 207-224.

25. Shug, A.L., Ferguson, S., Shrago, E. and Burlington,
    R.F. (1971): Changes in respiratory control and cyto-
    chromes in liver mitochondria during hibernation.
    Biochim. Biophys. Acta 226: 309-312.

26. South, F.E. (1958): Rates of oxygen consumption and
    glycolysis of ventricle and brain slices, obtained
    from hibernating and nonhibernating mammals as a
    function of temperature. Physiol. Zool. 31: 6-15.

27. South, F.E. (1960): Hibernation, temperature and
    rates of oxidative phosphorylation by heart mitochon-
    dria. Am. J. Physiol. 198: 463-466.

28. Verity, M.A. (1972): Cation modulation of synaptosomal
    respiration. J. Neurochem. 19: 1305-1317.

29. Willis, J.S. (1969): Cold adaptation of activities
    of tissues of hibernating mammals. In: Mammalian
    Hibernation III. Edited by Fisher et al., Oliver and
    Boyd, Edinburgh. III: 356-381.

BIOCHEMICAL ASPECTS OF CEREBRAL EDEMA

A. Baethmann, W. Oettinger, W. Rothenfusser
and R. Geiger

Institute for Surgical Research
Department of Surgery
University of Munich
Munich, W. Germany

Biochemical research on cerebral edema has certainly advanced the understanding of this complex condition. Biochemistry has been of great use in defining the different manifestations of this entity. Chemical investigations are probably more suitable than other methods to quantify the amount of edema in brain tissue, and to measure the effect of edema treatment. Since the pioneering work of Stewart-Wallace, Elliott and Jasper, and Pappius and co-workers, considerable material on the chemistry and biochemistry of brain edema has been published, supporting the existence of the two major prototypes described by Klatzo (10, 18, 30, 31, 43). Representative studies concern, for example: (1) the chemical composition of edematous tissue, (2) the edema fluid proper, (3) the distribution of edema in the tissue, and (4) the metabolic response of brain tissue to edema. Furthermore, investigations on chemical tissue or plasma factors promoting edema formation and persistence may be included in this subject. Obviously, a discussion of this material cannot be comprehensive within the scope of this brief introduction. Instead, a few aspects considered particularly significant will be emphasized.

a) CHEMICAL COMPOSITION OF EDEMATOUS BRAIN

First the water and electrolyte changes in various classic forms of either vasogenic or cytotoxic brain edema, shown in Figs. 1 and 2, may demonstrate the marked difference in edema relative to the underlying cause. Figures 1 and 2 make clear that the edema response of brain tissue to various insults is far from uniform.

Figure 1 shows the water, Na^+ and K^+ changes, measured in three different forms of vasogenic brain edema, which were produced by cortical freezing, tumor implantation and blunt skull trauma, respectively (13, 30, 36). Cold injury in the hands of Pappius and Gulati produced the most prominent rise in tissue water and, hence, swelling of the white matter (30). The gray matter was only minimally affected. Tumor implantation in rabbit brain studied by Herzog, Levy and Scheinberg induced edema amounting to 30%, while blunt skull trauma employed by Reulen, Hofmann and Baethmann in rats led to a small rise in brain water only (13, 36). However, the marked differences in edema illustrated by Figure 1 related not only to the type of injury, but also to the sampling technique used. Pappius and Gulati, and Herzog et al., studied tissue specimens sampled from discrete areas located relatively close to the focus, whereas Reulen et al. determined brain water and electrolytes of the total brain, i.e., edematous and normal tissue together. The swelling estimates calculated in the two former studies are thus representative only for the edematous areas sampled, while the latter study (36) permits evaluation of the total brain swelling, i.e., the extent of additionally occupied intracranial space.

The Na^+ and K^+ concentrations of the edema fluid in the white matter secondary to cold injury shown at the top of the left subset in Fig. 1 were calculated using the water and electrolyte changes (i.e. edema minus controls) on a dry weight basis (3). According to the data of Pappius and Gulati (30), the calculated Na^+ concentration of the edema fluid was 148.5 mM, that of potassium 14 mM. If these concentrations are compared with the data of Clasen and co-workers (Table I), who attempted to collect fluid of edematous white matter by centrifugation of the tissue, the agreement becomes obvious (7). Clasen et al found the edema fluid to contain 123 mM Na^+ and 15 mM K^+. The edema was again induced by cold injury. The ratios of the various fluid/serum concentrations were remarkably constant - ranging between 0.79 and 0.87 in the case of Na^+, chloride and albumin - except for potassium and Evans Blue. The data demonstrated the vasogenic origin of the edema but simultaneously indicated that the plasma filtrate entering the

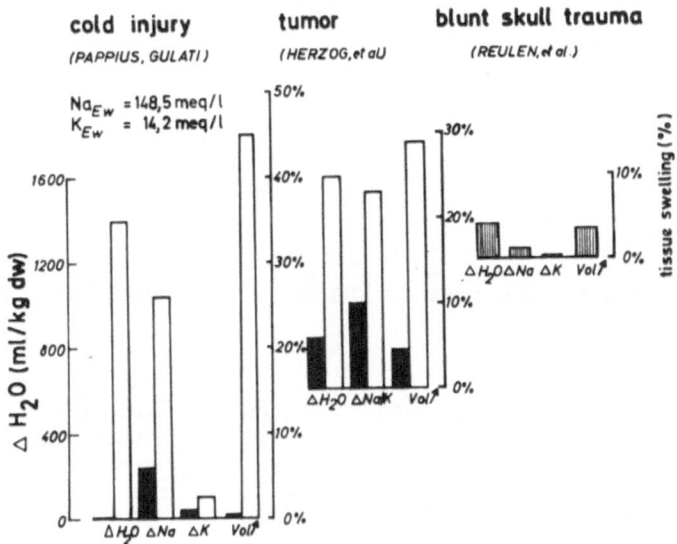

Fig. 1. Water and electrolyte changes (Δ: edema minus
        controls) on a dry weight basis in three forms
        of vasogenic brain edema. Tissue swelling
        (Vol↗) was calculated in percent according to
        (3, 10). The figures at the top of the subset of
        cold injury edema represent calculated Na^+ and
        K^+ concentrations in white matter (E_w) edema
        fluid (3). White and gray matter are symbolized
        by white and black columns, respectively. Water
        and electrolyte changes in blunt skull trauma are
        measured in total brain (shaded columns). From
        Baethmann and Schmiedek (3).

cerebral parenchyma is subject to modifications. The
rather high concentration of potassium was probably the
result of an artifact due to centrifugation of the
tissue; however, it is conceivable that in circumscribed
areas close to damaged tissue extracellular potassium
concentrations may increase.

Measurements of the protein content of edematous
brain have also confirmed the vasogenic nature of edema
produced by cold injury, or in brain tumors. In particular, the albumin content of edematous white matter rose
considerably as determined by disc-gel electrophoresis
(35).

Table I: Chemical composition of edema fluid*)

	Edema-Fl.*)	Serum	E/S**)	Edema-Fl.***) (computed)
Na$^+$ (mM)	123.4	143.0	0.86	148.5
K$^+$ (mM)	15.0	4.5	3.33	14.2
Cl$^-$ (mM)	86.7	110.0	0.79	---
Albumin	1.9g%	2.1g%	0.87	---
Evans Blue	16.4mg%	31.8mg%	0.52	---

*) According to Clasen et al. (7); **) E/S: edema/serum concentration ratio; ***) according to Pappius and Gulati (30).

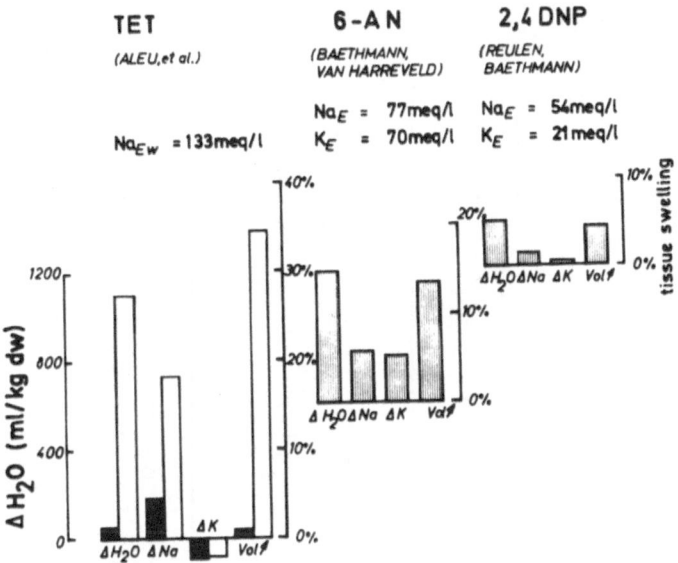

Fig. 2. Water and electrolyte changes in three forms of
cytotoxic brain edema induced by triethyl tin
(TET), 6-aminonicotinamide (6-AN) and dinitrophenol (2,4 DNP). For further explanation, see
Fig. 1. From Baethmann and Schmiedek (3).

In Fig. 2, three examples of cytotoxic brain edema
induced by metabolic inhibitors, such as triethyl tin
(TET), 6-aminonicotinamide (6-AN) and dinitrophenol
(2,4 DNP), are presented. The subset on the left shows
the data of Aleu et al., who studied triethyl tin, where
gray and white matter were separately analyzed (1). In
the graph, the changes in cerebral water and electrolytes and the respective tissue swelling are again shown.
Compared to the other agents, TET evidently is the most
effective compound to induce edema, with the swelling
amounting to 35%. This is 10% less than the white matter
swelling found after cold injury. The edema induced by
the nicotinamide analogue 6-AN led to brain swelling of
about 15%, while dinitrophenol produced tissue swelling
of only 5% (2, 37). Above the columns are the Na$^+$ and
K$^+$ concentrations of the edema fluid, which again were
calculated and not measured. Interestingly, the Na$^+$ concentration of the white matter edema after TET was quite
close to that of a plasma filtrate, although the blood-
brain barrier is known to remain intact and the fluid

Table II: Water, lipid content and tissue density
in perifocal brain edema *)

	brain water$_g$ (ml/100g fw)	brain water$_w$ (ml/100g fw)	total lipids$_w$ (g/100g fw)	density$_w$ (EMI-No.)
Brain-Edema	83.2 $\pm$ 0.9	78.8 $\pm$ 1.2	13.4 $\pm$ 1.3	12.1 $\pm$ 0.6
Controls	81.7 $\pm$ 1.0	69.1**)	19.1***)	p < 0.001 15.5

*) From Oettinger, et al. (25); **), ***) control data are taken
from Yates et al. (51) and McIlwain and Bachelard (21);
g: gray matter, w: white matter.

accumulated in spaces related to the intracellular rather than the extracellular compartment (1). The electrolyte concentrations of the edema fluid calculated in the 6-AN edema, or after dinitrophenol, were clearly different from a plasma filtrate, indicating an involvement of the intracellular compartment.

b) COMPUTER TOMOGRAPHY AND BRAIN EDEMA

Studies on the chemical composition of edematous brain may also help to understand computer-tomographical findings in patients suspected to be developing brain edema. Here, the question is how alterations in chemical brain composition influence the tissue density as determined in the CAT scan. The problem neurologists and neurosurgeons are particularly interested in is the threshold for the increase in brain-water where edema becomes visible in the tomogram. In this laboratory, the subject has been approached by measuring two components of brain tissue which on account of their abundancy in brain and their physical properties are major determinants of tissue density, i.e., water and lipids. Measurements were made in samples of perifocal edematous areas in patients with brain tumors, together with determinations of the corresponding density, numbers obtained in the computer-tomographical printout. The density figures, known as EMI numbers or Hounsfield units, were arbitrary, ranging from -500 (air) to +500 (bone), where water assumed a value of 0, fat of -50, normal white matter of +12 and gray matter of +18 (14).

Table II shows the results obtained in perifocal edema. Tissue density, given as EMI number, was determined only in the white matter, because gray matter density determinations may be less reliable on account of the adjacent bone structures (25). The perifocal gray and white matter was clearly edematous as evidenced by an increased tissue water content. The white matter lipids were reduced. The control data for brain water were taken from Yates et al. (51) and for cerebral fat content from McIlwain and Bachelard (21). The normal brain density numbers were obtained in 15 control patients without neurological symptoms. The reduction in tissue density shown in the right column in Table II is statistically significant ($p < 0.001$). Hence, the rise in tissue water is associated not only with a loss of cerebral lipids, but also with a loss in tissue density. The mean increase in white matter water content of approximately 10% corresponds to a decrease of mean tissue density of about 3 units. That means that the threshold in fluid uptake

Table III: Energy metabolism in perifocal brain edema *)

	Cr P (μM/gfw)	ATP/ADP	Lactate (μM/gfw)	ECP	L/P
Perifocal Edema (26)	3.53 ± 0.34	2.68 ± 0.22	4.18 ± 0.56	0.78	20 ± 3
Controls (6 dogs)	2.98 ± 0.15	4.00 ± 0.21	1.40 ± 0.13	0.85	4 ± 0.5

*) From Schmiedek et al. (41); Cr P: phosphocreatine;
ECP: energy charge potential; L/P = lactate/pyruvate ratio.

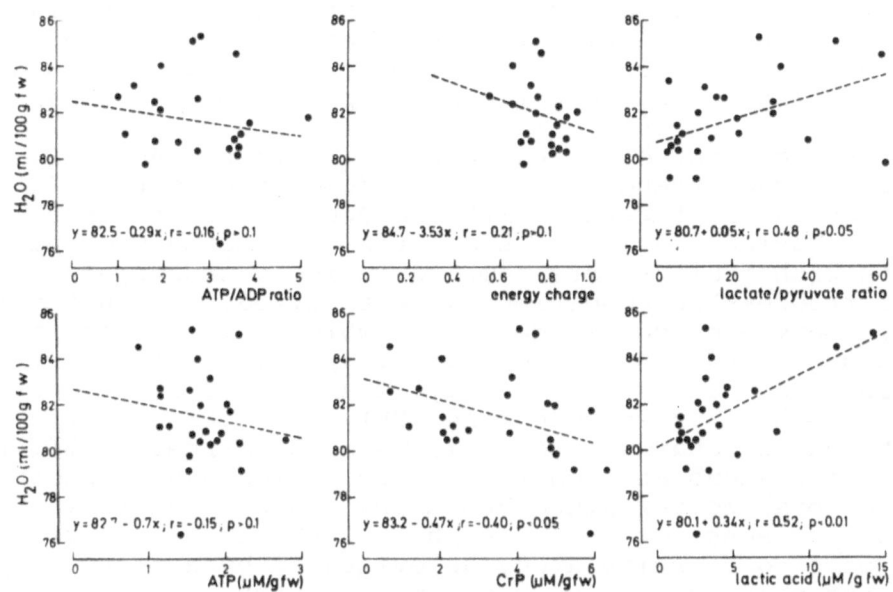

Fig. 3. Regression analyses of water content and labile metabolites in perifocal cortical areas around brain tumors in man. Only lactate and the lactate/pyruvate concentration ratio correlate significantly with water content of the respective tissue sample (r: 0.52, p< 0.01; r: 0.48, p< 0.05). From Schmiedek et al. (41).

in white matter leading to a change in density which is recognizable in the printout of the scanner is approximately 3%. It is thought that recognition of brain edema on the oscilloscope screen requires a water uptake 2-3 times higher. Since the tissue lipid content is inversely related to density, the decrease of lipids in edematous brain can be assumed to attenuate the effect of water uptake on x-ray absorption and consequently on brain edema recognition. This implies that computerized tomography would detect brain edema more sensitively if the lipid concentration remained unchanged.

c) BIOCHEMICAL RESPONSE OF BRAIN TISSUE TO EDEMA

With respect to the biochemical reaction of brain tissue to edema, cerebral energy metabolism may be considered briefly. When studying this subject in edematous

brain, one must be aware that alterations of metabolism
may not directly result from the edema, but from the
secondary effects such as impairment of tissue perfusion.
Various clinical and experimental approaches have been
used in study of energy metabolism in this condition,
such as measuring of global cerebral oxygen consumption,
labile tissue metabolites and the acid-base status. In
this laboratory together with the Department of Neurosurgery in Munich, perifocal brain tissue specimens of patients with brain tumors sampled during neurosurgery
under freeze-stop conditions were analyzed. For this purpose, a cryo-device was developed which made collection
of standardized tissue samples (41) possible. Table III
presents the data of this study. The parameters of energy
state, such as phosphocreatine (CP), the ATP/ADP ratio
or the energy charge potential, were not markedly different in edematous and control tissue, which was taken from
dog cerebral cortex since human brain was unavailable for
obvious reasons. On the other hand, the glycolytic activity seems to be stimulated, as evidenced by a 3- to 4-
fold increase in lactic acid concentration and the rise
of the lactic pyruvic acid ratio. Comparison of the individual metabolic tissue parameters with the brain water
content yielded regression analyses, which are shown in
Fig. 3.

In Fig. 3, the concentrations of ATP, CP and lactic
acid in the bottom row or the various metabolite ratios
in the top row were plotted against the water content on
the ordinate. No significant relations were found between
brain water and the parameters of the energy state, while
brain water appeared to correlate with lactate or the
L/P ratio. That means that the more edema present in the
tissue, the higher the glycolytic activity. Corresponding
observations were made by Frei et al. (12) and by Hjelm
et al. (15), who studied regional flow in perifocal edema
in relation to tissue glycolysis and ATP concentrations.
The regional flow correlated positively with lactic acid
except in one case with cerebral metastasis. On the other
hand, ATP appeared independent in respect not only to the
lactic acid, but also to the regional flow (15). Hence,
we may assume that brain edema surrounding brain tumors
or those after cold injury does not so much affect the
energy state, but rather stimulates glycolysis unless
tissue flow becomes severely curtailed.

    d) BRAIN EDEMA FACTORS

In this context it would be interesting to analyze
the effects of high lactic acid concentrations in brain

Table IV. Factors potentially associated with brain edema

Factors	Authors	Ref.
1. a) Lysosomes	Khattab	17
-enzymes	Bingham et al.	4
b) Proteolytic enzymes	Robert and	
(e.g., collagenase)	Godeau	38
2. Peptides	Sicuteri et al.	42
-kininogen-kinin	Blümel	5
system	Tzonos and Rana	46
3. Biogenic amines	Bulle	6
- Serotonin	Osterholm et al.	28
	Misra et al.	23
	Sachs	39
	Welch et al.	49
	Westergaard	50
	Costa et al.	8
	Klatzo	19
	Porta et al.	34
	Pausescu et al.	33
- norepinephrine	Osterholm	29
- histamine	Naftchi et al.	24
- dopamine		
4. Lipids		
- lipoperoxides	Suzuki and	
	Yagi	44
- prostaglandins	Jonsson and	
	Daniell	16
- free fatty acids	Sato et al.	40
5. Free radicals	Demopoulos et al.	9
	Ortega et al.	27

tissue on the process of edema formation and persistence. This point leads to the last part of this introduction, namely, the potential significance of chemical tissue or plasma factors which may be released or activated by the mechanisms causing edema. Such factors may be liable to a number of detrimental processes, e.g., enhancement, influx and spread of edema and inhibition of its resolution. They may affect various targets, e.g., the blood-brain barrier, microcirculation, or tissue metabolism. Table IV is a compilation of different candidates under discussion, such as lysosomal enzymes, studied by Khattab (17) in spinal cord compression-ischemia and Bingham et al. (4), in cold injury. Khattab found an increase in number and size of lysosomes dependent on the degree of tissue asphyxiation. In severe asphyxia, lysosomes seemed to lose their membranes (17). Bingham et al. provided evidence for a decrease in acid phosphatase and catepsin activity in cold injury edema, also suggesting that lysosomes were damaged, causing a release of the "suicide" enzymes.

Another interesting class of compounds are the biogenic amines, particularly serotonin, which has not lost its attractiveness since Bulle 20 years ago published his findings that serotonin injection into the brain leads to edema together with an enhanced barrier permeability (6). More recently it was shown by Westergaard that serotonin stimulates pinocytosis across cerebral blood vessels, a phenomenon which many consider highly significant in vasogenic brain edema (50). Other principles investigated in this context are lipoperoxides or free radicals, and of course the prostaglandins. While the role of free radicals in brain pathology, particularly cerebral edema, remains to be established, prostaglandins do not seem to play a major part (9, 27, 32).

The kallikrein-kinins aroused our interest because vasogenic edema is a condition where plasma constituents such as proteins make direct contact with the cerebral parenchyma. It is known from studies of Sicuteri et al. (42) that mere dilution of plasma with cerebrospinal fluid already leads to activation of the kallikrein-kinin system. On the other hand, glutamate merits attention on account of its potent pharmacological and toxic effects on nervous and glial cells (47) and because this compound abounds in the intracellular compartment in rather high concentrations, making massive leakage into the extracellular space in tissue damage very likely.

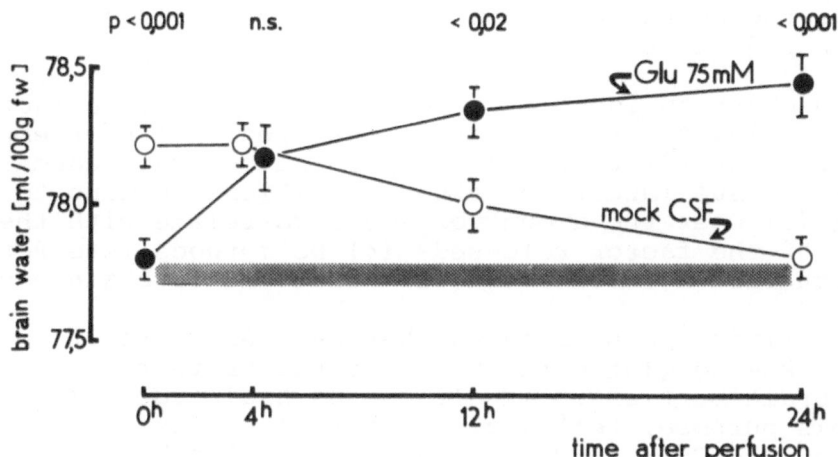

Fig. 4. Cerebral water content of rats on fresh weight
basis ($\bar{x}$ ± SEM) immediately and at various intervals after a 2 h perfusion period with isotonic,
buffered glutamate (75 mM) or mock CSF, respectively. The shaded bar at the bottom of the figure represents the range in brain water (± SEM)
of nonperfused normal rats. Time is given in
hours (h).

        Van Harreveld and Fifkova subjected rat cerebral
cortex to electrophoretic injections with glutamate for
1 hour. Histological and electronmicroscopic investigations of this material revealed that glutamate exposure
produced a halo of tissue necrosis surrounding the injection site with enormous swelling of glial elements
(47). Trubatch and van Harreveld calculated the glutamate threshold dose leading to this type of injury to be
about 8 mM (45). Assuming the intracellular glutamate
concentration to range from 12 to 15 mM, after correction
for a virtually glutamate-free extracellular compartment,
such damaging glutamate concentrations may conceivably
accumulate if many cells become destroyed in a cerebral
tissue insult. This in turn would induce further glutamate release, enhancing the process in a positive-feed-
back cycle (48). Sustained elevations of glutamate in the

permeability of cellular membranes, causing extra- to intracellular fluid shifts requiring active pumping to compensate for the enhanced influx. Eventually, exhaustion of the energy-producing capacity of cellular metabolism would ensue.

In order to identify a candidate as an edema factor, the following questions must be answered: (1) Does exposure of brain tissue to the factor induce brain edema? (2) Are tissue insults or other conditions leading to edema? (3) Does the amount of edema correlate with the amount of the factor released? (4) Do methods interfering with factor release benefit treatment of brain edema?

At first, we studied whether mere exposure of cerebral tissue to glutamate or plasma not protected by a blood-brain barrier in itself causes brain edema (26). For this purpose, ventriculo-cisternal perfusion was done in male Sprague-Dawley rats (250-300 g b.w.) with isotonic and buffered glutamate solutions (10 or 75 mM). Control animals were perfused with artificial CSF according to Merils (22). Figure 4 shows the cerebral water content of control and experimental animals perfused only with 75 mM glutamate immediately after perfusion, or at various intervals later, although 10 mM glutamate was found equally effective in inducement of brain edema. After perfusion with 75 mM glutamate, the cerebral Na^+ content rose from 207 ± 3 meq/kg dry weight in controls to 245 meq/kg dw ($p < 0.005$), while K^+ fell from 459 ± 5 meq/kg dw to 431 ± 10 meq/kg dw ($p < 0.05$). As seen in Fig. 4, control animals had a transient increase in cerebral water content immediately after perfusion, while in experimental animals edema started to develop after a certain interval. It is obvious from these data that the edema induced by exposure of central nervous tissue to glutamate is not as massive as that produced by other conventional methods. Nevertheless, the findings may serve as a first support of our concept that glutamate facilitates edema formation.

Similar experiments were conducted to test the kallikrein-kinin hypothesis. In Sprague-Dawley rats, ventriculo-cisternal perfusion with homologous plasma was employed to expose cerebral tissue to the kallikrein-kinin system (26). Kininogens, the precursors of kinins, are constituents of the plasma α-globulin fraction; in man, of the α_2-globulins (11).

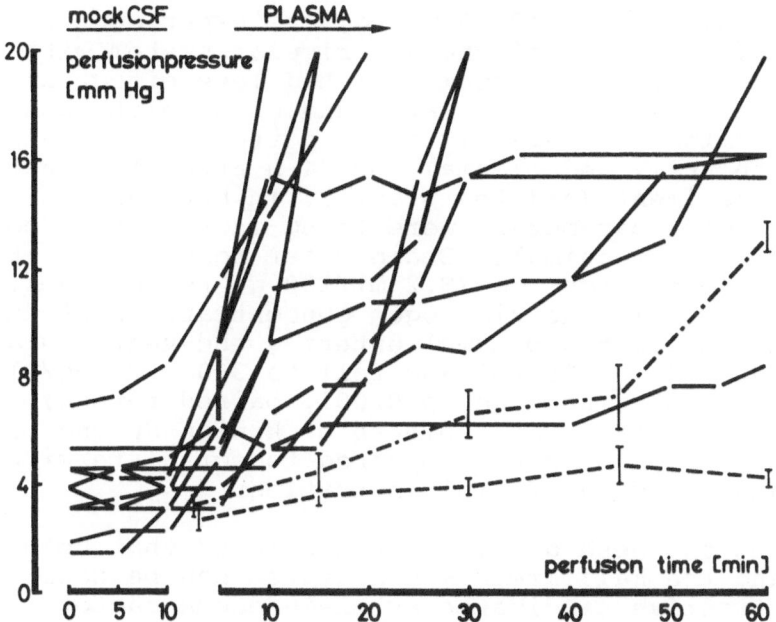

Fig. 5. Ventriculo-cisternal perfusion pressure manometrically recorded in rats perfused for 1 h
(flow: 5 ml/h) with undiluted (solid line) and
diluted plasma (-·-·-·-), and with artificial CSF
(----). Perfusion pressures of animals perfused
with undiluted plasma are shown as single curves;
those of the other two groups are shown as mean
values (x ± SEM). Ventricular perfusion with
mock CSF preceded perfusion with plasma.

We studied whether ventricular perfusion with plasma prevented from clotting also induces edema. Moreover,
the kininogen concentration of the in- and outflowing
perfusate was measured in order to assess release of
kinins during ventricular perfusion. The kininogen concentration was determined after conversion to kinins
using a biological test method according to Mann et al
(20).

Figure 5 demonstrates the manometrically recorded
ventricular perfusion pressure in control and experimental animals perfused with constant volumes of diluted
and undiluted plasma. Animals perfused with undiluted
plasma are shown as single curves, which in a considerable number demonstrate a marked rise upon changing of
the perfusion fluid from mock CSF to plasma. If the per-

fusion pressure reached 20 mm Hg, the experiment was dis-
continued. Perfusion of the ventricular system with diluted plasma (1:1 with mock CSF) had less effect on perfusion pressure. Cerebral water and electrolyte content
were determined immediately after perfusion (animals perfused with undiluted plasma), or 24 h after termination
of the experiment (diluted plasma). In both groups, the
chemical edema parameters were found increased as compared to control animals. Brain water rose from 77.6
ml/100 g fresh weight to 78.2 ml/100 g, respectively
($p < 0.01$; $p < 0.05$). The kininogen concentrations of the
perfusate (undiluted plasma) entering the ventricular
system were 486 $\pm$ 32ng/ml and fell to 360 $\pm$ 24 ng/ml
after ventricular passage ($p < 0.001$, paired t-test), reflecting a proportionate release of kinins during perfusion. The amount of edema appeared to correlate with the
amount and rapidity of kinin formation.

Based on these data, it is concluded that both glutamate and the kallikrein-kinin system can be considered
to be attractive candidates as brain edema factors. The
current and preliminary results make further attempts to
validate this concept worthwhile. Both hypotheses, if
proven, may influence current methods of brain edema
treatment.

ACKNOWLEDGMENT

Supported by DFG/Sonderforschungsbereich 51:
Medizinische Molekularbiologie und Biochemie, München.

REFERENCES

1. Aleu, F. P., Katzman, R. and Terry, R. D.: Fine structure and electrolyte analyses of cerebral edema induced by alkyl tin intoxication. J. Neuropathol. Exp.
   Neurol. 22: 403-413, 1963.

2. Baethmann, A. and van Harreveld, A.: Water and electrolyte distribution in gray matter rendered edematous with a metabolic inhibitor. J. Neuropathol. Exp.
   Neurol. 32: 407-423, 1973.

3. Baethmann, A., and Schmiedek, P.: Pathophysiology of
   cerebral edema: chemical aspects, in: Brain Edema,
   Cerebello-Pontine Angle Tumors, K. Schürmann, M. Brock,
   H. J. Reulen and D. Voth (eds), Advances in Neurosurger
   1: 5-18, Springer, Berlin-Heidelberg-New York, 1973.

4. Bingham, J. W., Paul, S. E. and Sastry, K. S. S.:
   Effects of cold injury on six enzymes in rat brain.
   Arch. Neurol. 21: 649-660, 1969.

5. Blümel, G.: Über die mögliche Beteiligung des Kinin-
   systems bei einem chirurgischen Zustandsbild, in:
   Haberland, G. L., and Matis, P. (eds), Neue Aspekte
   der Trasylol Therapie, 3: 125-130, F. K. Schattauer,
   Stuttgart-New York, 1969.

6. Bulle, P. H.: Effects of reserpine and chlorpromazine
   in prevention of cerebral edema and reversible cell
   damage. Proc. Soc. Exp. Biol. Med. 94: 553-556, 1957.

7. Clasen, R. A., Sky-Peck, H. H., Pandolfi, S., Laing, I.
   and Hass, G. M.: The chemistry of isolated edema fluid
   in experimental cerebral injury, in: Klatzo, I. and
   Seitelberger, F. (eds), Brain Edema, pp. 536-553,
   Springer, Wien, 1967.

8. Costa, J. L., Ito, U., Spatz, M., Klatzo, I. and
   Demirjian, C.: 5-Hydroxytryptamine accumulation in
   cerebrovascular injury. Nature 248: 2444: 135-136,
   1974.

9. Demopoulos, H. B., Milvy, P., Kakari, S. and Ransohoff,
   J.: Molecular aspects of membrane structure in cerebral
   edema, in: Reulen, H. J. and Schürmann, K. (eds),
   Steroids and Brain Edema, pp. 29-39, Springer, Berlin-
   Heidelberg-New York, 1972.

10. Elliott, K. A. C. and Jasper, H.: Measurement of ex-
    perimentally induced brain swelling and shrinkage.
    Amer. J. Physiol. 157: 122-129, 1949.

11. Frey, E. K., Werle, E. and Kraut, H.: Das Kallikrein-
    Kinin System und seine Inhibitoren, F. Enke, Stuttgart
    1968.

12. Frei, H. J., Wallenfang, T., Pöll, W., Reulen, H. J.,
    Schubert, R. and Brock, M.: Regional cerebral blood
    flow and regional metabolism in cold induced edema.
    Acta Neurochirurg. 29: 15-28, 1973.

13. Herzog, I., Levy, W. A. and Scheiberg, L. C.: Bio-
    chemical and morphologic studies of cerebral edema
    associated with intracerebral tumors in rabbits. J.
    Neuropathol. Exp. Neurol. 24: 244-255, 1965.

14. Hill, K. R. and Joyner, R. W.: Computerized X-ray tomography. Sci. Prog. (Oxford) 62: 237-262, 1975.

15. Hjelm, M., Overgaard, J., Lassen, N. A. and Tweed, W. A.: Brain tissue lactacidosis of hyperemic regions around brain tumors, in: Harper, A. M., Jennett, W. B., Miller, J. D. and Rowan, J. O. (eds), Blood Flow Metabolism in the Brain, pp. 1341-1342, Churchill Livingston, Edinburgh, 1975.

16. Jonsson, H. T. and Daniell, H. B.: Altered levels of PGF in cat spinal cord tissue following traumatic injury. Prostaglandins 11: 51-61, 1976.

17. Khattab, F. I.: Alterations in acid phosphatase bodies (lysosomes) in cat motoneurons after asphyxiation of the spinal cord. Exp. Neurol. 18: 133-140, 1967.

18. Klatzo, I.: Neuropathological aspects of brain edema. J. Neuropathol. Exp. Neurol. 26: 1-14, 1967.

19. Klatzo, I.: Pathophysiology of brain edema: pathological aspects in: Schürmann, K., Brock, M., Reulen, H. J. and Voth, D. (eds), Brain Edema, Cerebello-Pontine Angle Tumors, Advances in Neurosurgery, 1: 1-4, Springer, Berlin-Heidelberg-New York, 1973.

20. Mann, K., Geiger, R. and Werle, E.: A sensitive kinin liberating assay for kininogenase in rat urine, isolated glomeruli and tubules of rat kidney, in: Sicuteri, F., Back, N. and Haberland, G. I. (eds), Kinins: Pharmacodynamics and Biological Roles, pp. 65-73, Plenum, New York, 1973.

21. McIlwain, H. and Bachelard, H. S.: Biochemistry and the Central Nervous System, 4th ed., p. 309, Churchill Livingston, Edinburgh, 1971.

22. Merils, J. K.: The effect of changes in the calcium content of the cerebrospinal fluid on spinal reflex activity in the dog. Amer. J. Physiol. 131: 67-72, 1940.

23. Misra, S. S., Singh, K. S. P. and Bhargava, K. P.: Estimation of 5-hydroxytryptamine (5-HT) level in cerebrospinal fluid of patients with intracranial or spinal lesions. J. Neurol. Neurosurg. Psychiatry 30: 163-165, 1967.

24. Naftchi, N. E., Demeny, M., de Crescito, V., Tomasula, J. J., Flamm, E. S. and Campbell, J. B.: Biogenic amine concentrations in traumatized spinal cords of cats. Effect of drug therapy. J. Neurosurg. 40: 52-57, 1974.

25. Oettinger, W., Lanksch, W. and Baethmann, A.: The threshold of brain edema recognition by computerized tomography, in: Lanksch, W. and Kazner, E. (eds), Cranial Computerized Tomography, pp. 356-359, Springer, Berlin-Heidelberg-New York, 1976.

26. Oettinger, W., Baethmann, A., Rothenfusser, W., Geiger, R. and Mann, K.: Tissue and plasma factors in cerebral edema, in: Pappius, H. M. and Feindel, W. (eds), Dynamics of Brain Edema, pp. 161-163, Springer, Berlin-Heidelberg-New York, 1976.

27. Ortega, B. D., Demopoulos, H. B. and Ransohoff, J.: Effects of antioxidants on experimental cold-induced cerebral edema, in: Reulen, H. J. and Schürmann, K. (eds), Steroids and Brain Edema, pp. 167-175, Springer, Berlin-Heidelberg-New York, 1972.

28. Osterholm, J. L., Bell, J., Meyer, R. and Pyenson, J.: Experimental effects of free serotonin on the brain and its relation to brain injury. Parts I-III, J. Neurosurg. 31: 408-421, 1969.

29. Osterholm, J. L.: Noradrenergic mediation of traumatic spinal cord autodestruction. Life Sci. 14: 1363-1384, 1974.

30. Pappius, H. M. and Gulati, D. R.: Water and electrolyte content of cerebral tissues in experimentally induced edema. Acta Neuropathol. 2: 461-480, 1963.

31. Pappius, H. M.: Fundamental aspects of brain edema, in: Vinken, P. J. and Bruyn, G. W. (eds), Handbook of Clinical Neurology, Vol. 16, Tumors of the Brain and Skull, Part I, pp. 167-185, North Holland Publ. Co., Amsterdam, New York, 1974.

32. Pappius, h. M. and Wolfe, L. S.: Some further studies on vasogenic edema, in: Pappius, H. M. and Feindel, W. (eds), Dynamics of Brain Edema, pp. 138-143, Springer, Berlin-Heidelberg-New York, 1976.

33. Pausecu, E., Lugojan, R. and Pausescu, M.: Cerebral catecholamine and serotonin metabolism in post-hypothermic brain edema. Brain 93: 31-36, 1970.

34. Porta, M., Bareggi, S. R., Collice, M., Assael, B. M., Selenati, A., Calderini, G., Rossando, M. and Morselli, P. L.: Homovanillic acid and 5-hydroxyindole-acetic acid in the CSF of patients after a severe head injury. Eur. Neurol. 13: 545-554, 1975.

35. Rasmussen, L. E. and Klatzo, I.: Protein and enzyme changes in cold injury edema. Acta Neuropathol. 13: 12-28, 1969.

36. Reulen, H. J., Hoffmann, H. F. and Baethmann, A.: Die Beeinflussung des experimentellen traumatischen Hirnödems bei der Ratte mit einer Nikotinsäuretheophyllin-Verbindung. Zschr. ges. exp. Med. 138: 246-256, 1964.

37. Reulen, A. M. and Baethmann, A.: Das Dinitrophenolödem. Ein Modell zur Pathophysiologie des Hirnödems. Klin. Wschr. 45: 149-154, 1967.

38. Robert, A. M. and Godeau, G.: Action of proteolytic and glycolytic enzymes on the permeability of the blood-brain barrier. Biomedicine 21: 36-39, 1974.

39. Sachs, E.: Acetylcholine and serotonin in the spinal fluid. J. Neurosurg. 14: 22-27, 1957.

40. Sato, K., Jamaguchi, M., Mullan, S., Evans, J. P. and Ishii, S.: Brain edema. A study of biochemical and structural alterations. Arch. Neurol. 21: 413-424, 1969.

41. Schmiedek, P., Baethmann, A., Sippel, G., Oettinger, W., Enzenbach, R., Marguth, F. and Brendel, W.: Energy state and glycolysis in human cerebral edema. The application of a new freeze-stop technique. J. Neurosurg. 40: 351-364, 1974.

42. Sicuteri, F., Fanciullacci, M., Bavazzano, A., Franchi, G. and Del Bianco, P. L.: Kinins and intracranial hemorrhages. Angiology 21: 193-210, 1970.

43. Stewart-Wallace, A. M.: A biochemical study of cerebral tissue and of changes in cerebral .edema. Brain 62: 426-438.

44. Suzuki, O. and Yagi, K.: Formation of lipoperoxide in brain edema induced by cold injury. Experientia 30: 248, 1974.

45. Trubatch, J. and van Harreveld, A.: Spread of ionto-phoretically injected ions in a tissue. J. theoret. Biol. 36: 355-366, 1972.

46. Tzonos, T. and Rana, B.: Die Wirkung von Proteinasen und Fibrinolyse-Inhibitoren auf das experimentalle Hirnödem. Zschr. Neurol. 205: 61-70, 1973.

47. van Harreveld, A. and Fifkova, E.: Light- and elec-tron-microscopic changes in central nervous tissue after electrophoretic injection of glutamate. Exp. Molec. Pathol. 15: 61-81, 1971

48. van Harreveld, A. and Fifkova, E.: Effects of gluta-mate and other amino acids on the retina. J. Neuro-chem. 18: 2145-2154, 1971.

49. Welch, K. M. A., Meyer, J. S., Teraura, T., Hashi, K. and Shinmaru, S.: Ischemic anoxia and cerebral sero-tonin levels. J. Neurol. Sci. 16: 85-92, 1972.

50. Westergaard, E.: Enhanced vesicular transport of exog-enous peroxidase across cerebral vessels, induced by serotonin. Acta Neuropathol. 32: 27-42, 1975.

51. Yates, A. J., Thelmo, W. and Pappius, H. M.: Post-mortem changes in the chemistry and histology of normal and edematous brains. Amer. J. Pathol. 79: 555-564, 1975.

SIGNIFICANCE OF ADENYLATE CYCLASE IN THE REGULATION OF

THE PERMEABILITY OF BRAIN CAPILLARIES

F. Joó

Laboratory of Molecular Neurobiology
Institute of Biophysics
Biological Research Center
Szeged, Hungary

Certain enzymes found by histochemical methods to
be confined to the capillary wall have long been supposed
to take an active part in the regulation of brain-
capillary permeability. For example, in cases of
experimentally enhanced vascular permeability, an increase
of nonspecific alkaline phosphomonoesterase (EC 3.1.3.1)
has been reported (60). The capillary butyrylcholin-
esterase (EC 3.1.1.8) activity described originally by
Koelle (46) could be demonstrated at the light-microscopic
level in some species in those brain areas that were protected by the blood-brain barrier (19, 37). Recent results of Karcsú, Jancsó and Tóth (44) have shown, however,
that the capillaries in the area postrema, if studied
under the electron microscope, do manifest butyrylcho-
linesterase activity. "Extraneuronal" dopa-decarboxylase
(EC 4.1.1.26) has also been found to be confined to the
walls of brain capillaries, providing an enzyme trapping
mechanism for monoamine precursors (6, 7). A system that
is similar, although known in less detail, may operate
to prevent the entry of gamma-aminobutyric acid (GABA)
from the brain circulation. Namely, the strong GABA-
transaminase (EC 2.6.1.1.) positivity of brain vessels
has been interpreted - although no increase was observed
after inhibition with aminoacetic acid (18, 24) - in
relation to the regulation of capillary GABA transport
(22, 23, 25). Adenosine triphosphatase (ATPase) activity
(EC 3.6.1.3.), in striking contrast to that in the
capillaries of several organs (3, 31, 51) and in the
brain areas that are unprotected by the blood-brain

barrier, was reported by Torack and Barrnett (67) to be
localized mainly in the basal lamina of brain capillaries.
Inhibition of ATPase activity resulted in an increase of
macromolecular transport in brain capillaries (68),
accompanied by characteristic fine structural alterations
of the basal lamina (33). The role of the ATPase activity
in the basal lamina has been assumed to contribute
directly or indirectly to the maintenance of high
molecular organization of tropocollagen fibers (34).

In view of recent results (9, 35, 56, 59) empha-
sizing the primary importance of capillary endothelium
in the maintenance of the blood-brain barrier, it seemed
of interest to study whether the adenylate cyclase system
is involved, in a way similar to that reported by
Wagner et al. (69), in the mediation of effects of vasoactive substances. This paper shows that: (1) An increase
of pinocytotic activity together with the enhancement of
macromolecular transport could be elicited in brain
capillaries by dibutyryl cyclic adenosine monophosphate
treatment. (2) Adenylate cyclase activity could be
detected by histochemical methods in the walls of brain
capillaries. (3) The presence of histamine receptors
of H_1 and H_2 type could be revealed in a subcellular
fraction enriched in brain capillaries. (4) In view of
the above-mentioned findings, the effect of a nontoxic
histamine H_2-receptor blocker, metiamide, was investigated in experimental brain edema evoked by [90]yttrium
irradiation.

MATERIALS AND METHODS

Mice, weighing about 20-25 g, were each given
intraperitoneally a single dose (10 mg/kg) of N^6O^2-
dibutyryl cyclic 3',5'-adenosine monophosphate (dibu-
cAMP) dissolved in 0.1 ml saline solution. Control
mice were given 0.1 ml saline solution only. The
animals were killed by decapitation 5 or 20 minutes after
injection. Small cubes of the parietal cortex and the
cerebellar vermis were processed for electron microscopy.
Plates of the nonnuclear areas of the endothelial cells
to be measured were taken randomly from capillaries at
an original magnification of x 30,000. Portions of the
endothelial cytoplasm were outlined and measured planimetrically on prints of x 90,000 final magnification.
The pinocytotic and coated vesicles, both attached to
the luminal and abluminal surfaces and lying free in the
cytoplasm were counted on prints from the experimental
and control groups. The counts of vesicles per unit area

of endothelial cytoplasm in the experimental group were compared statistically to the control counts using the Student $\underline{t}$-test.

The permeability of brian capillaries was studied in adult rats by the sensitive fluorescence microscopic method (26) and after intravenous injections of ferritin (54).

The adenylate cyclase activity was demonstrated histochemically (57) in sections cut by Vibratome^R from paraformaldehyde-prefixed brain using 5'-adenylyl-imido-diphosphate, a highly specific new substrate.

Capillaries were isolated from rat cerebral cortex by a previously elaborated (38) procedure. Animals (10 rats in each experiment, 70 rats altogether) were perfused prior to homogenization with 0.9% sodium chloride through the aorta under ether anesthesia in order to eliminate the expected undesirable presence of blood corpuscles during enzyme determinations. Adenylate cyclase was determined biochemically by the method of Drummond and Duncan (15). Various concentrations of histamine and specific histamine-receptor blockers (8) and of noradrenaline were tested.

^{90}Yttrium cubes (approx. 4 mm^3) of varying strength (from 2.5 mCi to 0.1 mCi) were implanted into the surface of the parietal cortex on the right side in 4 adult dogs and 2 cats. For the quantitative expression of edema extent, the animals were given 2.5 ml/kg^{-1} of 1% Evans Blue 24 hours before investigation, the area of blue staining being measured planimetrically on symmetrical coronal slices (approx. 5 mm thick) obtained from the hemispheres of different animals. Corresponding areas of the right and left hemispheres from 3 control and 3 metiamide-treated animals were averaged, and the mean and S.D. were calculated. Half of the animals were treated with metiamide (generous gift of Dr. R. W. Brimblecombe from the Research Institute, Smith, Kline and French Laboratories Limited, Herts, England). Injections of metiamide were given intraperitoneally in a maintenance dose of 50 µg/kg^{-1}.

Fig. 1. Fluorescence of the trypan blue injected intravenously. The walls of capillaries show sharp
contours, which indicate the absence of macromolecular transport. x 700.

Fig. 2. Detail of the capillary wall under the electron
microscope. A few microvessels (thin arrow) and
tubular structures (thick arrow) are present in
the endothelial cell cytoplasm (E). In the cytoplasm of a glial cell (Gl) surrounding the capillary, a mitochondrion (M) is situated near the
basal lamina. x 14,000.

Fig. 3. Ferritin particles, as marker substances, cannot
penetrate brain capillaries in the normal condition. L = lumen, RBC = red blood corpuscle,
Gl = glial cell cytoplasm. x 70,000.

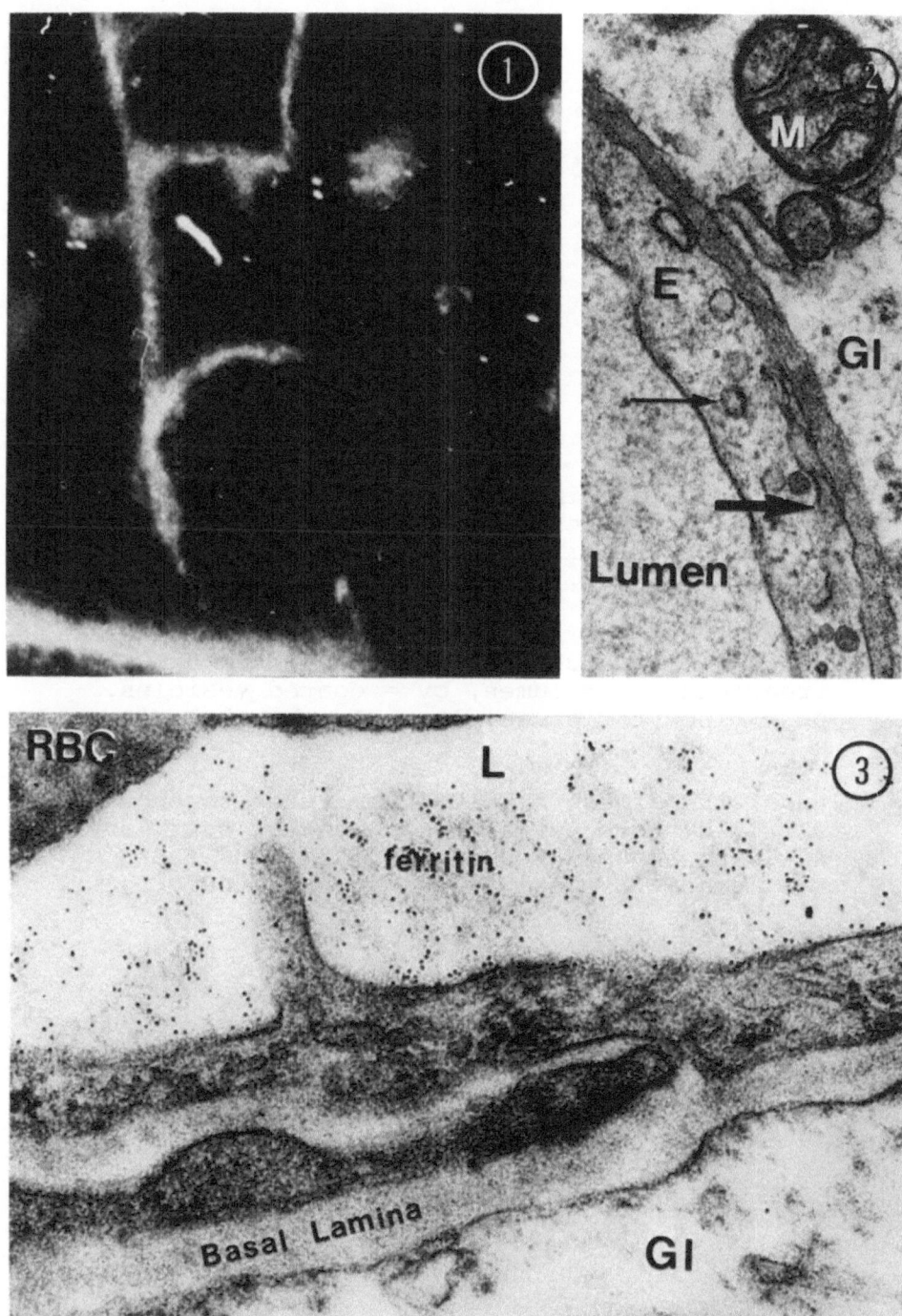

Fig. 4. The number of pinocytotic vesicles (pv) was in-
        creased 5 minutes after dibu-cAMP administration.
        L = lumen, M = mitochondria, Gl = glial cell cyto-
        plasm. x70,000.

Fig. 5. Tight junctions (arrow) between the endothelial
        cells remained closed 5 minutes after dibu-cAMP
        treatment. L = lumen, cv = coated vesicles,
        pv = pinocytotic vesicles, BL = basal lamina.
        x 130,000.

Fig. 6. Under the effect of dibu-cAMP, the exogenous
        ferritin has penetrated through the capillary
        wall and accumulated in the glial end feet (Gl).
        L = lumen, BL = basal lamina. x 45,000.

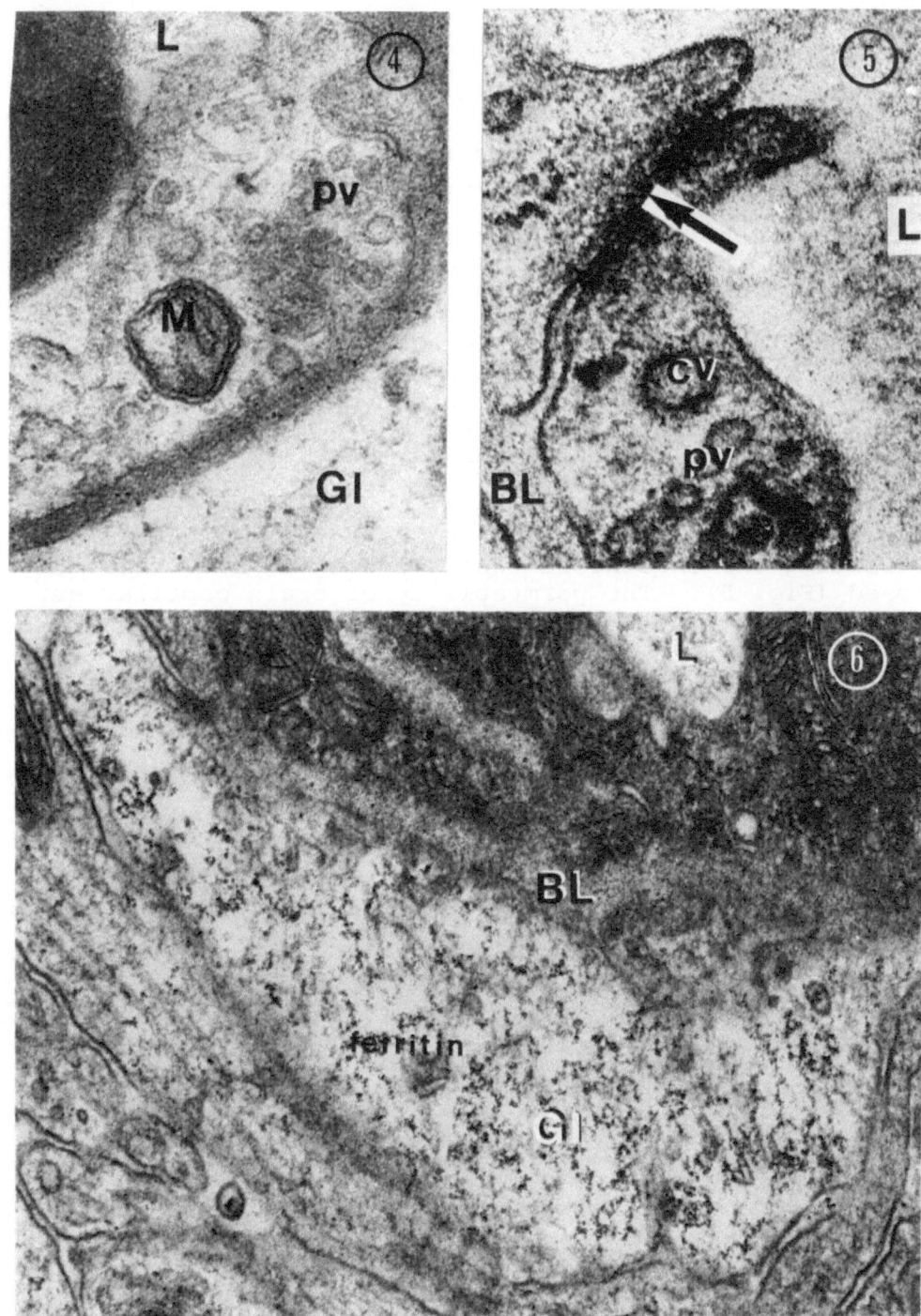

RESULTS

The effect of dibu-cAMP on the pinocytosis and permeability to macromolecules of brain capillaries

Following an intravenous injection of the barrier marker trypan blue, a wide range of capillaries could be seen under the fluorescence microscope (Fig. 1). The failure of penetration of macromolecules from the blood circulation into the brain parenchyma has already been reported by Reese and Karnovsky (56) related to the low rate of pinocytosis revealed by electron microscopy in the cytoplasm of capillary endothelial cells (Fig. 2). Figure 3 shows that, 30 minutes after an intravenous injection of ferritin, the electron-dense particles, though some have entered the endothelial cytoplasm but never penetrated the basal lamina, were mainly found in the lumen of capillaries. After dibu-cAMP treatment, however, there was a significant increase in the number of pinocytotic vesicles (Fig. 4). The tight junctions connecting the endothelial cells seemed to have remained closed (Fig. 5). The permeability of brain capillaries, judged by the penetration of ferritin, was found to be increased (Fig. 6). The effects of dibu-cAMP on the counts of different microvessels that are most likely involved in the evoked macromolecular transport is summarized below.

Number per unit area of microvessels after dibu-cAMP treatment (pv: pinocytotic vesicles; cv: coated vesicles):

	pv	cv	total counts
Control	7.09 ± 0.10	0.55 ± 0.21	7.89 ± 0.11
5 min	14.84 ± 0.18	1.51 ± 0.3	22.14 ± 0.24
P value	<0.001	<0.001	<0.001
20 min	15.37 ± 0.48	1.02 ± 0.52	21.15 ± 0.52
P value	<0.001	<0.02	<0.001

Histochemical localization of adenylate cyclase in brain capillaries

Figure 7 demonstrates the light-microscopic appearance of the adenylate-cyclase-positive structures in the cerebral cortex. In addition to the strong staining of capillaries it is to be noted that, as a result of avoiding freezing by Vibratome, the enzyme activity of delicate astrocytic processes of 2 to 5 μm diameter was also preserved. When the sections were incubated for a short time (Fig. 8), it was possible to confirm that, as in capillaries of epididymal fat (69), the adenylate cyclase activity could be inhibited by alloxan (Fig. 9). Noradrenaline in 10^{-5} M concentration did not seem to have any effect on the activity of capillary adenylate cyclase activity (Fig. 11). Under the electron microscope, the reaction product was observed in the luminal and basal membranes of capillaries (Fig. 12), as well as in the surface membranes of astrocytes (Fig. 13). The high enzyme reaction specificity was evidenced in control series (Fig. 14).

Characteristics of adenylate cyclase localized in brain capillaries

When chopped brain tissue was forced gently through nylon cloths of different pore sizes, abundant networks of capillaries with high structural integrity were found among the elements of neuronal and glial origin in the homogenate (Fig. 15). Profiles of capillaries were seen under the electron microscope in large number after differential and density gradient centrifugations (Fig. 16). The basal adenylate cyclase of the homogenate was 1387 + 331 pM/mg protein/5 minutes (mean + SE, n = 4), whereas that of the capillary-rich fraction was 10,245 + 994 pM/mg protein/5 min (mean + SE, n = 4). Activation of adenylate cyclase in the capillary-rich fraction was found with histamine but not with 10^{-5} M noradrenaline. Figure 17 shows the cAMP production in the presence of various concentrations of histamine and its receptor inhibitors chloropyramine and burimamide.

Therapeutic effect of metiamide treatment on brain edema of animals exposed to [90]yttrium irradiation

Irradiation with β-electrons from [90]yttrium is frequently used in neurosurgery, even though it induces severe brain edema in the white matter as an undesirable side effect. In our experiments, animals subjected to

Fig. 7. After incubation at 37°C for 60 minutes, adenylate cyclase was demonstrated histochemically
in the capillaries (arrows) and astrocytes.
x 900.

Fig. 8. After short incubations (5 minutes), capillary
adenylate cyclase (arrow) was much weaker.
x 600.

Fig. 9. Alloxan (in 5 mM concentration) completely
inhibited capillary adenylate cyclase if the
incubation was carried out for 5 minutes. x 600.

Fig. 10. Noradrenaline (in 10^{-5} M concentration) seemed
not to influence capillary adenylate cyclase
(arrow). x 600.

Fig. 11. Histamine (in 10^{-4} M concentration) seemed to
activate the adenylate cyclase of brain
capillaries (arrow). x 600.

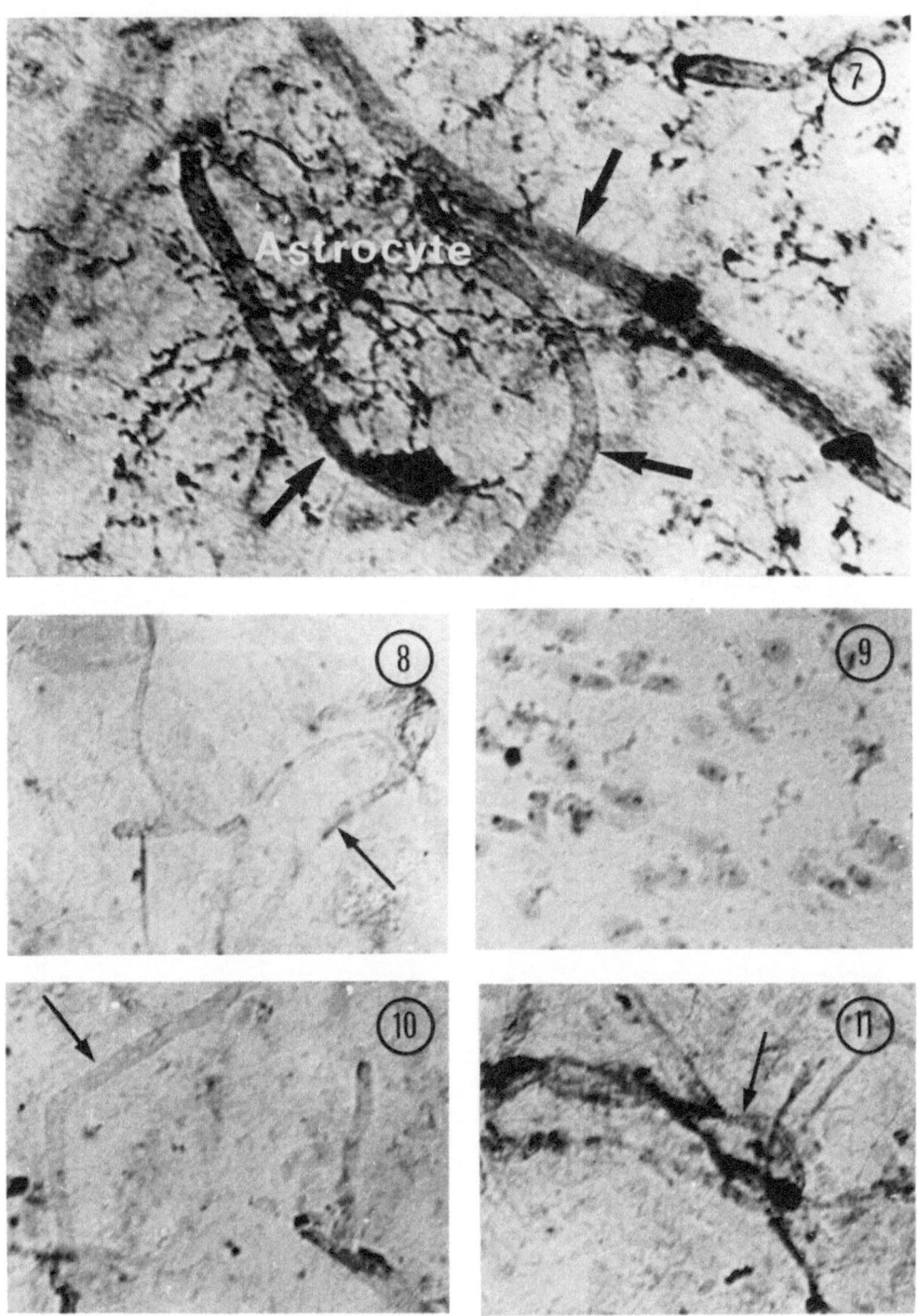

Fig. 12. Fine-structural localization of adenylate
         cyclase in the capillary wall. Luminal membrane
         and the basal lamina show (arrows) strong enzyme
         activity. L = lumen, M = inactive mitochondria.
         x 15,000.

Fig. 13. Adenylate cyclase positivity in the surface mem-
         brane of a glial end foot. Arrows point to the
         strong electron-dense precipitate, which indi-
         cates the localization of adenylate cyclase.
         L = lumen, M = mitochondrion. x 35,000.

Fig. 14. Incubation was carried out without the substrate.
         No electron-dense precipitate could be seen in
         the capillary wall. L = lumen, M = mitochondrion.
         x 30,000.

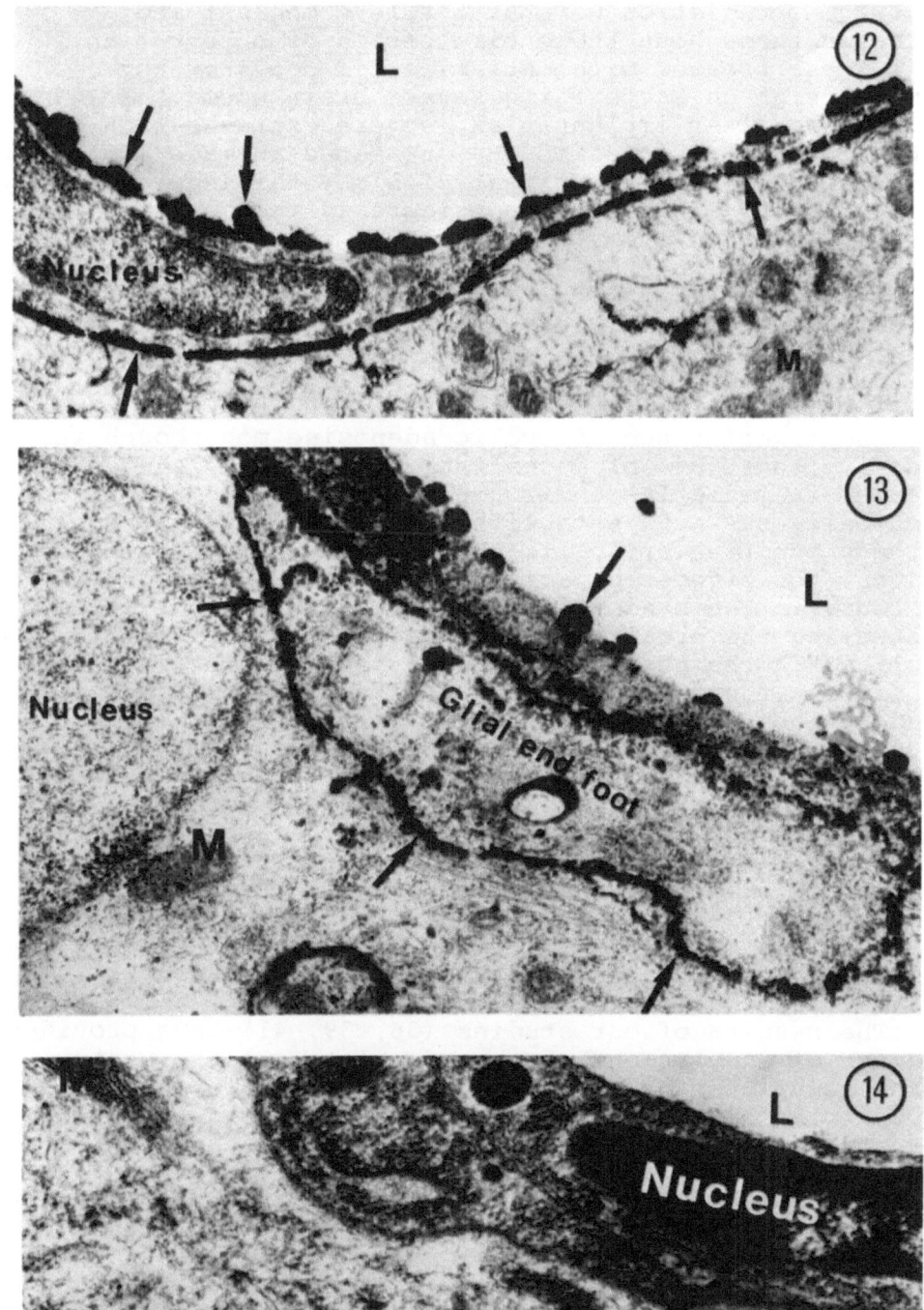

^{90}yttrium irradiation were as a rule somnolent and apathetic throughout the entire period of observation, while those treated with metiamide did not show any clinical sign characterizing severe brain edema. At 24 or 72 hours after implantation, severe extravasation of Evans Blue was observed in the untreated animals (Fig. 18, left side), whereas the extent of blue staining indicating the leakage of albumin was considerably reduced in metiamide-treated animals (Fig. 18, right side). The extent of edema in the control animals was found to be significantly greater (5.2 $\pm$ 2.5 cm^2) than that in metiamide-treated animals ($\overline{1}$.8 $\pm$ 1.0 cm^2).

DISCUSSION

The significance of cyclic adenosine monophosphate, an active substance produced from ATP by adenylate cyclase, in mediating environmental effects upon cell functioning has been established. In investigations of the production of cyclic AMP in different mammalian tissues, the highest capacity to synthesize cyclic AMP was found in the brain (11). Later, attention was focused on studying the problem of how cyclic AMP levels are regulated in the central nervous system. In brain slices, it has been found that formation of cyclic AMP was stimulated by incubation with putative neurotransmitters, e.g., noradrenaline (42, 43) and histamine (20, 63). Adenosine (61) and depolarizing agents, such as ouabain and veratridine (32), also resulted in an increase of cyclic AMP production. The activity of adenylate cyclase has been found in brain slices to be regulated by receptors linked to the membrane-bound enzyme. So far, alpha- and beta-receptors (12, 64), histamine (4, 63) and serotonin receptors (32) have been characterized, and a receptor for adenosine has been discovered (61).

The results of our studies (36, 39, 41) have provided evidence of involvement of the adenylate cyclase system in the regulation of the permeability of brain capillaries. At the same time, attention has been drawn to the fact that, in interpreting the findings obtained from the study of adenylate cyclase in brain slices, the presence of capillaries having enzymatic and receptor characteristics similar to, if not the same as, those of neuronal and glial elements must be taken into account. In regard to the cellular mechanism of enhanced protein transport resulting in vasogenic brain edema (45) two alternatives have been proposed: (1) tight junctions between the endothelial cells have been reported to be

widened reversibly in some cases (10, 55) and (2) the
transendothelial vesicular transport was found to be
stimulated (5, 35, 36, 48, 71). In pathological cases,
open junctions have been found in cerebral stab wound
(30), hypertension (21), cold injury (2), experimental
allergic encephalomyelitis (28) and lead encephalopathy
(47). On the other hand, transendothelial vesicular
transport was considered to be the primary leakage route
in hypertensive encephalopathy (17, 66, 73) and mercury
intoxication (70). Passage of horseradish peroxidase was
observed through channels, large vesicles and junctional
complexes (27). Later results (52, 53) have rendered
more likely the possibility of the transendothelial route
of protein transport. In regard to the significance of
micropinocytosis in transporting the plasma proteins,
some doubts have been raised recently (58), but it still
seems probable that, when an increase in micropinocytosis
is observed, random confluences of pinocytic vesicles,
similar to those reported by Simionescu, Simionescu and
Palade (65), can be established forming larger pores or
channels across the endothelial cells. An intracellular
system of tubules and cisternae of endoplasmic reticulum
has also been implicated in transcellular transport of
proteins (52).

In interpreting the possible significance of hista-
mine-sensitive adenylate cyclase in brain tissue, one
must be aware that, in addition to the capillaries, the
enzyme is localized in selective neurons (28) and glial
elements (13). Among the possible sources of endogenous
histamine and serotonin, the presence of mast cells (14,
16) in the central nervous system must also be emphasized.
It seems reasonable to assume that, when the cyclic AMP
level is increased in brain tissue as in ischemia (50),
histamine and serotonin can be released from mast cells
and contribute to the development of pathophysiological
changes. The possibility that hormone-sensitive receptors other than histamine are also confined to the
capillary wall cannot be ruled out. Baca and Palmer (1)
have shown that, in the 10,000g particulate fraction of
homogenized capillary-enriched fractions, the adenylate
cyclase was activated by norepinephrine, epinephrine
and dopamine. Westergaard (72) has reported that, after
perfusing through the cerebral ventricles, serotonin
produced an increase in transendothelial transport of
circulating exogenous peroxidase. In the interpretation
of earlier findings (49, 74) one has to take into account
the fact that, in addition to inhibiting Na^+-K^+-activated

Fig. 15. When a suspension is made by forcing brain tissue
        through nylon cloths of different pore sizes,
        the capillary beds (arrow), together with
        neuronal and glial elements, can be seen in
        large number. x 700.

Fig. 16. After differential and density gradient centrif-
        ugations, a fraction can be obtained that is
        enriched in capillaries retaining their fine
        structural characteristics. L = lumen, End =
        endothelial cytoplasm, pv = pinocytotic vesicles,
        BL = basal lamina. x 23,000.

Fig. 17. Effect of histamine and histamine-receptor block-
        ers (chloropyramine for H_1 receptors; burimamide
        for H_2 receptors) on capillary adenylate cyclase.

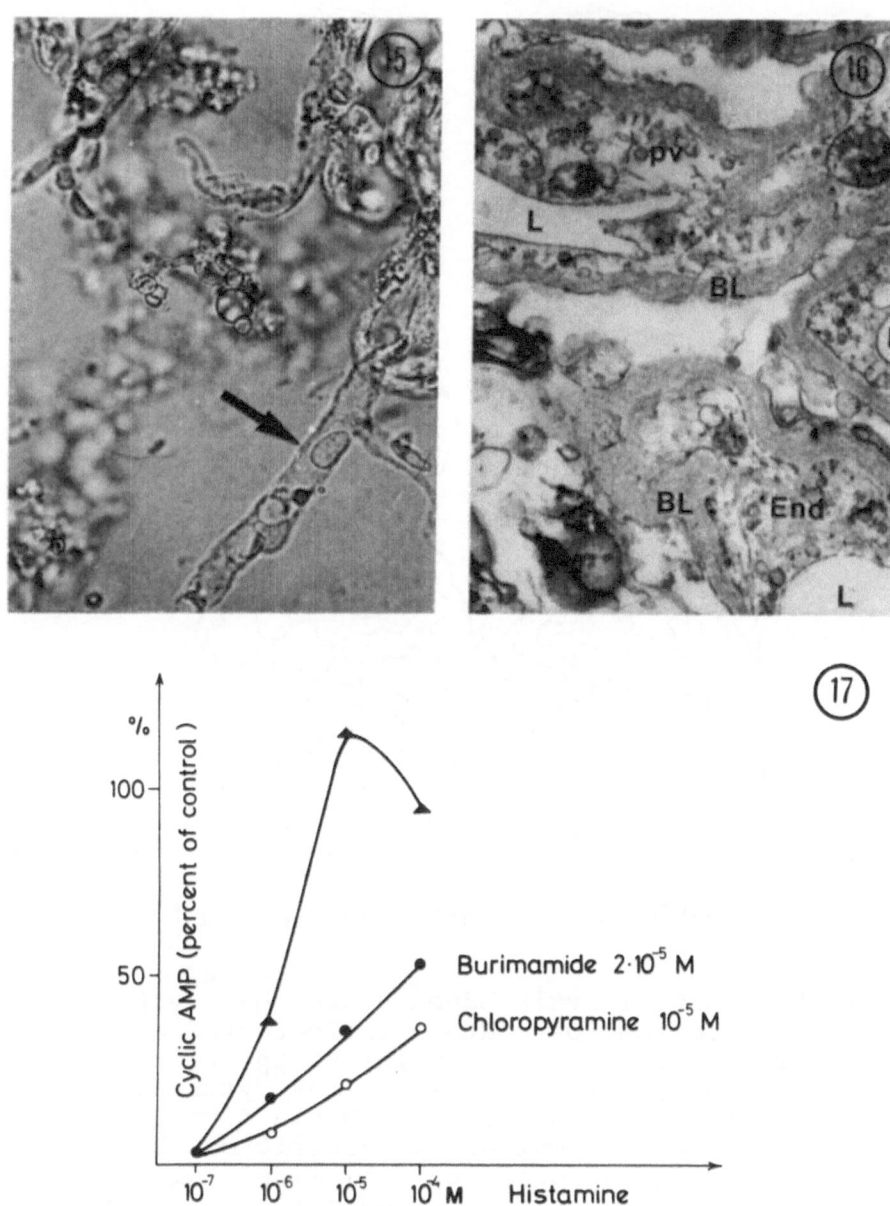

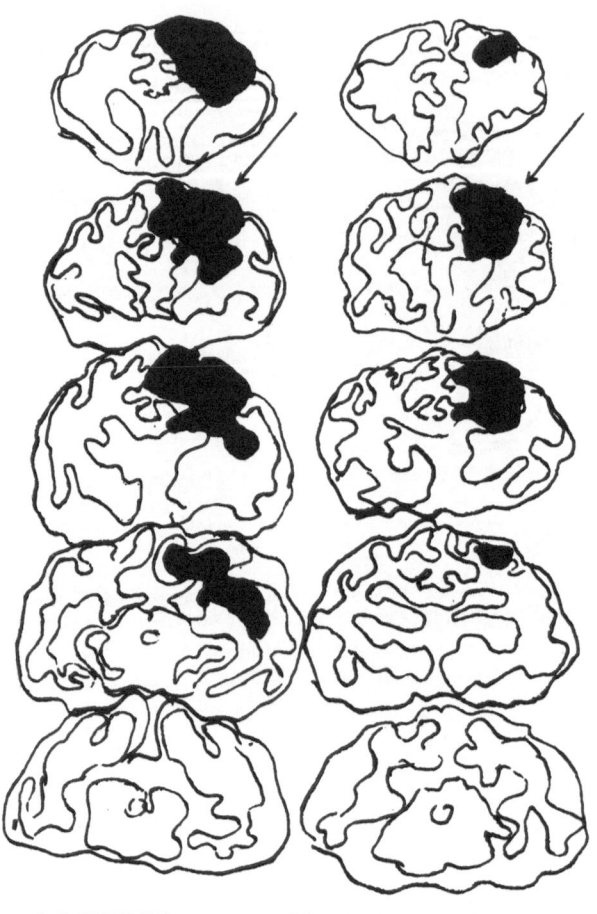

CONTROL   **Metiamide-TREATED**

Fig. 18. Expermental brain edema evoked by ^{90}yttrium
irradiation. Dark areas indicate the extent of
brain edema. Arrows point to the necrotic areas
caused by direct contact with ^{90}yttrium cubes.
These slices cut from the brain clearly show
that edema formation was restricted, by the
protection afforded by metiamide, to the portion
of the gyrus in direct contact with ^{90}yttrium.

ATPase, ouabain, after depolarizing the membrane, can interact indirectly with the adenylate cyclase system (62).

Our results (40) have shown that metiamide is effective in reducing the extent of brain edema. This finding clearly indicates that, following ^{90}yttrium implantation, H_2 receptors were indeed involved in the development of brain edema. As metiamide is known to be unable to cross the blood-brain barrier, it is conceivable that in our experiments, the prevention of brain edema was effected by the metiamide-induced insensitivity of brain capillaries to histamine released from ^{90}yttrium-irradiated tissue.

It still remains to be elucidated whether the non-toxic H_2 receptor blockers have therapeutic effect on other types of brain edemas.

ACKNOWLEDGMENTS

The author is grateful to collaborators, with whom certain parts of this paper have already been published I am thankful to Miss Gabriella Gazdagh, Mrs. Zsuzsanna Horváth, Mrs. Krisztina Mohácsi, Mrs. Gizella Rubicsekné-Kereki, Mr. József Farkas and Mr. János Szeles for their skilled technical assistance. The manuscript was prepared during the tenure of a sabbatical leave spent on a fellowship from the Max-Planck Society at the Max-Planck-Institut für biophysikalische Chemie, Göttingen-Nikolausberg, F.R.G.

REFERENCES

1. Baca, G.M. and Palmer, G.C. (1975): Presence of hormonally-sensitive adenylate cyclase receptors in capillary-enriched fractions from rat cerebral cortex. Blood Vessels (in press).

2. Baker, R.N., Cancilla, P.A., Pollock, P.S. and Frommes, S.P. (1971): The movement of exogenous protein in experimental cerebral edema. An electron microscopic study after freeze-injury. J. Neuropathol. Exp. Neurol. 30: 668.

3. Bartoszewitz, W. and Barrnett, R.J. (1964): Fine structural localization of nucleoside phosphatase activity in urinary bladder of the toad. J. Ultrastruct. Res. 10: 599-609.

4.   Baudry, M., Martres, M.P. and Schwartz, J.C. (1975):
     H_1 and H_2 receptors in the histamine-induced
     accumulation of cyclic AMP in guinea pig brain
     slices. Nature 253: 362-364.

5.   Beggs, J.L. and Waggener, J.D. (1976): Transendothe-
     lial vesicular transport of protein following
     compression injury to the spinal cord. Lab. Invest.
     34: 428-439.

6.   Betler, A., Falk, B. and Rosenberg, E. (1964): The
     direct demonstration of a barrier mechanism in the
     brain capillaries. Acta Pharmacol. Toxicol. 20:
     317-321.

7.   Betler, A., Falk, B., Owman, Ch. and Rosengreen, E.
     (1966): The localization of monoaminergic blood-
     brain barrier mechanisms. Pharmacol. Rev. 18:
     369-385.

8.   Black, J.W., Duncan, W.A.M., Durant, C.J., Ganellin,
     C.R. and Parson, E.M. (1972): Definition and
     antagonism of histamine H_2-receptors. Nature 236:
     385-390.

9.   Bodenheimer, T.S. and Brightman, M.W. (1968): A
     blood-brain barrier to peroxidases surrounded by
     perivascular spaces. Am. J. Anat. 122: 249-268.

10. Brightman, M.W., Hori, M., Rapaport, S.I., Reese, T.S.
     and Westergaard, E. (1973): Osmotic opening of
     tight junctions in cerebral endothelium. J. Comp.
     Neurol. 152: 317-326.

11. Butcher, R.W. and Sutherland, E.W. (1962): Adenosine
     3',5'-phosphate in biological materials. I. Purifi-
     cation and properties of cyclic 3',5'-nucleotide
     phosphodiesterase and use of this enzyme to character-
     ize adenosine 3',5'-phosphate in human urine. J.
     Biol. Chem. 237: 1244-1250.

12. Chasin, M., Rivkin, I., Mamrak, F., Samaniego, S.
     and Hess, S.M. (1971): α and β-adrenergic receptors
     of accumulation of cyclic adenosine 3',5'-monophos-
     phate in specific areas of guinea pig brain. J.
     Biol. Chem. 246: 2037-2041.

13.   Clark, R.B. and Perkins, J.P. (1971): Regulation
      of adenosine 3',5'-cyclic monophosphate concentra-
      tion in cultured human astrocytoma cells by
      catecholamines and histamine. Proc. Natl. Acad.
      Sci. U.S.A. 68: 2757-2760.

14.   Dropp, J.J. (1972): Mast cells in the central
      nervous system of several rodents. Anat. Rec.
      174: 227-238.

15.   Drummond, G.I. and Duncan, L. (1970): Adenyl
      cyclase in cardiac tissue. J. Biol. Chem. 245:
      976-983.

16.   Edvinsson, L., Owman, C. and Sjöberg, N.O. (1976):
      Autonomic nerves, mast cells, and amine receptors
      in human brain vessels. A histochemical and
      pharmacological study. Brain Res. 115: 377-393.

17.   Eto, T., Omae, T. and Yamamoto, T. (1971): An
      electron microscope study of hypertensive encephal-
      opathy in the rat with renal hypertension. Arch.
      Histol. Jap. 33: 133.

18.   Fisher, M.A., Hagen, D.Q. and Colvin, R.B. (1966):
      Aminooxidacetic acid: interactions with gamma-
      amino-butyric acid and the blood-brain barrier.
      Science 153: 1668-1670.

19.   Flumerfelt, B.A., Lewis, P.R. and Gwyn, D.G. (1973):
      Cholinesterase activity of capillaries in the rat
      brain. A light and electron microscopic study.
      J. Histochem. 5: 67-77.

20.   Forn, J. and Krishna, G. (1970): Effect of
      norepinephrine, histamine and other drugs on cyclic
      3',5'-AMP formation in brain slices of various
      animal species. Pharmacology 5: 193-204.

21.   Giacomelli, F., Wiener, J. and Spiro, D. (1970):
      The cellular pathology of experimental hypertension.
      V. Increased permeability of cerebral arterial
      vessels. Am. J. Pathol. 59: 133-159.

22.   Van Gelder, N.M. (1965): The histochemical
      demonstration of gamma-aminobutyric acid metabolism
      by reduction of a tetrazolium salt. J. Neurochem.
      12: 231-237.

23. Van Gelder, N.M. (1965): A comparison of gamma-
     amino-butyric acid metabolism in rabbit and mouse
     nervous tissue. J. Neurochem. 12: 239-244.

24. Van Gelder, N.M. (1966): The effect of aminoacetic
     acid on the metabolism of gamma-aminobutyric acid
     in brain. Biochem. Pharmacol. 15: 533-539.

25. Van Gelder, N. M. and Elliott, K.A.C. (1958):
     Disposition of gamma-aminobutyric acid administered
     to mammals. J. Neurochem. 3: 139-143.

26. Hamberger, A. and Hamberger, B. (1966): Uptake of
     catecholamines and penetration of trypan blue after
     blood-brain barrier lesions. Z. Zellforsch. 70:
     386-392.

27. Hansson, H.A., Johansson, B. and Blomstrand, C.
     (1975): Ultrastructural studies on cerebrovascular
     permeability in acute hypertension. Acta Neuropathol.
     32: 187-198.

28. Hegstrand, L.R., Kanof, P.D. and Greengard, P. (1976):
     Histamine-sensitive adenylate cyclase in mammalian
     brain. Nature 260: 163-165.

29. Hirano, A., Dembitzer, H.M., Becker, N.H., Levine, S.
     and Zimmerman, H.M. (1970): Fine structural
     alterations of the blood-brain barrier in experimental
     allergic encephalomyelitis. J. Neuropathol. Exp.
     Neurol. 29: 432-440.

30. Hirano, A., Becker, N.H. and Zimmerman, H.M. (1969):
     Pathological alterations in the cerebral endothelial
     cell barrier to peroxidase. Arch. Neurol. 20: 300-
     308.

31. Hoff, H.F. (1968): A comparison of the fine-
     structural localization of nucleoside phosphatase
     activity in large intracranial blood vessels and
     the thoracic aorta of rabbits. Histochemie 13:
     183-191.

32. Huang, M., Shimizu, H. and Daly, J.W. (1972):
     Accumulation of cyclic adenosine monophosphate in
     incubated slices of brain tissue. 2. Effects of
     depolarizing agents, membrane stabilizers, phospho-
     diesterase inhibitors, and adenosine analogs. J.
     Med. Chem. 15: 462-466.

33. Joó, F. (1968): The effect of the inhibition of adenosine triphosphatase activity on the fine structural organization of brain capillaries. Nature 219: 1378-1379.

34. Joó, F. (1969): Changes in the molecular organization of the basement membrane after inhibition of adenosine triphosphatase activity in rat brain capillaries. Cytobios 3: 289-301.

35. Joó, F. (1971): Increased production of coated vesicles in the brain capillaries during enhanced permeability of the blood-brain barrier. Brit. J. Exp. Pathol. 52: 646-649.

36. Joó, F. (1972): Effect on N^6O^2-dibutyryl cyclic 3',5'-adenosine monophosphate on the pinocytosis of brain capillaries of mice. Experientia 28: 1470.

37. Joó, F. and Csillik, B. (1966): Topographic correlation between the hematoencephalic barrier and the cholinesterase activity of brain capillaries. Exp. Brain Res. 1: 147-151.

38. Joó, F. and Karnushina, I. (1973): A procedure for the isolation of capillaries from rat brain. Cytobios 8: 41-48.

39. Joó, F., Rakonczay, Z. and Wollemann, M. (1975): cAMP-mediated regulation of the permeability in the brain capillaries. Experientia 31: 582-583.

40. Joó, F., Szücs, A. and Csanda, E. (1976): Metiamide-treatment of brain oedema in animals exposed to ^{90}yttrium irradiation. J. Pharm. Pharmacol. 28: 162-163.

41. Joó, F., Tóth, I. and Jancsó, G. (1975): Brain adenylate cyclase: its common occurrence in the capillaries and astrocytes. Naturwissenschaften 8: 397.

42. Kakiuchi, S. and Rall, T.W. (1968): The influence of chemical agents on the accumulation of adenosine 3',5'-phosphate in slices of rabbit cerebellum. Mol. Pharmacol. 4: 367-378.

43.   Kakiuchi, S. and Rall, T.W. (1968):   Studies on
      adenosine 3',5'-phosphate in rabbit cerebral
      cortex. Mol. Pharmacol. 4: 379-388.

44.   Karcsú, S., Jancsó, G. and Toth, L. (1977):   Butyryl-
      cholinesterase in fenestrated capillaries of the
      rat area postrema. Brain Res. 120: 146-150.

45.   Klatzo, I. (1967):   Presidential Address: Neuro-
      pathological aspects of brain edema. J. Neuropathol.
      Exp. Neurol. 26: 1-13.

46.   Koelle, G.B. (1954):   The histochemical localization
      of cholinesterase in the central nervous system of
      the rat. J. Comp. Neurol. 100: 211-235.

47.   Lampert, P., Garro, F. and Pentschew, A. (1967):
      Lead encephalopathy in suckling rats. In Brain
      Edema, Klatzo, I. and Seitelberger, F. (eds.),
      p 207, Springer, New York.

48.   Laursen, H., Schrøder, H. and Westergaard, E. (1975):
      The effect of portocaval anastomosis on the per-
      meability to horseradish peroxidase of cerebral
      vessels of the rat. Acta Pathol. Microbiol. Scand.
      83: 266-268.

49.   Lowe, D., Schieweck, Chr., Meier-Ruge, W., Bangerter,
      D. and Wolff, J.R. (1975):   The effect of ouabain
      on the ultrastructure of cerebral arterioles and
      surrounding tissue studied by a cannulation of a
      cerebral artery. Res. Exp. Med. 166: 97-114.

50.   Lust, W.D., Mršulja, B.B., Mršulja, B.J., Passonneau,
      J.V. and Klatzo, I. (1975):   Putative neurotrans-
      mitters and cyclic nucleotides in prolonged ischemia
      of the cerebral cortex. Brain Res. 98: 394-399.

51.   Marchesi, V.T. and Barrnett, R.J. (1964):   The
      localization of nucleotide phosphatase activity in
      different types of small blood vessels. J.
      Ultrastruct. Res. 10: 103-115.

52.   Møllgard, K. and Saunders, N.R. (1975):   Complex
      tight junctions of epithelial and of endothelial
      cells in early foetal brain. J. Neurocytol. 4:
      453-468.

53. Møllgard, K., Malinowska, D.H. and Saunders, N.R.
     (1976): Lack of correlation between tight junction
     morphology and permeability properties in developing
     choroid plexus. Nature 264: 293-294.

54. Raimondi, A.J., Evans, J.P. and Mullan, S. (1962):
     Studies of cerebral edema. III. Alterations in
     the white matter: an electron microscopic study
     using ferritin as a labeling compound. Acta
     Neuropathol. 2: 177-197.

55. Rapaport, S.I. (1970): Effect of concentrated
     solutions on blood-brain barrier. Am. J. Physiol.
     219: 270-274.

56. Reese, T.S. and Karnovsky, M.J. (1967): Fine
     structural localization of a blood-brain barrier
     to exogenous peroxidase. J. Cell Biol. 34: 207-217.

57. Reik, L., Petzold, G.L., Higgins, J.A., Greengard, P.
     and Barrnett, R.J. (1970): Hormone-sensitive adenyl
     cyclase: cytochemical localization in rat liver.
     Science 168: 382-284.

58. Rippe, B., Kamiya, A. and Folkow, B. (1977): Is
     capillary micropinocytosis of any significance for
     the transcapillary transfer of plasma proteins?
     Acta Physiol. Scand. 100: 258-260.

59. Rodriquez, L.A. (1955): Experiments on the histologi-
     cal locus of the haematoencephalic barrier. J.
     Comp. Neurol. 102: 27-45.

60. Samorajski, T. and McCloud, J. (1961): Alkaline
     phosphomonoesterase and blood-brain permeability.
     Lab. Invest. 10: 492-501.

61. Sattin, A.W. and Rall, T.W. (1970): The effect of
     adenosine and adenine nucleotides on the cyclic
     adenosine 3',5'-monophosphate content of guinea pig
     cerebral cortex slices. Mol. Pharmacol. 6: 13-23.

62. Shimizu, H., Creveling, C.R. and Daly, J.W. (1970):
     Stimulated formation of adenosine 3',5'-cyclic
     phosphate in cerebral cortex: synergism between
     electrical activity and biogenic amines. Proc.
     Natl. Acad. Sci. U.S.A. 65: 1033-1040.

63.   Schultz, J. and Daly, J.W. (1973): Adenosine
       3',5'-monophosphate in guinea pig cerebral cortical
       slices: effects of α- and β-adrenergic agents,
       histamine, serotonin and adenosine. J. Neurochem.
       21: 573-579.

64.   Schultz, J. and Daly, J.W. (1973): Accumulation of
       cyclic adenosine 3',5'-monophosphate in cerebral
       cortical slices from rat and mouse: stimulatory
       effect of α- and β-adrenergic agents and adenosine.
       J. Neurochem. 21: 1319-1326.

65.   Simionescu, M., Simionescu, N. and Palade, G.E.
       (1975): Permeability of muscle capillaries to
       small hemepeptides. J. Cell Biol. 64: 584-607.

66.   Sonkodi, S., Joó, F. and Maurer, M. (1970): The
       permeability state of the blood-brain barrier in
       relation with the plasma renin activity in early
       stage of experimental renal hypertension. Brit.
       J. Exp. Pathol. 51: 448-452.

67.   Torack, R.M. and Barrnett, R.J. (1964): The fine
       structural localization of nucleotide phosphatase
       activity in the blood-brain barrier. J. Neuropathol.
       Exp. Neurol. 23: 46-59.

68.   Várkonyi, T. and Joó, F. (1968): The effect of
       nickel chloride on the permeability of the blood-
       brain barrier. Experientia 24: 452.

69.   Wagner, R.C., Kreiner, P., Barrett, R.J. and Bitensky,
       M.W. (1972): Biochemical characterization and
       cytochemical localization of catecholamine-sensitive
       adenylate cyclase in isolated capillary endothelium.
       Proc. Natl. Acad. Sci. U.S.A. 69: 3175-3179.

70.   Ware, R.A., Chang, L.W. and Burkholder, P.M. (1974):
       An ultrastructural study on the blood-brain barrier
       dysfunction following mercury intoxication. Acta
       Neuropathol. 30: 211-224.

71.   Westergaard, E. (1974): Transport of protein tracers
       across cerebral arterioles under normal conditions.
       In: Pathology of Cerebral Microcirculation. J.
       Cervos-Navarro (ed.), pp 218-227, W. de Gruyter & Co.
       Berlin, New York.

72.   Westergaard, E. (1975):   Enhanced vesicular
      transport of exogenous peroxidase across cerebral
      vessels, induced by serotonin.   Acta Neuropathol.
      32: 27-42.

73.   Westergaard, E., van Deurs, B. and Brøndstedt, H.E.
      (1977):   Increased vesicular transfer of horseradish
      peroxidase across cerebral endothelium, evoked by
      acute hypertension.   Acta Neuropathol. 37: 141-152.

74.   Wolff, J.R., Schieweck, Chr., Emmenegger, H. and
      Meier-Ruge, W. (1975):   Cerebrovascular ultra-
      structural alterations after intra-arterial
      infusions of ouabain, scilla-glycosides, heparin
      and histamine.   Acta Neuropathol. 31: 45-58.

BRAIN MICROVESSELS: GLUCOSE METABOLIZING ENZYMES IN ISCHEMIA AND SUBSEQUENT RECOVERY

B.M. Djuričić and B.B. Mršulja

Laboratory for Neurochemistry
Institute of Biochemistry
Faculty of Medicine, Belgrade, Yugoslavia

Hypoxia (arterial pO2 reduction) and ischemia (a critical degree of the brain tissue hypoperfusion) can deprive central nervous system cells of the oxygen supply which they require for normal function. In addition, ischemia deprives the brain of glucose, amino acids and other blood constituents passing the blood-brain barrier. Among other organs, the brain is unique in requiring a constant glucose and oxygen supply. With the restriction of the blood supply to the brain, the main sources of energy production are ATP, creatine phosphate, and glycogen (17). However the concentration of these brain metabolites falls rapidly following the onset of ischemia as produced in gerbils (13, 16, 18, 22).

There are at least three morphologically distinct brain compartments: neurons, glial cells and endothelial cells, which at the same time represent different metabolic compartments (5, 9, 25, 28). These compartments seem to react differently to hypoxia and ischemia (9, 25). The neurons are without any doubt, the elements most sensitive to injury in ischemia.

Recently, we have shown a significant decrease in specific capillary uptake of ^3H -2-deoxy-D-glucose in gerbils subjected to ischemia for 6 min., but not for 1 and 3 min. (31), indicative of an energy-dependent uptake of the glucose analogue. Generally, it has been accepted that under physiological conditions the glucose and glucose analogues are transported from the blood to the brain by a facilitated carrier-mediated process which requires no energy (3). However, a decreased unidirectional entry

239

of the hexose into the brain was observed under pathologic conditions (1, 2). Since we have suggested that
some enzymes involved in glucose metabolism (e.g., capillary hexokinase and glucose-6 phosphatase might be par-
ticipating in glucose transport through the capillary
endothelium (5), the behavior of glucose-metabolizing
enzymes in brain microvessels and parenchyma in ischemia
was examined further. Also, the beneficial effect of
dihydroergotoxine (a combination of dihydroergocornine,
dihydroergocristine and dihydroergocriptine in equal
proportions: Redergin[R], LEK, Ljubljana) upon the enzymic
changes, particularly in brain parenchyma, are demonstrated during ischemia and subsequent recovery from it.
All experiments were performed on groups of Mongolian
gerbils because the gerbil provides a good experimental
model for the study of ischemia and the post-ischemic
period (12).

METHODS

Global forebrain ischemia was produced in groups of
Mongolian gerbils by bilateral common carotid-artery occlusion for 1 and 3 min. (and 9 min. for the LDH-isoenzyme
study). The cerebral circulation was reestablished for 5
min. in animals subjected to 3 min. ischemia only. Sham-
operated gerbils, without occluded carotid arteries, were
used as controls for each experimental period. Dihydroergotoxine methane sulfonate (DHE, 1 mg/kg body weight, i.p.)
was administered 90 min. prior to ischemia. Non-treated
gerbils were used for the determination of the enzymes
base line activity in capillaries and parenchyma. The
cerebral capillaries were separated from the non-vascular
tissue of gerbil forebrain according to our recently described technique (23), Ringer's solution containing 1%
of albumin was replaced by the homogenizing medium as it
is more suitable for enzymic studies (4). Following the
separation of the capillary fraction the remainder was
centrifugated at 105,000 x g for 60 min., and the enzyme
activities were assayed in the appropriate fraction (pellet or supernatant), depending on their intracellular
localization. The enzymatic activities of hexokinase (HK),
phosphofructokinase (PFK), pyruvate kinase (PK), lactate
dehydrogenase (LDH), NAD^+-dependent isocitric dehydrogenase (ICDH-NAD), $NADP^+$-dependent isocitric dehydrogenase
(ICDH-NADP), glucose-6-P dehydrogenase (G-6-PDH) and
glucose-6-phosphatase (G-6-Pase) were measured using the
established spectrophotometric or fluorimetric techniques.
LDH isoenzymes were revealed after scanning acrylamide
gels that were stained for LDH activity (34).

Table 1. Enzymatic activities in isolated Mongolian gerbil forebrain microvessels and brain parenchyma.
Activity ratio: (brain microvessels activity)/
(brain parenchyma activity).

Enzyme	Activity in microvessels	in parenchyma	Ratio
1. Hexokinase [a]	22 ± 2	109 ± 4	0.20
2. Phosphofructokinase [a]	10 ± 1	56 ± 4	0.18
3. Pyruvate kinase [a]	893 ± 28	4972 ± 388	0.18
4. Lactate dehydrogenase [a]	625 ± 12	2844 ± 107	0.22
5. Glucose 6-P dehydrogenase [b]	120 ± 16	592 ± 41	0.20
6. Isocitrate dehydrogenase-NAD[+] dependent [b]	12 ± 1	39 ± 1	0.31
7. Isocitrate dehydrogenase-NADP[+] dependent [b]	315 ± 52	415 ± 18	0.76
8. Glucose 6-phosphatase [b]	432 ± 32	20 ± 1	21.50

Enzyme activities were determined at pH 7.1 using either
triethanolamine-hydrochloride or imidazole-hydrochloride
buffers, 50 mM, at constant temperture of 37°C. Values
represent the mean ± S.E.M. for 4-6 assays done in dupli-
cate. Each enzymatic assay was linear regarding protein
concentration and incubation time.

[a] The activities were given as nMoles of substrate/min/mg
protein, 37°C.

[b] The activities were given as nMoles of substrate/hour/mg
protein, 37°C.

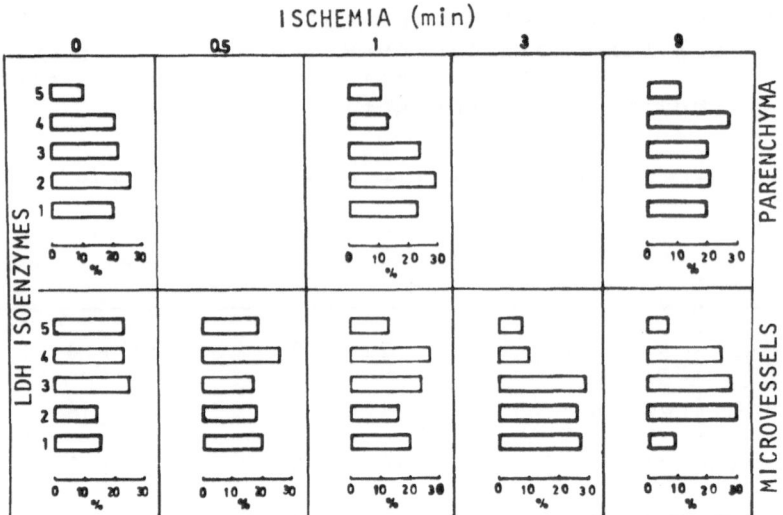

Fig. 1. Schematic representation of the lactate dehydrogenase (LDH) isoenzymatic patterns after various
periods of ischemia in brain parenchyma and
isolated brain microvessels of Mongolian gerbils.
Inserted abscissas represent the isoenzymes, relative participation in the total activities.

CAPILLARY/PARENCHYMA RATIO OF GLUCOSE-METABOLIZING ENZYMES

Among the eight investigated enzymes of glucose
metabolism only G-6-Pase was enriched in the gerbil forebrain microvessel fraction (Table 1), which is in agreement with the previous finding in the rat brain capillaries (5). However, this enzyme was found enriched, when
compared with parenchyma, in an extent indicative for its
almost exclusive localization in capillaries. Also, in
capillaries relatively high activity of ICDH-NADP was
noted. Other enzymes of glucose metabolism in microvessels
showed activities of about 20-30% of the parenchyma ones.

Presence of G-6-Pase in brain microvessels leads us
to the hypothesis that this enzyme, in concordance with
HK, may play important role in glucose transport from
blood to parenchyma, enabling the glucose-6-P splitting
formed in HK reaction, and glucose release in adjacent
glial cells (5). The fact that glucose blood to brain
transport is oxygen dependent process (31) increases our

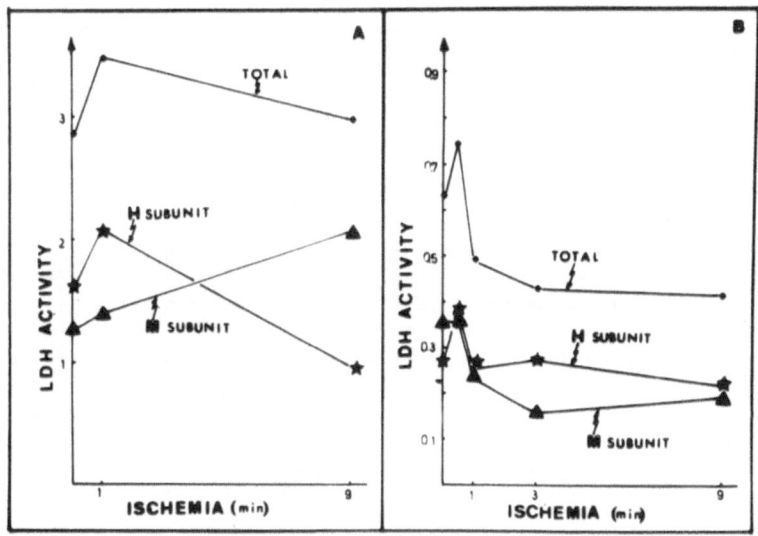

Fig. 2. Subunit composition of lactate dehydrogenase (LDH
in brain parenchyma (A) and isolated brain microvessels (B) of Mongolian gerbils after various
periods of ischemia. LDH activities are represented as μMoles of pyruvate/min/mg protein, 37°C,
ph 7.1.

interest to study both enzymes under the ischemic conditions.

    LDH in brain microvessels displayed fairly high
activity (Table). Regarding its isoenzymic composition
(predominance of the M subunits typical for the tissues
tolerating decrease of pO_2 during certain periods, e.g.
muscle), microvessels were clearly oposed to the parenchyma. LDH isoenzymic pattern in parenchyma was typical
for the aerobic tissue (Fig. 1). LDH_1 (HHHH tetramer)
and LDH_2 (HHHM tetramer) represents minor part in microvessels in contrast to parenchyma. Quantitatively, 60%
of total activity in microvessels was due to the M subunits, and reversal is true for the parenchyma (Fig.2).
Microvessel LDH possibly plays some role in the process
(es) of the lactate removal from the brain, especially
in the conditions of its accumulation; the possibility
that lactate may play some role in the genesis of cerebral edema was suggested (14). Certain kinetic properties

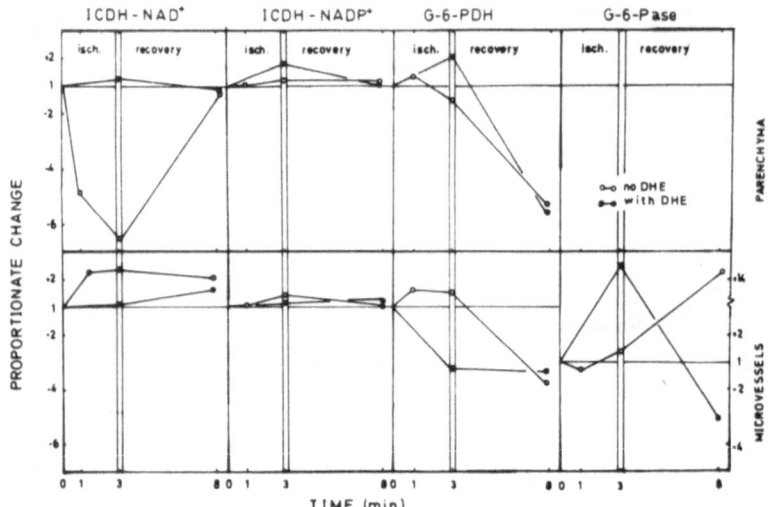

Fig. 3. The effect of global cerebral ischemia and
       subsequent recovery in the activities of
       hexokinase (HK), phosphofructokinase (PFK),
       pyruvate kinase (PK) and lactate dehydrogenase
       (LDH) in brain parenchyma and isolated brain
       microvessels of Mongolian gerbils. DHE = dihydro-
       ergotoxine. Proportionate changes which give
       equal weight to negative and positive changes,
       were calculated using the following expression:

$$X \left(\frac{\text{experimental value}}{\text{control value}} \right) X \quad , \text{ where } X = +1 \text{ for}$$

       experimental values equal or greater than the
       controls, and X = -1 for experimental values
       less than control values.

of the rat brain microvessel LDH favor this hypothesis
(5). On the other hand, LDH isoenzymic pattern of brain
microvessels indicates that endothelial cells are able
to withstand pO_2 decrease to the greater extent than
the parenchymal cells which might explain the findings
that [³H]-deoxy-D-glucose uptake was not changed in
brain capillaries during 1 and 3 minutes of global
ischemia in gerbils (31).

     Another interesting property of the microvessel
enzymatic organization is relatively high ICDH-NADP
activity. This enzyme is known to be localized in cytosol

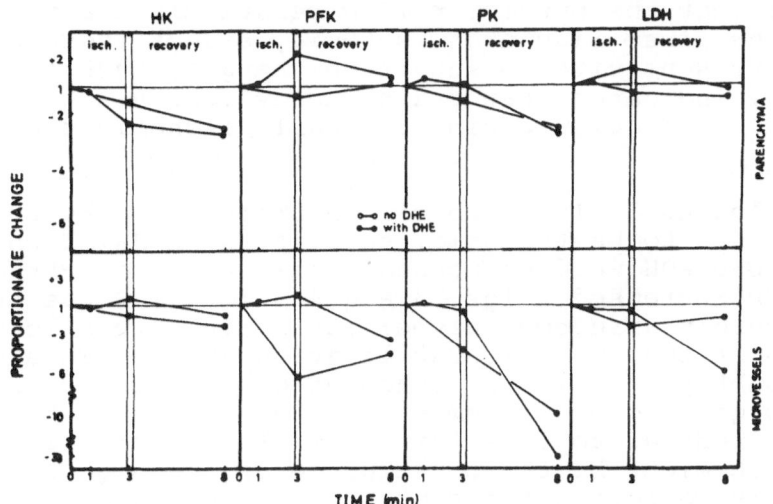

Fig. 4. The effects of global cerebral ischemia and
subsequent recovery on the activities of NAD^+
- dependent and $NADP^+$ - dependent isocitrate
dehydrogenases (ICDH-NAD and ICDH-NADP),
glucose-6-dehydrogenase (G-6-PDH) and glucose-
6-Pase (G-6-Pase) in brain parenchyma and isolated brain microvessels of Mongolian gerbils.
DHE = dihydroergotoxine. Proportionate changes
were calculated as described in Fig. 3.

(19), contrary to ICDH-NAD, the key enzyme of tricarboxylic acid cycle. ICDH-NADP probably plays role in biosyn-
thetic rather than in catabolic processes (15).

CHANGES IN ISCHEMIA

One-minute ischemia resulted in dramatic drop of
ICHD-NAD activity and significant increase in that of
brain parenchyma G-6-PDH (Fig.3). On the other hand,
ICDH-NAD activity in microvessels was significantly
increased, while increase in G-6-PDH activity was
greater than in parenchyma (Fig. 3). The activity of
the other enzymes of glucose metabolism both in parenchyma and microvessels remained practically the same
after the period of ischemia (Figs. 3 and 4). The same
period of global ischemia in gerbils was characterized
by great reduction of brain energy metabolites (13).

The fact that the enzymes remain relatively unaltered
during the first minute of ischemia, except those known
to be very sensitive to substrate changes, such as ICDH-
NAD (19), indicates that brain energy reserves can sup -
port ATP production during the early ischemia stages
(20).

Following 3 min. ischemia in brain parenchyma only
PK was found to be the same; activities of HK, ICDH-NAD,
LDH and G-6-PDH were decreased, while those of PFK and
ICDH-NADP increased (Figs. 3 and 4). ICDH-NAD was the
most severely affected enzyme; its activity was less
than 1/6 of the initial. Facilitation at the PFK step
was undoubtly the sign of enhanced glycolysis occuring
at the beginning of ischemia (17, 20). However, HK activity was reduced to less than half of its initial activity which might be the consequence of the high glucose-
6-P levels following the short term ischemia (33). Ischemia reduced the particle-bound HK to 40% of its normal
activity, but soluble HK only to 53% (Djuričić and
Mršulja, unpublished). Hence, 2-fold increase in PFK
activity might be able to compensate the fall in HK
activity since it was suggested that cellular compartmentation of HK is functional compartmentation also (8,
32). After 9 min. of ischemia HK was not changed further,
but PFK and PK activities were markedly reduced which is
indicative for the glycolysis swich off (5).

Endothelial cells of brain microvessels seem to be
more resistant to ischemia with regard to the glucose
metabolizing enzymes (Figs. 3 and 4). Following 3 min.
of the circulation cessation ICDH-NAD and G-6-PDH remained highly active but activities of PFK, ICDH-NADP
and G-6-Pase increased for about 40-50% while LDH activity was slightly reduced. Such results suggest that
glycolysis still persisted in microvessels.

LDH isoenzymic pattern differed in brain parenchyma,
which exhibits typical aerobic pattern, from that in
microvessels where M subunits dominated (Figs. 1 and 2).
When the changes in isoenzymic pattern were followed during the ischemia they appeared to be greater than the
changes in overall LDH activity. Half-minute ischemia
resulted in great increase in H subunits activity in
microvessels; the subunit at this moment participated
with more than 50% of the total activity (Fig. 2 B) most
probably due to the activation of the presynthetized
protein (29). Although following 1 min. ischemia both
H and M subunits were reduced, M subunits were decreased

as compared to the control levels. This shift in isoenzymes towards anodic forms was more emphasized following 3 min. ischemia (Fig. 1). Nine min. ischemia produced in capillaries changes most likely resulted by
random subunits reassociation. On the other hand, short-
term ischemia resulted in increase of H but not M subunits in parenchyma, while significant loss of H subunits and increase of M subunits were found following
9 min. ischemia (Figs. 1 and 2). These findings are
similar to those of Penney et al. (27) in heart of rats
suffering from sideropenic anemia. However, the relationship between isoenzymic changes and ischemia remains to
be eluciated when all the factors influencing association
and reassociation of LDH subunits become clearer.

CHANGES IN RECOVERY

Normalisation of the brain metabolism during short
term ischemia is taken for granted. However, evidences
have accumulated that post-ischemic period is actually
new pathophysiologic situation (13, 22, 26). Our data
following 5 min. of recovery from 3 min. ischemia strongly support this hypothesis. Practically, only ICDH-NAD
and PFK activities in parenchyma returned towards the
normal values, while HK, G-6-PDH and LDH were further
reduced (Figs. 3 and 4). PK, the enzyme unaltered during
ischemia, had markedly reduced activity during recovery
(Fig. 3). Similarly, neither adenylate cyclase nor Na^+-
K^+-activated ATPase changed during ischemia but their
activities was decreased during the post-ischemic period
(30).

Response of glucose metabolizing enzymes upon the
recirculation in microvessels was striking (Figs. 3 and 4).
Only isocitric dehydrogenases activities turned towards
normalization, PFK, PK and LDH, which were either increased or decreased during ischemia were found strongly
reduced. HK activity was also reduced in postischemic
period although during ischemia its activity was enhanced;
simmilarly G-6-PDH activity decreased in recovery but
much more than that of HK. On the other hand, G-6-Pase
activity was about 14-fold enhanced during post-ischemia
(Fig. 4). This enzyme was thought to participate in glucose transport through the endothelial cells (5). Whether
or not the enhancement of G-6-Pase activity in postischemia is in any conection with the increased $[^3H]$-de-
oxy-D-glucose uptake in cerebral capillaries in recovery
from global ischemia (31) has to be established. In addition to G-6-Pase, PK was the most severe affected enzyme;

its activity during recovery being reduced to about 1/30 of its initial activity (Fig. 3). PK activity strongly depends on potassium concentration (6) which is shown to be changed in ischemia (10) but is not rapidly restored upon recirculation (7).

THE EFFECT OF DIHYDROERGOTOXINE

Using hypovolemic oligemia and biphasic ischemia in cat brains to simulate conditions of cerebral metabolic insufficiencies, Meier-Ruge et al. (21) found that resulting disturbances in brain electrical activity and lactate and glucose metabolism were improved by DHE. Also, we were able to show that DHE had benefitial effect on parenchymal enzymic disturbances when administered to the gerbils prior the ischemia (25). According to our data, DHE in general expressed the oposite effects on enzymic disturbances in brain parenchyma and capillaries (Figs. 3 and 4). DHE had no effect upon the enzymic activities in normal animals but supressed the changes of glucose metabolizing enzymes in ischemia in parenchymal cells while it potentiated in endothelial cells of brain capillaries. However, it is of particular importance that DHE suppressed the ischemic changes both in parenchyma and capillaries of ICDH-NAD (Fig. 4); this enzyme is the one of the key enzymes of the tricarboxylic acid cycle, main generator of ATP from glucose. On other hand, DHE-treated gerbils exposed to 3 min. ischemia showed 14-fold increased G-6-Pase activity in microvessels which was essentially the same as after 5 min. of recovery from 3 min. ischemia in animals not treated with DHE (Fig. 4). In contrast to the benefitial effects of DHE in brain parenchyma, we were not able to find any pronounced effect of the drug during subsequent recovery period with respect to the glucose metabolizing enzymes (Figs. 3 and 4) and enzymes of glutamate metabolism (data not presented). It is proposed that DHE has a tendency towards normalizing or "economizing" brain metabolism (21). According to our data this is true only in situations of restricted oxygen and glucose supply to the brain. However, in brain microvessels, DHE prevented the post-ischemic changes in G-6-Pase, PK and LDH (Figs. 3 and 4). The fact that DHE expressed the effect in recovery is indicative that endothelial cells in this period of recirculation are far from normalizing their metabolism, probably due to the outflow of catabolic end-products formed during ischemia in the tissue (lactate, ammonia, etc.), which can alter cell metabolism. DHE seemed to prevent such changes.

COMMENT

The changes of glucose metabolizing enzymes observed in parenchyma and microvessels during ischemia and subsequent post-ischemic period, once again enhanced the view of postulated metabolic differences between these two brain compartments (5, 25). The responsivness to the drug (DHE) gave further evidence supporting such assumption. It is clear that post-ischemic period is not just a reversal of the events occuring in ischemia. As shown by the treatment with DHE parenchyma reacts to blood deprivation earlier than the endothelial cells of brain microvessels, but also demonstrated earlier tendency towards the metabolic events normalization. This post - poned endothelial cells reaction to ischemia, which seems to occur during the recovery period, and may be related to the "maturation" phenomenon (11, 24). The results of the study imply that despite apparent metabolic recovery during the post-ischemic period, enzymic changes that may be important for the quality of recovery are occuring.

ACKNOWLEDGEMENT

These studies were partly supported by grant from The Union of Medical Scientific Institutions (ZMNU) to Dr Bogomir B. Mršulja. DHE (Redergin[R]) was the gift of LEK, Ljubljana.

REFERENCES

1. Berson, F.G., Spatz,M. and Klatzo,I. (1975): Effect of oxygen saturation and pCO_2 on brain uptake of glucose analogues in rabbits, Stroke, 6: 691-696.

2. Betz, A.L., Gilboe,D.D. and Drewes, L.R. (1974): Effect of anoxiy on net uptake and undirectional transport of glucose into the isolated dog brain, Brain Res., 67: 307-316.

3. Crone, C. (1965): Facilitated transfer of glucose from blood into brain tissue. J. Physiol. (Lond.) 181: 103-113.

4. Djuričić, B.M. and Mršulja, B.B. (1977): Enzymic activity in the brain: microvessels vs. total forebrain homogenate, Brain Res. (in press).

5. Djuričić, B.M., Rogač, Lj., Spatz,M., Rakić,Lj.M. and Mršulja, B.B. (1977): Brain microvessels. I.

Enzymic activities. In: Proc. Inter. Erwin Riesch
Symposion on the Pathology of Cerebrospinal Micro-
circulation, Berlin - to be published by Raven Press
New York.

6. Dyson, R.D., Cardenas, J.M. and Barsotti, R.J. (1975):
   The reversibility of skeletal muscle pyruvate kinase
   and an assessment of its capacity to support glyco-
   genogenesis. J. Biol. Chem., 250: 3316-3321.

7. Gilboe, D.D., Drewes, L.R. and Kintner,D. (1976):
   Edema formation in the isolated canine brain: anoxia
   vs. ischemia. In: Dynamics of Brain Edema (ed. by
   H.M. Pappius and W. Feindel, Springer Verlag, Berlin-
   Heidelberg-New York, pp. 228-235.

8. Gots, R.E. and Bessman, S.P. (1974): The functional
   compartmentation of mitochondrial hexokinase. Arch.
   Biochem. Biophys. 163: 7-14.

9. Hamberger, A. and Hydén, H. (1963): Inverse enzymatic
   changes in neurons and glia during increased function
   and hypoxia. J. Cell. Biol. 16: 521-525.

10. Hossmann, A.-K. (1976): Development and resolution
    of ischemic brain swelling. In: Dynamics of Brain
    Edema, ed. by H.M. Pappius and W. Feindel, Springer
    Verlag, Berlin-Heidelberg-New York, pp. 219-227.

11. Ito,U., Spatz,M., Walker, T.J., Jr. and Klatzo,I.
    (1975): Experimental cerebral ischemia in Mongolian
    gerbils. I. Light microscopic observations. Acta
    Neuropathol. (Berl.), 32: 209-223.

12. Kahn,K. (1972): The nature course of experimental
    cerebral infarction in the gerbil. Neurology (Minn.)
    25, 510-515.

13. Kobayashi, M., Lust, W.D. and Passonneau, J.V.(1977):
    Concentrations of energy metabolites and cyclic
    nucleotide during and after bilateral ischemia in
    the gerbil cerebral cortex. J. Neurochem. 29: 53-59.

14. Laborit, H.(1969): Neurophysiologie, Masson et Cie,
    Paris.

15. Lehninger, A.L. (1972): Biochemistry, Worth Publ. Inc.
    New York.

16. Levy, D.E. and Duffy, T.E. (1977): Cerebral energy metabolism during transient ischemia and recovery in the gerbil. J. Neurochem. 28: 63-70.

17. Lowry, O.H., Passonneau,J.V., Hasselberger, F.K. and Schultz, D.W. (1964): Effect of ischemia on known substrates and cofactors of the glycolytic pathway in brain. J. Biol. Chem. 239: 18-30.

18. Lust, W.D., Mršulja, B.B., Mršulja, B.J.,Passonneau, J.V. and Klatzo,I. (1975): Putative neurotransmitters and cyclic nucleotides in prolonged ischemia of the cerebral cortex. Brain Res. 98: 394-399.

19. Mahler, H.R. and Cordes, E.H. (1966): Biological Chemistry, Harper and Raw Publ. New York - London.

20. Maker, H.S. and Lehrer, G.M. (1971): Effect of ischemia, In: Handbook of Neurochemistry, ed. A. Lajtha, vol. 6. Plenum Press, New York, 267-310.

21. Meier-Ruge, W., Enz. A. Gygax,P., Hunziker,O., Iwangoff,P. and Reichlmeier,K. (1975): Experimental pathology in basic research of the aging brain. In Aging, ed. Gershon, S. and Raskin,A., vol.2, Raven Press, New-York, 55-126.

22. Mršulja, B.B., Lust, W.D., Mršulja, B.J., Passonneau, J.V., and Klatzo,I. (1976): Post-ischemic changes in certain metabolites following prolonged ischemia in the gerbil cerebral cortex. J. Neurochem. 26: 1099-1lo3.

23. Mršulja, B.B., Mršulja, B.J., Fujimoto,T., Klatzo,I. and Spatz, M. (1976): Isolation of brain capillaries: a simplified technique. Brain Res. 110: 361-365.

24. Mršulja, B.B., Lust, W.D., Mršulja, B.J.,Passonneau, V.J. and Klatzo,I (1976): Brain glycogen following experimental cerebral ischemia in gerbils (Meriones unguiculatus), Experientia, 32: 732-733.

25. Mršulja, B.B., Djuričić, B.M., Mršulja, B.J.,Rogač, Lj., Spatz, M. and Klatzo,I. (1977): Brain microvessels II. The effect of ischemia and dihydroergotoxine on the enzymic activities. In: Proc. Inter. Erwin Riesch Symposion on the Pathology of Cerebrospinal Microcirculation, Berlin, to be published by Raven Press, New York.

26. Mršulja, B.B. (1977): Some new aspects in the patho-
    chemistry of the post-ischemic period. In: Proc.
    Inter. Symposium on the Pathophysiology of Cerebral
    Energy Metabolism. Belgrade.

27. Penney, D.G., Bugiasky, L.B. and Mieszala, J.R.(1974):
    Lactate dehydrogenase and pyruvate kinase in rat
    heart during sideropenic anemia. Biochim. Biophys.
    Acta. 334: 24-30.

28. Rose, S.P.R. (1976): Functional biochemistry of
    neurons and glial cells. In: Perspectives in Brain
    Res., ed. M.A. Corner and D.F. Swaab, Progress in
    Brain Research, vol. 45, Elsvier, North-Holland
    Biomedical Press, Amsterdam, 67-82.

29. Schimke, R.T. (1973): Control of enzyme levels in
    mammalian tissues, In: Advances in Enzymology, ed.
    A. Meister, vol. 37. John Wiley and Sons, New York,
    135-187.

30. Schwartz, J.P. Mršulja, B.B., Mršulja, B.J., Passo-
    nneau, J.V. and Klatzo,I. (1976): Alterations of
    cyclic nucleotide-related enzymes and ATPase during
    unilateral ischemia and recirculation in gerbil
    cerebral cortex. J. Neurochem. 27: 101-107.

31. Spatz,M., Mršulja, B.B., Mršulja, B.J. and Klatzo,I.
    (1977): Ischemic and post-ischemic effects on the
    2-deoxy-D-$[^3H]$-glucose uptake in cerebral capillaries,
    Brain Res. 120: 141-145,

32. Teichgräber,P. and Biesold,D. (1968): Properties of
    membrane-bound hexokinase in rat brain. J. Neurochem.
    15: 979-989.

33. Watanabe, H. and Passonneau, J.V. (1974): The effect
    of trauma on cerebral glycogen and related metaboli-
    tes and enzymes. Brain Res. 66: 147-159.

34. Wilkinson, J.H. (1965): Isoenzymes. E. and F.N. Spon
    Ltd. London.

NON-INVASIVE MEASUREMENTS OF REGIONAL CEREBRAL

BLOOD FLOW IN PATIENTS WITH CEREBROVASCULAR DISORDERS

J.S. Meyer, H. Naritomi and F. Sakai

Department of Neurology, Baylor College of
    Medicine
                        and
Baylor-Methodist Center for Cerebrovascular
    Research, Houston, Texas 77030, U.S.A.

The recently developed ^{133}Xe inhalation method (3, 7, 8) has certain advantages compared to the carotid injection method which may be listed as in Table I. The method is noninvasive for measuring regional cerebral blood flow (rCBF), with little or no discomfort to patients or volunteers during the measurement. Serial measurements of rCBF may thus be correlated with the clinical condition of the patient as well as with other diagnostic tests, such as computer-assisted tomography and the EEG. Regional blood flow measurements may be made within the vertebrobasilar arterial distribution by means of a special helmet with placement of detectors over the brainstem and cerebellum (4), as well as over both cerebral hemispheres in the carotid arterial distribution. These advantages of the ^{133}Xe inhalation method are particularly useful in the clinical investigation of cerebrovascular disease in both carotid and vertebrobasilar arterial territories.

The purpose of the present communication is to summarize results of such measurements of rCBF in different groups of patients with transient ischemic attacks (TIAs) in the carotid and vertebrobasilar arterial distribution, patients with cerebral infarction and finally in patients with recent subarachnoid hemorrhage.

Regional CBF was measured by a modification of the ^{133}Xe inhalation method described by Obrist et al. (7)

Table I

ADVANTAGES OF ^{133}Xe INHALATION rCBF METHOD

---

1. Measurements are noninvasive with minimal discomfort, suitable for patients and volunteers.

2. Procedure is safe.

3. Measurements may be made on an out-patient or in-patient basis.

4. Measurements may be repeated frequently.

5. Multiple regions of both cerebral hemispheres and, according to Meyer et al. (3, 5), the brainstem, cerebellum and total CBF may be measured simultaneously.

6. Measurements may be combined with EEG, EMG, EOG, BP, pulse, respiration, temperature and end-tidal pO_2, pCO_2 and ^{133}Xe.

7. Measurements may be made with the subject awake, asleep or engaged in standard mental activity, including speech, motor and sensory tests, problem-solving and psychological tasks.

8. Measurements may be combined with evoked responses.

9. Measurements may be made with the subject sitting, standing or lying, and during postural tilting to test autoregulation.

10. Measurements may be made while the subject is hyperventilating or inhaling CO_2 mixtures.

11. Pharmacological effects of medications on cerebral blood flow may be tested.

12. The natural history of disease and/or the effects of medical or surgical treatment may be evaluated.

13. Measurements may be correlated with neurological, psychiatric, neuropsychological, biochemical and CT evaluations.

14. Measurements may be used for screening populations at risk from cerebrovascular disease.

---

which has been reported elsewhere (3). In brief, ^{133}Xe gas mixed with room air (5-6 mCi/liter was inhaled for 1 minute through a close-fitting face mask, and the ^{133}Xe clearance curve from the head and end-tidal air was recorded throughout a 10-minute interval by means of a PDP 11-5 computer.

Sixteen properly collimated sodium iodide crystal scintillation detectors were mounted in place over both hemispheres as well as over the brainstem-cerebellar regions by means of a specially designed helmet. The brainstem-cerebellar probes were placed below and behind the mastoid process on a line connecting the internal auditory meatus and the inion looking slightly upward in a manner designed to record desaturation curves derived from the midbrain, pons and overlying cerebellum. The relationship of the probes to the skull was confirmed by roentgenograms of the head with the probes in place. Each probe was mounted perpendicular to the skull in a port fitted with an adjustable thumb screw. The output from each detector was fed in parallel to separate discriminators adjusted to accept pulses between 67.5 and 94.5 kEV for γ activity and between 23.5 and 38.5 kEV for X activity. The X-ray curves were available for correction for extracerebral contamination of the γ curves by the use of the spectrum-subtraction technique. However, rCBF values obtained from γ curves alone are not significantly different from values obtained with γ -X curves (4).

First-compartmental flow and weight (F_1 and W_1), second-compartmental flow and weight (F_2 and W_2) and mean flow (MF) were printed out by the computer on the basis of two-compartmental analysis as well as the initial slope index after 2.5 minutes of desaturation ($ISI_{2.5}$) described by Risberg et al. (8). The computer also prints out regional flow values placed in appropriate anatomical areas of a brain map together with the normal values for that region plus two standard deviations obtained from a series of normal volunteers measured in this laboratory under similar circumstances (Fig. 1). It was determined from values obtained from these normal volunteers that if a regional F_1 value exceeded the mean hemispheric F_1 value by 22%, that value was abnormal. This derived from the interregional coefficient of variation plus two standard deviations for F_1 values (3).

End-tidal partial pressures for carbon dioxide ($PECO_2$), oxygen (PEO_2) and ^{133}Xe were recorded from the face mask during the rCBF measurements along with the

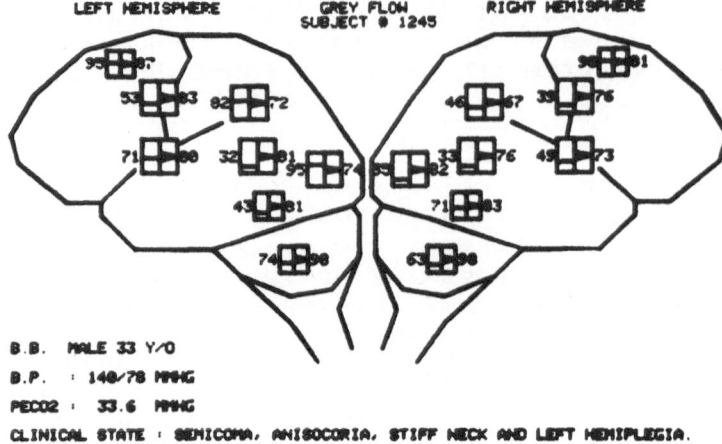

Fig. 1. In the right half of each square a triangle with
        its base on the center line is drawn with a num-
        ber printed at its apex denoting the standard
        (normal) value for the parameter being displayed.
        The width of the base of the triangle represents
        the standard deviation of the normal values on
        the scale just established. In the left half of
        the square a horizontal line is drawn at a height
        which represents the measured value for that probe
        based on the scale established from the normal
        value. Also, at the left side of the square, the
        measured value is printed.

blood pressure, pulse, EEG and skin temperature. It has
been shown previously that the reproducibility of serial
runs carried out in normal volunteers is on the order of
−2.5% for F_1.

MEASUREMENTS OF rCBF IN PATIENTS WITH TIAs IN THE VERTEBROBASILAR ARTERIAL TERRITORY

Regional CBF in both cerebral hemispheres and the
brainstem-cerebellar regions was measured simultaneously
in 36 patients with vertebrobasilar arterial insufficiency
(VBI) and in 3 patients with recent brainstem infarction.
Their mean age was 59.0 ± 10.0 years. All 36 patients with
VBI tested had recurrent episodes of transient ischemic

Table II SUMMARY OF CLINICAL OBSERVATIONS IN PATIENTS WITH VERTEBROBASILAR DISEASE

Classification of Patients	Group A VBI	Group B VBI	Brainstem Infarction
Mean Age in Years	58.8 ± 8.9	59.7 ± 7.1	51.0 ± 10.4
Total Cases	N = 26	N = 10	N = 3
ANGIOGRAPHIC OBSERVATIONS OF VERTEBRAL BASILAR ARTERIAL SYSTEM			
Normal	1	1	0
Minor Atherosclerotic Irregularities	3	1	0
Marked Stenosis or Occlusions	13	5	2
Not examined	9	3	1
RISK FACTORS			
Hypertensive Cardio-vascular disease	21	7	1
Rheumatic Heart Disease	2	1	1
Cervical Spondylosis	8	4	0
Hiperlipidemia (Type IV)	5	3	1
Diabetes Mellitus	5	2	0
DURATION OF VBI SYMPTOMS			
1 Year or Less	13	3	
More than 1 Year	13	7	*)
INTERVAL BETWEEN LAST ISCHEMIC EPISODE AND rCBF MEASUREMENT			
3 Weeks or less	14	1	
More than 3 Weeks	12	9	**)

Group A = Untreated with vasoactive drugs; Group B = treated with papaverine 360 mg daily. *) Symptoms of VBI preceded brainstem infarction for 1 and 5 months in two cases. **) Brainstem infarction occurred 1, 4 and 5 weeks preceding rCBF measurements respectively.

attacks (TIAs) referable to the vertebrobasilar arterial
territory. The majority of patients had demonstrable
atherosclerotic changes of the vertebral and basilar
arteries demonstrated by angiography. Two of the three
patients with recent brainstem infarction showed complete occlusion of the basilar artery. Recurrent TIAs
preceded the complete brainstem infarction in two of
these patients. A summary of the clinical findings in
these 39 patients is listed in Table II. In the 36 patients with VBI, 26 had received no vasoactive drugs
(Group A), but 10 of the patients had been treated with
oral papaverine hydrochloride in a dose of 360 mg daily
(Group B) for 2 weeks to 4 years preceding the rCBF measurements. In the papaverine-treated group, transient ischemic symptoms were reported to be improved and occurred
less frequently than prior to institution of therapy.

Regional CBF was measured before and after the induction of orthostatic hypotension by means of postural
tilting in the head-up position by 30° in order to test
cerebral autoregulation. The results obtained were compared with those in 15 age-matched normal controls (mean
age = 53.8 ± 9.8 yr) measured in an identical manner.

In the horizontal position, mean arterial blood
pressure (MABP) was significantly higher in patients with
VBI compared to age-matched normal controls. The regional
F_1 values for both cerebral hemispheres and the brainstem-
cerebellar regions in untreated patients with VBI were
not significantly different from those measured in age-
matched normals. This was true whether the rCBF measurements were made within or after a 3-week interval following the last transient ischemic symptoms. This finding
differs from rCBF measurements made in patients with TIAs
in the carotid arterial distribution (carotid TIAs), who
showed persistent reduction of rCBF for 3 weeks after
the TIA. This is taken to indicate that transient ischemic
symptoms in patients with VBI are not regularly caused by
thromboembolism, as appears to be the case in patients
with carotid TIAs. However, it should be noted that regional F_1 levels in the 26 untreated patients with VBI
showed significant correlation with spontaneous blood
pressure levels both for brainstem-cerebellar regions
($r = 0.73$, $p < 0.01$) and for the cerebral hemispheres
($r = 0.51$, $p < 0.01$), indicating a state of dysautoregulation. In the 15 normal controls there was no correlation of rCBF with the blood pressure levels, indicating
intact autoregulation.

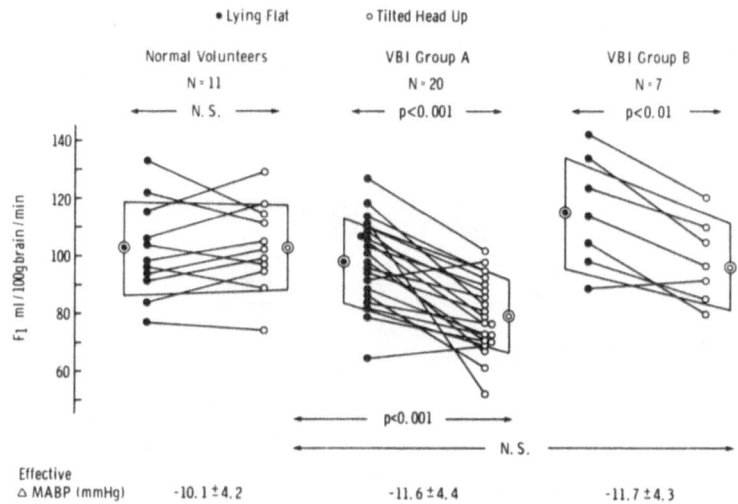

Fig. 2. Effect of testing autoregulation on regional
        brainstem-cerebellar F₁ values in normals
        compared to untreated (Group A) and treated
        (Group B) patients with vertebrobasilar in-
        sufficiency.

In papaverine-treated patients with VBI, regional F_1
values for both cerebral hemispheres were significantly
increased compared to values in age-matched normals, and
compared to untreated patients with VBI, regional F_1
values were significantly increased both for the brainstem-cerebellar regions and for the cerebral hemispheres.

In the three patients with recent brainstem infarction, brainstem-cerebellar F_1 values were all reduced.
When rCBF measurements were repeated following orthostatic
tilting by 30°, effective MABP was decreased by 10-12 mm
Hg both in patients with VBI and in normal controls, although $PECO_2$ did not change significantly. Figure 2 com-
pares the effect of induced orthostatic hypotension on
brainstem-cerebellar F_1 values in normals and in Group A
and B patients. In normal subjects with intact autoregulation, brainstem-cerebellar F_1 values did not change
significantly, following the reduction of effective MABP.
Unlike the values in normal subjects, brainstem-cerebellar
F_1 values became significantly reduced in both untreated
and treated patients with VBI during induced orthostatic
hypotension, indicating impaired autoregulation in the
brainstem-cerebellar regions in all patients with VBI.

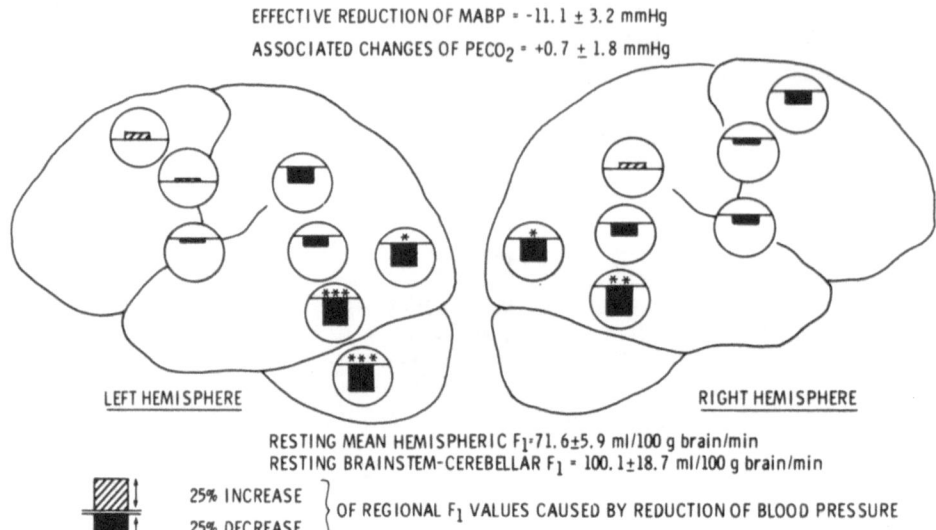

EFFECTIVE REDUCTION OF MABP = -11.1 ± 3.2 mmHg

ASSOCIATED CHANGES OF $PECO_2$ = +0.7 ± 1.8 mmHg

LEFT HEMISPHERE   RIGHT HEMISPHERE

RESTING MEAN HEMISPHERIC F_1·71.6±5.9 ml/100 g brain/min

RESTING BRAINSTEM-CEREBELLAR F_1 = 100.1±18.7 ml/100 g brain/min

25% INCREASE } OF REGIONAL F_1 VALUES CAUSED BY REDUCTION OF BLOOD PRESSURE

25% DECREASE

* $P < 0.05$ ** $P < 0.01$ *** $P < 0.005$

Fig. 3. Testing regional cerebral autoregulation in pa-
    tients with vertebrobasilar insufficiency pres-
    ent for 1-12 months (N = 10).

During induced postural hypotension F_1 values were reduced
compared to normals in untreated VBI patients (Group A),
but in the patients treated with papaverine (Group B),
the F_1 values were not reduced significantly below normal
values despite the dysautoregulation.

    Dysautoregulation in patients with VBI remains re-
gional and restricted to the vertebral, basilar and posterior cerebral arterial distribution unless they suffered recurrent transient ischemic symptoms for longer
than 1 year. Figure 3 illustrates the regional pattern
of dysautoregulation in untreated patients with VBI of
1-12 months' duration. In Fig. 3, any change of regional
F_1 values during induced postural hypotension was compared to the steady state and expressed as a percentage
of resting F_1 values for each region. Significant decreases of regional F_1 values were restricted to the
brainstem-cerebellar, occipital and inferior temporal
regions in the distribution of vertebral, basilar and
posterior cerebral arteries, indicating regional dysautoregulation in the distribution of these arteries. However, in patients with long-standing VBI of 18-60 months'

duration, dysautoregulation was more widespread and involved the frontal, parietal, sylvian-opercular regions in carotid arterial distribution as well as the territory supplied by the vertebrobasilar arterial system.

In summary, CBF in patients with VBI was usually normal in the steady state under resting blood pressure conditions, but regional dysautoregulation was demonstrated in the vertebrobasilar arterial distribution during mild orthostatic hypotension by postural tilting. Dysautoregulation is presumably caused by the initial thromboembolic event occurring within the vertebrobasilar arterial territory, which predisposes the patient to repeated episodes of ischemia when there are alterations in perfusion pressure.

MEASUREMENTS OF rCBF IN PATIENTS WITH TIAs IN THE CAROTID ARTERIAL TERRITORY

Regional CBF was also measured in 19 patients with TIAs occurring within one carotid arterial territory (carotid TIAs). All patients had atherosclerotic occlusive changes of ipsilateral (N = 16) or bilateral (N = 3) carotid arteries demonstrated by angiography. Unlike the values in patients with VBI, F_1 values in patients with carotid TIAs were significantly reduced in both the ischemic and nonischemic hemispheres provided that rCBF measurements were made within 3 weeks after the ischemic episode, diaschisis (6). After 3 weeks, F_1 values returned to normal in both hemispheres. Regional F_1 values during the first 3 weeks after the TIA were maximal in the ischemic middle cerebral arterial (MCA) territory, and in the nonischemic hemisphere, regional F_1 values were reduced in the homologous zones, showing a mirror image of regional flow pattern in the ischemic hemisphere. This symmetrical pattern of flow reduction is characteristic in diaschisis, as mentioned later. Cerebral autoregulation was also tested in patients with carotid TIAs by postural head-up tilting. Unlike the case in patients with VBI, autoregulation was usually normal in patients with carotid TIAs, and only a few patients showed impaired autoregulation within ischemic MCA territory.

MEASUREMENTS OF rCBF IN PATIENTS WITH UNILATERAL CEREBRAL INFARCTION

The 32 patients with unilateral cerebral infarction were classified according to the severity of neurological deficit as Grade 2 (N = 3), Grade 3 (N = 17) and Grade 4 (N = 12) (4). According to this clinical classification

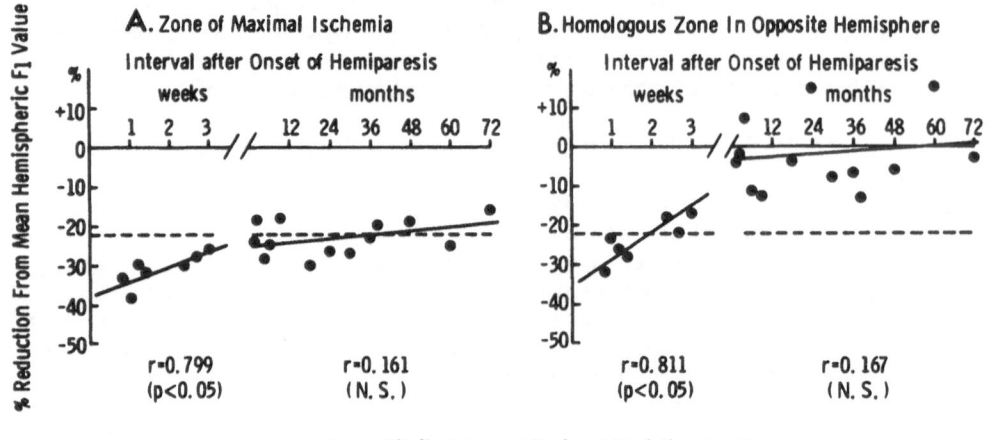

Fig. 4. Grade 2 and 3 unilateral cerebral infarction
       (N = 20).

(4), a deficit persisting for less than 24 hours was
classified Grade 1, which was already described as carotid TIAs; a deficit clearing within 3 weeks was classified Grade 2; a persistent deficit lasting longer than
3 weeks but with partial improvement was classified Grade
3; and a severe and persistent deficit with little or no
recovery was classified as Grade 4.

     In general, the more severe the neurological defi-
cit, the greater in degree and the longer in duration
was the local reduction of rCBF. Furthermore, the more
severe the neurological deficit, the more remarkable was
the diaschisis. Figure 4 illustrates the time course of
reduction of regional F_1 values in the zones of maximal
ischemia (panel A), which were all located in the infarcted MCA territory, and in the homologous zones in
the nonischemic hemisphere (panel B) observed in patients
with Grade 2 and 3 cerebral infarction. During the first
3 weeks after the onset of stroke there was bilateral
and symmetrical flow reduction, as shown in the maximal
ischemic zone and in the homologous zones in the non-
ischemic hemisphere. Such a symmetrical flow pattern of
diaschisis is likely to be caused by a transcallosal
mechanism as hypothesized by Kempinsky (2). After 3 weeks,
regional F_1 values returned to normal in the nonischemic

hemisphere but remained reduced in the ischemic hemisphere. In the brainstem-cerebellar regions, on the contrary, regional F_1 values were significantly increased during the first 3 weeks after the stroke provided that the patients were alert. However, if the patients were drowsy or stuporous due to the cerebral infarctation with cerebral edema, brainstem-cerebellar F_1 values were markedly reduced.

TESTING CEREBROVASCULAR FUNCTIONAL RESERVE BY MULTIPLE PSYCHOPHYSIOLOGICAL ACTIVATION IN CEREBROVASCULAR DISEASE

Cerebrovascular functional reserve was tested by measuring rCBF at rest in quiet darkness and during standard multiple psychophysiological activation. The activation procedure included counting, conversation, music and observing movements during the rCBF measurements.

In normals, the cerebral activation increased F_1 values in the entire brain, with the greatest increase being in brainstem-cerebellar regions - by 20% compared to the resting state.

In patients with carotid occlusive disease, F_1 values increased only in the cerebral hemisphere supplied by the intact carotid artery, remaining unchanged in the cerebral hemisphere supplied by the occluded carotid artery. In patients with dysphasis due to occlusive disease of the left middle cerebral artery, F_1 values did not increase in either hemisphere with a -12% mean reduction in the motor speech area. The degree of F_1 change in the speech area correlated with prognosis for ultimate recovery of speech. If F_1 increased in areas homologous to the speech zones in the right hemisphere, prognosis for speech recovery was better.

MEASUREMENTS OF rCBF IN PATIENTS WITH SUBARACHNOID HEMORRHAGE (SAH)

Single or serial measurements of rCBF were found to be useful in the management of patients with recent SAH from aneurysm (N = 9) or arteriovenous malformation (N = 1). Regional F_1 values were reduced in a pattern which correlated well with the anatomical locus of vasospasm demonstrated by angiography and/or the intracerebral hematoma demonstrated by computer-assisted tomography as shown in Fig. 1. The degree of reduction of the

F_1 values also correlated well with the localization of the neurological deficits and with the overall status of the patient as graded by the classification of Hunt and Hess (1) as Grade 2 ($N = 2$), 3 ($N = 5$) or 4 ($N = 3$), in which Grade 2 = severe headache, nuchal rigidity with no neurological deficit other than cranial nerve palsy; Grade 3 = drowsiness, confusion, or mild focal deficit; Grade 4 = stupor, moderate to severe hemiparesis, early decorticate rigidity.

In patients with vasospasm, reduction of regional F_1 values was evident after the fourth day of SAH and showed gradual recovery after the second week of SAH, resulting in a return to normal levels by the fifth week of SAH. Surgical intervention appeared to be safe and may be recommended when reduced regional F_1 values exceed the level of 48 ml/100 g brain per min.

When regional F_1 values were decreased below 38 ml/100 g brain per min or mean flow (MF) values were decreased below 23 ml/100 g brain per min, a permanent neurological deficit ensued. This critical level appears to hold for patients suffering either SAH or occlusive cerebrovascular disease due to thromboembolism.

Regional CBF measurements also proved to be helpful in two patients who developed normal pressure hydrocephalus (NPH) following SAH. In these two patients, the decrease of intracranial pressure by removal of 15 ml of cerebrospinal fluid markedly increased F_1 values as well as W_1 values, indicating the presence of NPH. Both patients recovered after the ventriculoperitoneal shunt or the serial removal of CSF.

ACKNOWLEDGMENTS

This work was supported by USPHS Grant NS 09287 and by the Cooper Laboratories, Cedar Knolls, New Jersey.

Send reprint requests to John Stirling Meyer, M.D., Baylor College of Medicine, Department of Neurology, 1200 Moursund, Houston, Texas, 77030, U.S.A.

REFERENCES

1. Hunt, W.E. and Hess, R.M. (1968): Surgical risk as related to time intervention in the repair of intracranial aneurysms. J. Neurosurg. 28: 14-20.

2. Kempinsky, W.H. (1958): Experimental study of distant effects of acute focal brain injury: A study of diaschisis. Arch. Neurol. Psychiatry 79: 376-389.

3. Meyer, J.S., Ishihara, N., Deshmukh, V.D. et al. (1978): An improved method for noninvasive measurement of regional cerebral blood flow and blood volume by ^{133}Xe inhalation. Part I: Description of the method and normal values obtained in healthy volunteers. Stroke 9: 195-205.

4. Meyer, J.S., Kanda, T., Fukuuchi, Y. et al. (1971): Clinical prognosis correlated with hemispheric blood flow in cerebral infarction. Stroke 2: 383-394.

5. Meyer, J.S., Sakai, F., Naritomi, H. et al. (1978): Normal and abnormal patterns of cerebrovascular reserve tested by ^{133}Xe inhalation. Arch. Neurol. (in press).

6. Meyer, J.S., Shinohara, Y., Kanda, T. et al. (1970): Diaschisis resulting from acute unilateral cerebral infarction. Quantitative evidence for man. Arch. Neurol. 23: 241-247.

7. Obrist, W.D., Thomson, H.K. Jr., Wang, H.S. et al. (1975): Regional cerebral blood flow estimated by ^{133}Xe inhalation. Stroke 6: 245-256.

8. Risberg, J., Ali, Z., Wilcox, E.M. et al. (1975): Regional cerebral blood flow by ^{133}xenon inhalation. Stroke 6: 142-148.

ENERGY METABOLISM IN FOCAL SEIZURES

W. F. Caveness

Laboratory of Experimental Neurology
National Institutes of Neurological and
Communicative Disorders and stroke
Bethesda, Maryland 20014

INTRODUCTION

The present day concept of Epilepsy has evolved
from astute clinical observations and painstaking laboratory investigations. Flashes of insight have been separated by long periods of unimaginative thinking.
Hippocrates taught in 400 B.C. that"if a man be struck
on one side of the head, he may develop convulsions on
the opposite side of the body" (26). This not only took
this disorder out of the realm of "sacred diseases," but
also implied a distinct organization within the brain.
Hughlings Jackson, in the latter part of the 19th cen-
tury, in his truly remarkable concept of the hierachy
of the central nervous system recognized the importance
of focal seizures in the study of localization of function within the nervous system. Jackson's statememt,
most often quoted, is that "a convulsion is but a symptom and implies only that there is an occasional, an
excessive, and a disorderly discharge of nerve tissue
on muscles" (16). This has been interpreted by many to
mean a chaotic disorder within the brain. However,
Jackson continues, "It is to be insisted on that a convulsion, in any case, is nothing whatever else than a
sudden excessive and rapid development of the normal
movements which the center, suddenly and excessively
discharging, represents" (15). In the succeeding years,
it has become increasingly apparent that a convulsive
seizure is a "highly organized process made up of
individual events which interlock with one another in
a variety of ways, even though their exact relationship

267

may still sometimes be obscure" (22). In the search for
insight into this organization, the cerebral cortex has
received the greatest attention in man and in the experimental animal (6, 21), with secondary interest in the
cerebellar cortex (9, 17) and thalamus (19). In recent
years, this attention has been focused on the microcir-
cuitry in small aggregates of cells, recognizing the
interplay of excitatory and inhibitory postsynaptic potentials (1, 12, 14). These studies have significantly
advanced our knowledge, but their meaning will be made
even clearer when the results are more carefully inte-
grated with overall CNS function. In this presentation,
we will consider relatively large neuronal aggregates,
connected by long pathways, that may be involved in experimental focal seizures induced by injecting penicillin
into the face-hand area of the monkey cerebral cortex.
Prior to setting forth the experimental procedure and
results, it may be worthwhile to review some of the
better known macrocircuits utilized by the sensorimotor system in carrying out normal functions.

ESTABLISHED CIRCUITRY WITHIN THE SENSORIMOTOR
SYSTEM OF THE MONKEY

In the cerebral cortex, there are the immediate
and profuse pathways between Brodmann's area 4 and areas
3-1-2 and 6, as well as connections with the supplementary motor cortex. From these cortical areas there are
the direct cortico-spinal pathways and cortico-rubral-
spinal pathways to the motoneurons of the spinal cord
that supply flexor muscles.

A less direct but highly significant pathway is the
circuit from the cortex through striatum, pallidum, and
thalamus, VL and VA, back to the cortex. With a potent
influence on the sensorimotor cortex, its basic function
is the modulation of large movements and posture (7).
This loop is enhanced by the addition of to-and-from
connections with the substantia nigra and subthalamic
nucleus (3). Further connections between the basal ganglia and thalamus, e.g., striatum to DM, and substantia
nigra to CM, VL and VA of thalamus (24), undoubtedly
add to the flexibility of this system in modulating motor function. Implied throughout the description of this
macrocircuitry is the interaction of excitation and inhibition.

There is a powerful projection from the sensorimotor cortex to the pontine nuclei (2). The projection
from the pontine nuclei to the contralateral cerebellum,

as mossy fibers, make up part of a loop that extends to
the dentate nucleus to VL of the thalamus and then back
to the cerebral cortex, enhancing the activity in the
latter. A similar circuitry involves the interpositus
nucleus, VA of the thalamus, and an additional connection
to the red nucleus.

Quite a different effect is brought about by the input from the inferior olivary nucleus, via the climbing
fibers. The activated Purkinje cells of the cerebellar
cortex inhibit the activity in the cerebellar nuclei,
withdrawing their excitatory effect on the thalamus and
in turn on the cerebral sensorimotor cortex (11). In
this regard, it is pertinent to remember the spino-oli-
vary input from the proprioceptors of contracting muscles
(20). Obviously, the cerebellum is an integral part of
another system, whose microstructure is best suited to
the modulation of fine movements. Another pathway from
the dorsal column of the cord through n. cuneatus and
VPL of the thalamus to the sensorimotor cortex is described by Eccles as being entirely inhibitory in its influence (10).

Within the realm of the neurotransmitters, cells
and pathways, the basal ganglia have received considerable
attention. Current thinking places a significant number
of gabanergic neurons within the pallidum, with their
projections acting upon dopamine cells in the substantia
nigra which in turn influence acetylcholine neurons in
the striatum. The latter, through more complicated pathways, influence the gabanergic neurons. Further evidence
is accumulating for connections between cerebellar nuclei
as well as the cerebral cortex with the nigro-striatal
system (13).

Against this background we would like to share with
you our most recent findings regarding macroneuronal
aggregates utilized in the propagation of focal fits,
using an expression of energy metabolism for their iden-
tification and the extent of their activity.

MATERIALS AND METHODS

Ten monkeys, Macaca mulatta, 3.0 to 3.5 kg in weight,
were prepared as follows: Under halothane anesthesia, the
abdominal aorta and inferior vena cava were catheterized
via the femoral artery and vein in preparation for the
metabolic studies. By means of a right scalp incision,
trephine of the skull, and reflection of the dura, the

face-hand area of the right motor cortex was identified.
The area was protected with Gelfoam, and the scalp temporarily closed. Two hours after completion of the surgical procedure and the end of the general anesthesia,
the experimental observations were conducted with the
animal fully alert, in a prone positon, the head fixed,
and the limbs loosely restrained. Discomfort from the
head clamps was minimized by topical anesthesia.

The focal seizures were induced by stereotaxically
controlled injection of 25,000 units of crystalline potassium penicillin G in 0.025 ml of distilled water,
into area 4 of the face-hand area of the right motor cortex at a depth of 2.5 mm. The overflow from the needle
tract was immediately washed from the pial surface of the
cortex with normal saline.

Bipolar electroencephalographic (EEG) recording was
obtained with six Beckman scalp electrodes arrayed bilaterally in three pairs over the frontal, temporal and
occipital regions, respectively. Eight pairs of needle
electrodes for electromyographic (EMG) recording were
placed in the right and left masseter and orbicularis
oris muscles and the flexors of the four extremities.
The EEG, EMG, and an electrocardiogram were obtained with
a Grass 8-18 electroencephalograph. Clinical phenomena
were monitored by visual observations as well as by
electromyography.

For the determination of metabolic changes we have
elected a global appraisal of the whole brain for local
glucose utilization. We use the glucose utilization as
a criterion of local neuronal metabolism. The method
was devised by Sokoloff et al (25) and uses $2[^{14}C]$deoxy-
glucose uptake and phosphorylation to the rate of glucose phosphorylation in the tissue of the brain.   It
permits a quantitative determination of the rate of glucose consumption in the structural and functional units
of the brain (18).

The combination of these techniques was as follows:
Following the intracortical penicillin injection and
after the development of ipsilateral electrographic
spikes and the beginning of the contralateral clinical
expression, ^{14}C-labeled deoxyglucose, 100 μCi/kg, was
injected (in a bolus) by vein. After 30 minutes, during
which time arterial blood samples were obtained, the
animal was decapitated, and the head immediately immersed
in Freon chilled by liquid nitrogen to -100°C. Subse-

quently the brain, in the skull, was serially sectioned at 30 μm with a PMV cryomicrotome at a temperature of -20°C. Approximately 100 sections from each brain were dried at 60°C and placed on blue-sensitive X-ray film for macroautoradiographs. From determination of the concentration of [^{14}C]deoxyglucose and glucose in arterial plasma and the concentration of [^{14}C] in the autoradiographs, the actual local glucose utilization value, in milligrams per gram per minute, was calculated from the formula devised by Sokoloff et al. (25). For the present study the glucose utilization was determined, from the average of 10 densitometric measurements each, for 48 bilateral structures within and outside the sensorimotor system.

RESULTS

CLINICAL: The time course of the observations was dictated by the demands of the [^{14}C]deoxyglucose method, i.e., 30 minutes for the necessary conversion of [^{14}C]-deoxyglucose to [^{14}C]deoxyglucose-6-phosphate that is trapped in the tissues. During this 30-minute period, not one but a series of focal seizures took place. While the number varied from monkey to monkey, with a mean of 12, the characteristics of each episode were quite similar: a rapid buildup, seconds in duration, sustained clonic movements for a minute or so, followed by an abrupt cessation. The latter gave every indication of being an active process, rather than one of fatigue or exhaustion, e.g., the suddenness of the termination, with no loss in awareness of the surroundings, and with minimal lapse in time before the buildup of the precise contractions of the next episode. The movements were flexion contractions with little if any tonic component. The extent of the contractions was confined to the contralateral face in four monkeys, the contralateral face and upper extremity in four monkeys, and in a single monkey the spread of activity involved the ipsilateral face and upper extremity, as well. A single monkey, without penicillin injection, served as a control. (This series is being expanded to include three additional controls and six additional experimental preparations.) The electroencephalographic expression consisted of high-amplitude spike-and-slow-wave complexes, at 3 to 4 Hz, confined to the ipsilateral hemisphere, with the spikes coincident with the contralateral flexion contractions. In the monkey with the bilateral clinical expression, there were concomitant bilateral spike-and-wave complexes.

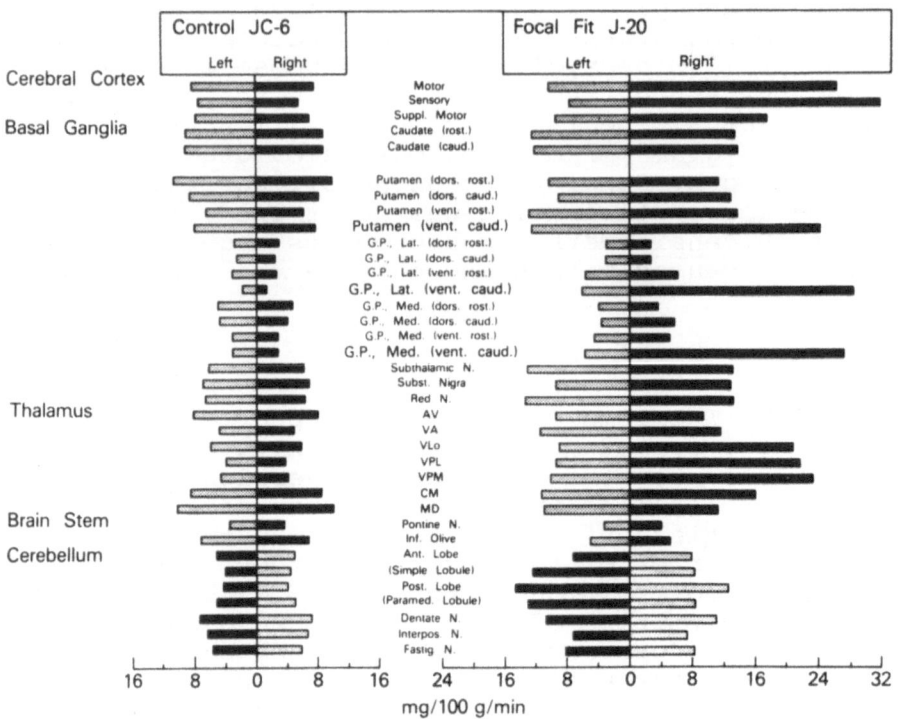

Fig. 1. Cerebral glucose utilization

GLUCOSE UTILIZATION: When a representative seizure monkey, with the clinical expression in the contralateral face and upper extremity, was compared to a control, the patterns of local glucose utilization were seen to be quite different (Fig. 1). In the seizure monkey, there was both an overall increase and a marked ipsilateral (to the side of the penicillin injection) predominance in the motor, sensory and sup. motor cortex, ventral caudal putamen, ventral caudal globus pallidus, in both medial and lateral segments, VL, VPL, and VPM, of the thalamus, and the contralateral simple and paramedian lobules of the cerebellum. The differential magnitude of the unilateral differences is shown in Table I. This was quite remarkable in the cerebral cortex, the right (ipsilateral) predominance being 318% in the postcentral gyrus.

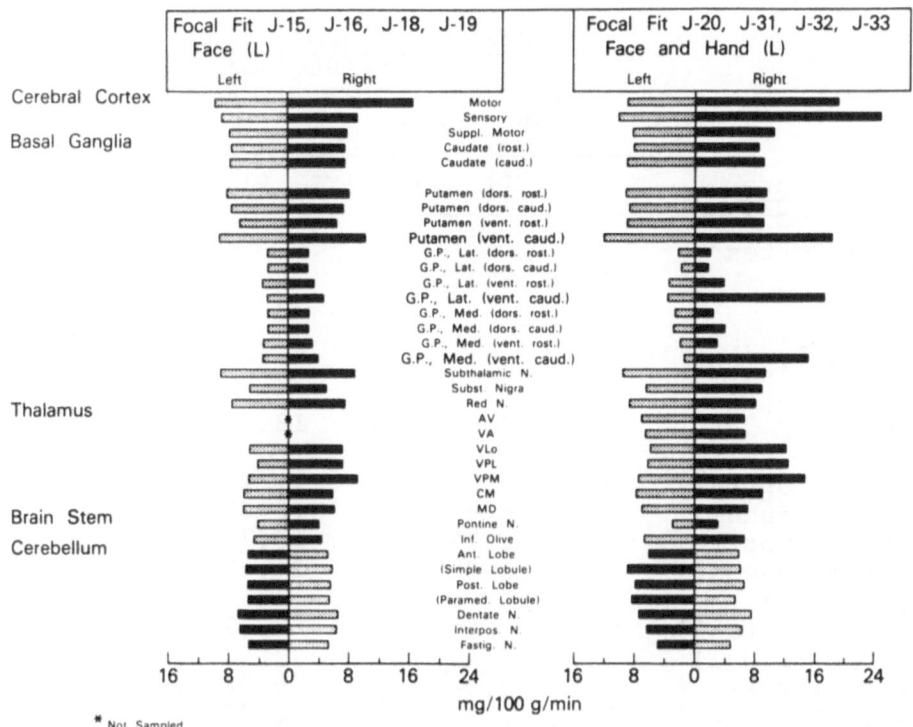

Fig. 2. Cerebral glucose utilization

In the striatum, the most prominent right-to-left difference was in the ventral caudal putamen, 93%. But in the globus pallidus there was a difference, greater than that in the cerebral cortex, in the ventral caudal aspect of both the lateral and medial divisions, of 368 and 371%, respectively. The substantia nigra showed a modest 36% difference. In the thalamus, VL, VPL and VPM, the right-to-left difference was 127, 130 and 131%, respectively, with CM showing a difference of 41%. In the cerebellum there were moderate left-to-right (contralateral) differences in the simple and paramedian lobules, 52 and 60%, respectively. There was no appreciable unilateral increase in structures outside the sensorimotor system, e.g., the prefrontal, occipital and temporal cortices and samples from the forebrain white matter.

Table I   DIFFERENTIAL MAGNITUDE OF THE UNILATERAL DIF-
          FERENCES IN GLUCOSE UTILIZATION

	CONTROL			FOCAL SEIZURE		
	Cerebral Glucose Utilization (mg/100g/min)					
	Monkey JC-6			Monkey J-20		
Structure	Rt.	Lt.	%(R-L)/L	Rt.	Lt.	%(R-L)/L
Cerebral Cortex						
Motor	7.67	8.12	-6	25.38	10.37	145
Sensory	5.69	7.22	-21	31.82	7.61	318
Sup. Motor	7.04	7.49	-6	17.42	9.45	84
Basal Ganglia						
Caudate						
(caudal)	8.93	8.93	0	13.74	12.21	13
Putamen						
(dorsal rostral)	10.10	10.56	-4	11.29	10.37	9
(dorsal caudal)	8.30	8.48	-22	12.82	10.07	27
(ventral rostral)	6.32	6.32	0	13.74	12.82	7
(ventral caudal)	7.85	7.85	0	24.16	12.52	93
G.P., Lateral						
(dorsal rostral)	3.07	2.80	10	2.71	3.02	-10
(dorsal caudal)	2.71	2.44	11	2.71	3.02	-10
(ventral rostral)	2.80	2.98	-6	6.08	5.78	5
(ventral caudal)	1.63	1.63	0	28.45	6.08	368
G.P., Medial						
(dorsal rostral)	4.87	4.87	0	3.63	3.94	-8
(dorsal caudal)	4.15	4.69	-12	5.78	3.63	59
(ventral rostral)	2.98	3.07	-3	5.16	4.55	13
(ventral caudal)	2.98	3.07	-3	27.22	5.78	371
Subth. N.	6.32	6.14	3	13.13	13.13	0
Subst. Nigra	7.04	6.86	3	12.82	9.45	36
Red N.	6.50	6.50	0	13.13	13.44	-2
Thalamus						
AV	8.12	8.12	0	9.30	9.45	-2
VA	5.05	4.87	4	11.44	11.60	-1
VL	5.96	5.96	0	20.79	9.15	127
VPL	3.97	3.97	0	21.71	9.45	130
VPM	4.33	4.69	-8	23.24	10.07	131
CM	8.48	8.48	0	15.89	11.29	41
MD	10.10	10.29	-2	11.29	10.98	3
Brain Stem						
Pont. N.	3.70	3.52	5	3.94	3.32	19
Inf. Olive	6.95	7.22	-4	5.16	5.16	0

Table I - cont.

Structure	CONTROL			FOCAL SEIZURE		
	Cerebral Glucose Utilization (mg/100g/min)					
	Monkey JC-6			Monkey J-20		
	Rt.	Lt.	%(R-L)/L	Rt.	Lt.	%(R-L)/L
Cerebellum						
Ant. Lobe	5.05	5.23	4	7.92	7.31	-8
(Simple Lobule)	4.51	4.15	-8	8.23	12.52	52
Post. Lobe	4.15	4.33	4	12.52	14.66	17
(Paramed. Lobule)	5.14	5.14	0	8.23	13.13	60
Dentate N.	7.22	7.40	2	10.98	10.68	-3
Interpos. N.	6.77	6.41	-5	7.31	7.31	0
Fastig. N.	5.87	5.69	-3	8.23	8.23	0

It is only when the focal seizure spreads from the contralateral face to the contralateral face and upper extremity that the unilaterality in glucose utilizaton is clearly evident. This is demonstrated when two representative groups of four animals each are compared (Fig. 2). In the single animal in which the flexion contraction extended to the ipsilateral side, although remaining stronger on the contralateral side, the overall glucose utilization was appreciably increased throughout the structures in the sensorimotor system, but with a remaining unilateral predominance in the motor, sensory and supplementary motor cortex, ventral caudal putamen, ventral caudal globus pallidus, both lateral and medial divisions, VL, VPL, and VPM of the thalamus and, less prominent, the contralateral anterior and posterior lobes of the cerebellum (Fig. 3).

DISCUSSION

The determination of local glucose utilization shows the location and extent of the metabolic energy requirements of the experimental focal seizure. A striking feature is the unilateral increase in appropriate elements of the sensorimotor system with propagation of the seizure from face to upper extremity, and the added bilateral increase when the propagation includes a clinical expression on both sides of the body. The second notable feature

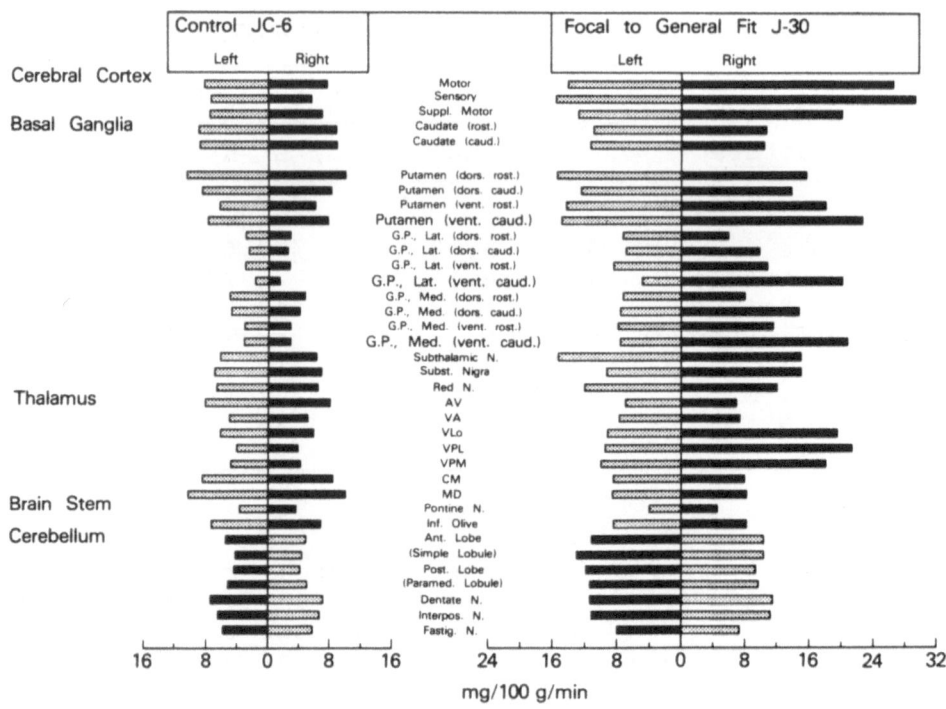

Fig. 3. Cerebral glucose utilization

is the graded expenditure of energy from the extraordinary
expression in the sensory and motor cortices (3-1-2 and 4)
and the ventral caudal parts of the putamen and globus
pallidus, with somewhat less expression in motor and sensory thalamic relay nuclei, and the least expression in
parts of the cerebellar cortex. In interpreting these findings, one must consider the continuing influence from the
penicillin remaining at the site of the injection in area
4, and the recognized importance of the cerebral cortex
in the development and propagation of the focal seizure.
Equal in degree of glucose utilization were the very discrete, somatotopically related areas (3) in the putamen
and globus pallidus, along with VA of the thalamus. This
strongly suggests an essential role for the cortico-
striato-pallido-cortical loop. This loop is thought of
as modulating posture and gross movements, and may well

play such a part in the stereotyped clonic movements of the focal fit. The lack of an overt increase in glucose utilization in the pontine and contralateral cerebellar nuclei, and the relatively modest increase in the cerebellar cortex, were unexpected. The function of the cerebellum in seizures arising in the cerebral cortex, though not fully defined, is well recognized (5, 8). In the model at hand, during previous observations of simultaneous electrographic recordings from the surface of the cerebral and cerebellar cortices, the paroxysmal expression was found to be similar in wave form, somatotopically linked in location and coincident in time, as displayed by ink-writing equipment. But the average voltage of the abnormal activity from the cerebellar cortex was only one fourth that from the cerebral cortex. In one of these monkeys, with the progression of the focal seizure over time, the paroxysmal activity from the cerebral cortex increased in amplitude from 810 μV to 1650 μV, while that from the cerebellar cortex was negligible, from 190 μV to 220 μV (4). The previous findings are consistent with the current observations of differences in energy metabolism between the cerebral and cerebellar cortices. The lack of a more prominent glucose utilization may be due to the more efficient function of the cerebellum with less expenditure of energy, its circuitry permitting a rapid in-out handling of signals (11) without the capability of self-sustained paroxysmal activity. Further, there may be less demand upon its function of modulating fine movements by gross contractions of the focal fit. The afferent input from the proprioceptors of the contracting muscles, as well as the input from the cortico-pontine-cerebellar pathways, must be considerable. Perhaps the cerebellum handles this by "resetting its threshold," as suggested by Szentágothai (27). The prominent glucose utilization by the sensory relay nuclei of the thalamus, VPL and VPM, suggests that there are powerful afferent volleys acting on the cerebral cortex, if not on the cerebellum.

The temporal characteristics of the clinical and electrographic phenomena must be taken into consideration when interpreting the metabolic findings. The observations are of glucose utilizaition in a series of recurrent focal seizures over a 30-minute period, not of glucose utilization during a sustained uniform activity over 30 minutes. Excitatory and inhibitory phenomena are strongly suggested by the recurrent build-up and abrupt cessation of the focal seizures, each episode lasting 1 or 2 minutes. On a faster time scale, the electrographic spike and wave, coincident with contralateral clonic movements of face and upper extremity, reflect an alternating excitatory and inhibitory activity (23), occurring about four times a second. These external expressions of excitation and inhibition on two time scales must be only the most obvious expressions of multiple and interlocking modulating systems within the sensorimotor system that are subserving the seizure. It is apparent that we have identified, with some degree of accuracy in location and quantitation of extent, cortical and subcortical macrostructures within the sensorimotor system that are metabolically active in focal seizures without differentiating whether that action is excitatory, inhibitory, or both.

ACKNOWLEDGMENT

This chapter contains unpublished work in collaboration with M. Kato, S. Hosokawa, B. L. Malamut, S. Wakisaka and R. R. O'Neill in the Laboratory of Experimental Neurology, NINCDS, National Institutes of Health, Bethesda, Maryland.

REFERENCES

1. Ajmone-Marsan, C. (1969): Acute effects of topical epileptogenic agents. In Jasper, H. H., Ward, A. A., Jr., and Pope, A. (eds.). Basic Mechanisms of the Epilepsies. Boston, Little, Brown and Company, 299-328.

2. Brodal, A. (1969): Neurological Anatomy in Relation to Clinical Medicine. Second edition, New York, Oxford University Press.

3. Carpenter, M. B. (1976): Anatomical organization of the corpus striatum and related nuclei. In Yahr, M. D.(ed.). The Basal Ganglia, New York, Raven Press.

4. Caveness, W. F., Kosaka, K., Hosokawa, S. and O'Neill, R. R. (1977): Cerebral-cerebellar paroxysmal activity in experimental focal seizures. Ann. Neurol. 1: 287-289.

5. Cooper, I. S.,, Riklan, M., and Snider, R. S. (1974): The Cerebellum, Epilepsy, and Behavior. New York, Plenum Press.

6. Creutzfeldt, O. D. (1973): Synaptic organization of the cerebral cortex and its role in epilepsy. In Brazier, M. A. B. (ed.). Epilepsy: Its Phenomena in Man. New York, Academic Press, 11: 12-29.

7. Denny-Brown, D. (1967): The fundamental organization
    of motor behavior. In Yahr, M. D., and Purpura, D. P.
    (eds.). Neurophysiological Basis of Normal and Abnormal
    Motor Activities. Hewlett, N.Y. Raven Press, 415-444.

8. Dow, R. S. (1965): Extrinsic regulatory mechanisms of
    seizure activity. Epilepsia 6: 122-140.

9. Dow, R. S., Fernandes-Guardiola, A. and Manni, E.
    (1962): The influence of the cerebellum on experi-
    mental epilepsy. Electroencephalogr. Clin. Neuro-
    physiol. 14: 383-398.

10. Eccles, J. C. (1969): The Inhibitory Pathways of the
    Central Nervous System. Springfield, Illinois, Charles
    C. Thomas.

11. Eccles, J. C., Ito, M. and Szentágothai, J. (1967):
    The Cerebellum as a Neuronal Machine. Heidelberg,
    Berlin, Göttingen and New York, Springer-Verlag.

12. Fisher, R. S. and Prince, D. A. (1977): Spike-wave
    rhythms in cat cortex induced by parenteral penicillin
    II. Cellular features. Electroencephalogr. Clin.
    Neurophysiol. 42: 625-639.

13. Glowinski, J. (1977): Some properties of the ascending
    dopaminergic pathways. Interaction of the nigrostria-
    tal dopaminergic system with other neuronal pathways.
    NRP Meeting, June 20 - July 1, 1977, Boulder, Colorado.

14. Herz, A. and Zielgansberger, W. (1972): Changes of
    focal potential by iontophoretic application of glu-
    tamic acid and gamma-amino-butyric acid. In Petsche,
    H. and Brzier, M. A. B. (eds.). Synchronization of
    EEG Activity in Epilepsies. New York, Springer-Verlag,
    141-153.

15. Jackson, J. H. (1958): A study of convulsions. In
    Taylor, J., Holmes, G. and Walshe, F. M. R. (eds.).
    Selected Writings of John Hughlings Jackson. New
    York, Basic Books, Inc., 354.

16. Jackson, J. H. (1958): A study of convulsions. In
    Taylor, J., Holmes, G. and Walshe, F. M. R. (eds.).
    Selected Writings of John Hughlings Jackson, New
    York, Basic Books, Inc., 8.

17. Julien, R. M. (1974): Experimental epilepsy: cerebro-
    cerebellar interaction and antiepileptic drugs. In
    Cooper, I. S., Riklan, M. and Snider, R. S (eds.).
    The Cerebellum, Epilepsy and Behavior. New York,
    Plenum Press, 97-118.

18. Kennedy, C., Des Rosiers, M. H., Sakurada, O., Shino-
    hara, M., Reivich, M., Jehle, J. W. and Sokoloff, L.
    (1976): Metabolic mapping of the primary visual sys-
    tem of the monkey by means of the autoradiographic
    [^{14}C]deoxyglucose technique, Proc. Natl. Acad. Sci.
    U.S.A. 73: 4230-4234.

19. Matsumoto, H. (1964): Intracellular events during the
    activation of cortical epileptiform discharges.
    Electroencephalogr. Clin. Neurophysiol. 17: 294-307.

20. Oscarsson, E. (1967): Functional significance of in-
    formation channels from the spinal cord to the cere-
    bellum. In Yahr, M. D., and Purpura, D, O, (eds.).
    Neurophysiologic Basis of Normal and Abnormal Motor
    Activities. New York, Raven Press, 93-117.

21. Penfield, W. and Jasper, H. (1954): Epilepsy and the
    Functional Anatomy of the Human Brain. Boston, Little
    Brown and Co.

22. Petsche, H. (1976): Pathophysiological apsects of
    epileptic seizures. In Birkmayer, W. (ed.). Epileptic
    Seizures - Behavior - Pain. Baltimore, University
    Park Press, 11-31.

23. Prince, D. A. (1968): Inhibition in "epileptic"
    neurons. Exp. Neurol. 21: 207-321.

24. Purpura, D. P. (1976): Physiological organization of
    the basal ganglia. In Yahr, M. D. (ed.). The Basal
    Ganglia. New York, Raven Press, 91-114.

25. Sokoloff, L., Reivich, M. Kennedy, C., Des Rosiers, M.
    H., Patlak, C. S., Pettigrew, K. D., Sakurada, O. and
    Shinohara, M. (1977): The [^{14}C]deoxyglucose method for
    the measurement of local cerebral glucose utilization:
    Theory, procedure and normal values in the conscious
    and anesthetized albino rat. J. Neurochem. 28: 897-916.

26. Souqeus, A. (1936): Etapes de la neurologie dans l'
    antiquité" grecque. Paris, Masson et Cie., 246.

27. Szentágothai, J., University of Budapest, Hungary,
    Personal communication.

SOME BIOCHEMICAL ASPECTS OF ELECTROCONVULSIVE SEIZURE

Lj. Rakić, R. Mileusnić, Lj. Rogač and R. Veskov

Institute of Biochemistry, Faculty of Medicine
   and
Institute for Biological Research
Belgrade, Yugoslavia

Among experimental procedures used in neurochemical research on seizure, electroconvulsive shock (ECS) has been the object of numerous and extensive studies (29, 52, 57, 89, 95). The seizure, as a general expression of the sudden onset of an intense, rapidly repetitive focal or generalized electrical discharge in the brain, is not only the expression of epilepsy as prototypical disease, but also a common symptom of numerous neurological diseases. Experimentally, it can be induced by different methods. Past neurochemical studies of convulsions induced by different methods categorically refute the existence of a single chemical agent which triggers the abnormal electrical discharge with consequent behavioral manifestations in all cases, although the electrical and behavioral manifestations in convulsions caused by different mechanisms are very similar in their general pattern of expression. Measurements of biochemical parameters in convulsions are mainly limited to the phenomena accompanying or following convulsive attacks. ECS has been used as a model for different neurobiological studies (e.g., excitation, convulsions, stress, affective behavior, learning and memory, sleep) with the intention of interpreting its central effects. ECS has several clinical therapeutic effects, e.g., antidepressive, antimanic, anticonfusional, antipsychotic and above all diminution of memory disturbances (22, 73). Experimentally induced seizure in animals produces a remarkable behavioral effect--retrograde amnesia (61) and blocking of the rebound of REM sleep after chronic sleep deprivation (14). Seizure markedly enhances oxidative

metabolism (23, 47, 48, 74, 88) and reduces the high-
energy phosphates in the brain by elevating lactic acid
content, thus implying that energy utilization during
seizure exceeds its production (18, 32, 49). Despite the
increased metabolism, hypoxia of the brain does not appear
so long as the arterial blood pressure and the oxygen
tension in the arterial blood are not allowed to fall below
the normal levels (41). ECS studies concerned with the
correlation between behavioral effects and biochemical
alterations in the context of the molecular organization
of the central nervous system point out the controversy
related to the role of seizure itself in creating the
corresponding behavioral and biochemical disturbances.
While brain seizure activity was suggested to be necessary
for the induction of amnesia (91), there are some opinions
stating that neither behavioral convulsions nor cortical
seizure activities appear to be important for amnestic
effect (25, 90). Several studies concerned with the latter
consideration have pointed out the significance of the
relationship between electroshock current and retrograde
amnesia (21, 28, 56, 80). In another study it was found
that the biochemical effects of ECS on inhibition of amino
acid incorporation into the brain proteins were linearly
correlated with the current used in the electroshock, and
the presence or absence of convulsions did not disturb
this correlation (24). In the numerous neurochemical
studies of ECS, particular attention has been given to a
few aspects. These are, in addition to brain energy
metabolism, RNA and protein synthesis (42), electroshock
effects on cerebral electrolytes (29), on the metabolism
of acetylcholine as well as of biogenic amines, and on the
blood-brain barrier and its permeability (82).

     Clinical observations that the therapeutic effective-
ness of convulsive therapy was related to the number of
seizures (8, 64) led to experimental studies on brain and
behavior with the basic aim of discovering brain biochemical changes associated with repeated ECS that might be
related to the persisting behavioral consequences of the
treatment. Repeated ECS produces an increase in brain
weight (76), increases the activity of monoaminoxidase (77)
as well as of acetylcholinesterase (76) and tyrosine
hydroxylase (68), and increases the levels of serotonin
(30), norepinephrine (46) and deaminated metabolites of
norepinephrine (52).

     With these findings in mind, and aiming for better
comprehension of the mechanisms of the biochemical changes
in ECS, we made it our goal to examine some key aspects of

the problem which could be, in our experience, involved in the structural and metabolic regulation at different levels, e.g., at the blood-brain barrier and in the motility transport system of the cells.

MATERIALS AND METHODS

Adult male Lewis strain rats (200-250 g body weight) were used. The animals were afforded unlimited access to standard laboratory chow and water until the time of the experiment. Electroshock convulsions were induced by stimulation of each animal through alligator clips attached to the ears, at parameters which consistently elicited seizure in freely moving animals. Control animals were sham-treated. Repeated convulsions were carried out with the same procedure in two ways: (1) ECS was repeated 6 times during one day, with an interval of 1 hour between convulsions, and the animals were sacrificed immediately after the last seizure (daily repeated convulsions); and (2) repeated ECS was applied during 5 days (6 times daily with an interval of 1 hour between convulsions), and the animals were also sacrificed after the last seizure (several days' repeated convulsions). Only animals having full and developed seizures were used for the experiments. A developed generalized epileptiform seizure is characterized by certain consistency, duration and intensity of the convulsive phases. The electroshock seizure pattern in rats begins with a so-called seizure spasm and initial clonus which is characterized by short-duration extension, as well. This phase is followed by the tonic phase, then the clonic phase, again ending with automatisms, but sometimes with coma. Only animals exhibiting this pattern were used for biochemical studies. Total duration of the developed ECS was 40-60 sec. The tonic patterns were more intense than the clonic patterns. Repeated electroshock stimulations usually increased the ECS threshold.

Rat brain microvessels were isolated using the technique described by Mršulja et al. (66, 67). Rats were decapitated, and the forebrains were quickly removed and diced with scissors into 5 vols of an ice-cold homogenizing medium of the following composition (final concentrations): 58 mM NaCl, 8 mM KCl, 0.9 mM $CaCl_2$, 0.5 mM NaH_2PO_4, 0.4 mM $MgSO_4$, 2.8 mM glucose and 14 mM HEPES; final pH 7.4. Tissue was homogenized in the cold with a Teflon-pestle Potter-Elvehjem homogenizer, and centrifuged at 1500 g for 10 min (0.4°C). The pellet was resuspended in 5 vols of ice-cold 0.32 M sucrose (pH 7.1) and overlayed on a preformed discontinuous 1.0-1.8 M sucrose gradient. Centrifugation for

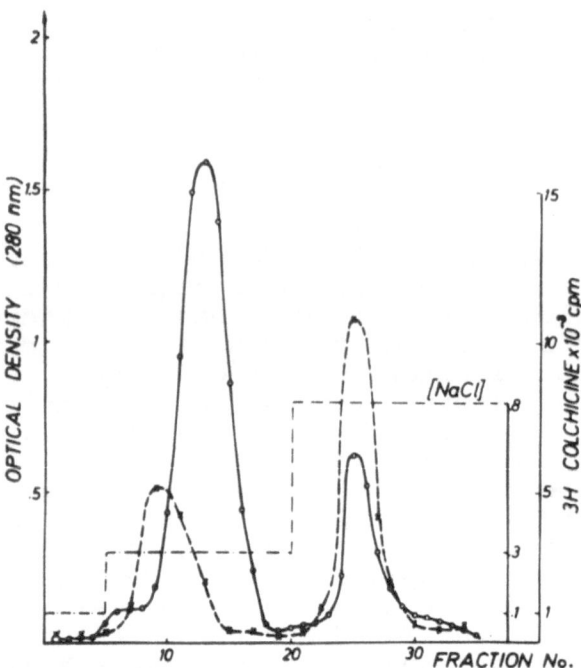

Fig. 1.   Chromatography of the soluble proteins on DEAE-
          Sephadex A-50.

30 min at 58000 g (0-4°C) in a Beckman SW27 rotor yielded
a pure microvessel fraction (67).   The microvessel pellet
was treated with 0.5% (vol/vol) Triton X-100, and the
suspension was allowed to stand in the cold for 30 min;
the resulting suspension was the enzyme source (60).   Less
than 10% activity of the membrane-bound enzymes was found
in the Triton X-100 pellet, but no activities of the cytosol or the intramitochondrial enzymes.   The whole brain
enzymatic activities were measured in the supernatants that
were obtained from the forebrain homogenates in 10 mM Tris-
(hydroxymethyl)-aminoethane HCl buffer (pH 7.1) with 0.5%
(vol/vol) Triton X-100 added, which were allowed to stand
for 30 min in the cold and centrifuged at 15000 g for 45
min (0-4°C).   Protein contents were determined according
to Lowry and Passonneau (58), with bovine serum albumin as
the standard; values were corrected for the change caused
by the presence of Triton X-100.   All the spectrophotometric enzymic assays were performed at 37°C in 50 mM tri-
ethanolamine/HCl buffer, pH 7.1, except that of alkaline
phosphatase, which was assayed at pH 9.8.   Acetylcholine

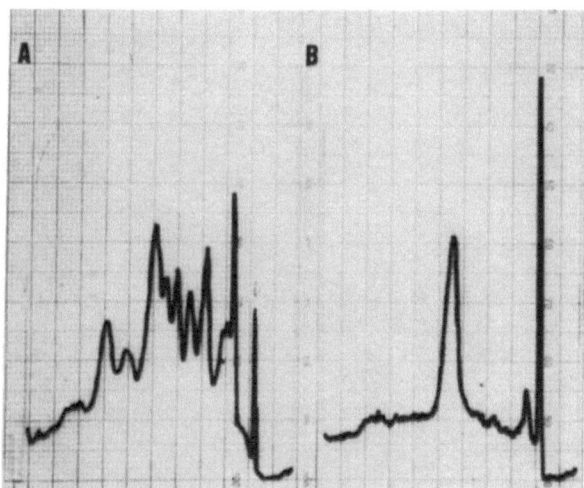

Fig. 2.   Acrylamide gel electrophoresis of soluble pro-
          teins (A) and purified tubulin (B).

esterase (AChE) activity was assayed with acetylcholine
iodid as substrate and 1 μM tetraisopropylpyrophosphor-
amide (isoOMPA) as the selective inhibitor of butyrylcholine esterase (BuChE) (85), which in turn was assayed with
butyrylcholine iodide as substrate and 30 μM 1:5-bis-4-
allyl dimethylammonium phenylpentan-3-one dibromide (BW
284C51) as the selective inhibitor of AChE (85).

     Identification and quantification of rat brain tubu-
lin were done employing the following procedures: electrophoresis, isolation of colchicine-binding protein and
filter-binding assay. Gel electrophoresis was performed
on 10% acrylamide gels containing 0.1% SDS, according to
Laemmli and Favre (54). The samples were run on the
Shandon electrophoresis apparatus at a current of 1 mA/gel
until the tracking dye reached the end of the gel (about
6-7 hours). The running buffer was 25 mM Tris-glycine
buffer (pH 8.6). The gels were stained overnight in 0.02%
Coomassie brilliant blue. For the isolation of colchicine-
binding protein, brains were placed in a 20 mM Na-phosphate
buffer (pH 6.8) containing 5 mM MgCl$_2$, 0.24 mM sucrose
(P-Mg buffer) in a beaker packed in ice. All subsequent
operations were performed in the cold. The blood vessels
and meninges were removed (92). The following brain regions were dissected: frontal cortex cerebri (Cx), n.
caudatus (N.Cd), thalamus (Th), hippocampus (Hippo),

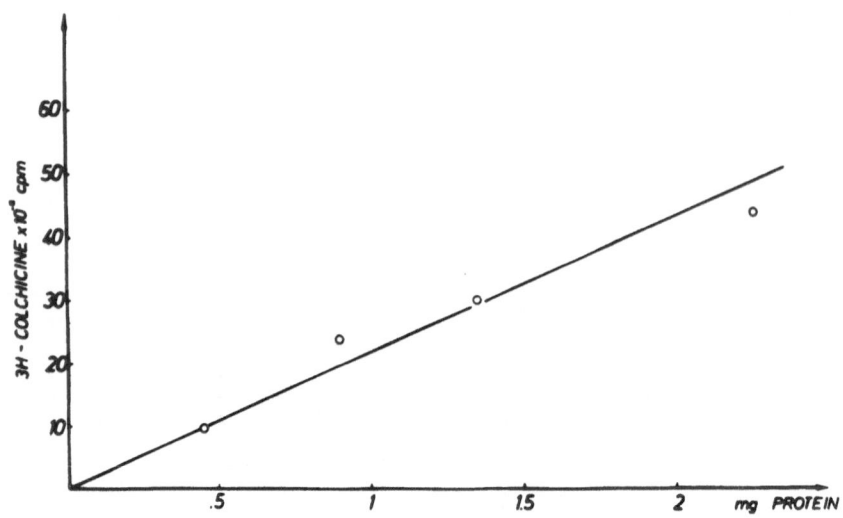

Fig. 3.   Determination of relationship between tubulin
           and ^{3}H-colchicine

cerebellum (Cer) and medulla + pons (MP). The tissue was
washed in the same buffer. Homogenization was done in 2
vol of the P-Mg buffer containing 10^{-4} M GTP. The homogenate was centrifuged in a Beckman J-21 centrifuge at
16,000 g for 30 min. The supernatant was then centrifuged
in a Beckman-Spinco G centrifuge at 105,000 g for 1 hour.
Soluble tubulin was identified by filter assay for colchicine-binding. Assays were performed by Weisenberg and
Timasheff procedures (93). The fraction of soluble proteins after centrifugation at 105,000 g was incubated for
60 min at 37°C with ^{3}H-colchicine (final concentration
2.5 x 10^{-6} M, 0.1 μCi), then, cooled to 0°C. About 1 mg
of soluble proteins were emptied directly on DEAE-cellulose
chromedia paper (DE 81, Whatman Co.) disks. The tubes were
rinsed with 1 ml 10^{-4} M nonradioactive colchicine; 8 ml of
P-Mg buffer were added to the filter funnel. Filtration
took approximately 10 min. The disks were washed by 4
additions of 10 ml of P-Mg buffer-containing fluid (92)
and the radioactivity was counted in a Nuclear Chicago
scintillation counter. The tubulin content was measured
by chromatography on DEAE-Sephadex A50 (2.5 x 6 cm) column
eluted with P-Mg buffer (pH 6.8) according to Brajan and
Wilson (10).

     A typical elution profile from DEAE-Sephadex is shown
in Fig. 1. Soluble proteins isolated from rat brain were

incubated with 2.5×10^{-6} colchicine for 1 hour at 37°C
and quickly cooled to 0°C. The majority of proteins did
not absorb to the ion exchanger. Only the last peak represented protein-bound colchicine. Acrylamide gel electrophoresis of the purified tubulin showed a band representing 80% of the protein in the gel (Fig. 2). In our
experiments the tubulin content was determined by two
methods: filter assay for colchicine-binding and electrophoresis in SDS polyacrylamide gels. Probably due to some
overlapping of the proteins in the gel or to nonidentical
degradation of tubulin subunits in minor components with
a molecular weight of about 33,000, very similar but nonidentical values were obtained from the colchicine-binding
assays and SDS polyacrylamide gels.

The linear relationship between increased amounts of
soluble proteins and colchicine (2.5×10^{-6} M; ^{3}H-colchi-
cine) is shown in Fig. 3. Effects of ECS on tubulin content and enzymatic activity as well are represented as the
proportionate changes (exp. = experimental, cont. = control):

$$\left(\frac{\text{exp. value}}{\text{cont. value}} \right) x \quad ;$$

x = 1 when exp. value is greater than cont. value; x = -1
when exp. value is less than cont. value.

The GTP used in the purification procedure was 95%
pure product obtained from Sigma Biochemical Co. Radioac-
tive ^{3}H-colchicine was obtained from New England Nuclear.
Unlabeled colchicine was obtained from Fisher Scientific
Co. The standard buffer used in experiments was 20 mM
Na-phosphate (pH 6.8) containing 5 mM $MgCl_2$ (P-Mg buffer),
10^{-4} M GTP or 0.24 M sucrose.

RESULTS

A) <u>Microvessel/parenchyma enzyme activity ratios in
ECS</u>

The preservation of neuronal integrity requires the
proper function of the specific control in the transport
of many materials in the central nervous system. This
special transport mechanism is called the blood-brain
barrier (BBB) and has been attributed to cerebral microvessels (16). A chemical flow occurs from neurons to glial
cells to capillaries and back again. Evidence that the
permeability of the BBB is increased by ECS has been
accumulated in a number of studies (4, 13, 27, 37, 55).
Better understanding of the mechanisms regulating BBB
permeability will be provided by more precise research

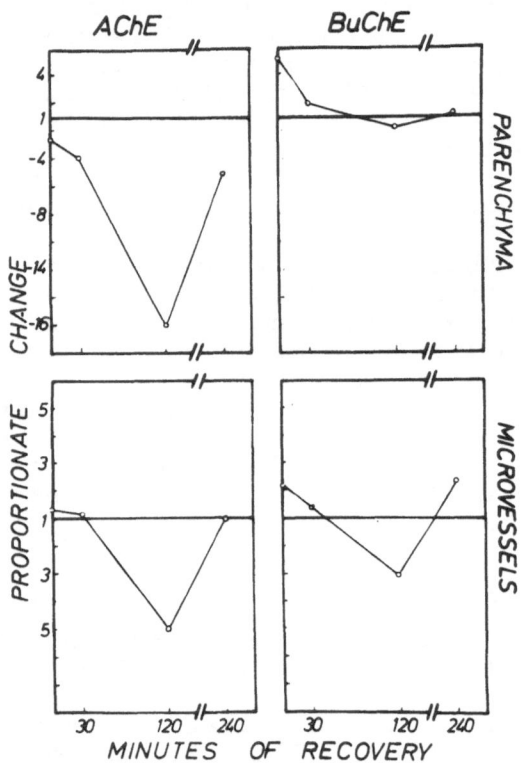

Fig. 4.   AChE and BuChE activity in and after a single ECS.

on the metabolic processes in microvessels, taking into
consideration that transport of many substances through
the BBB includes stereospecific carrier-mediated transport systems (72). The isolation of cerebral capillaries
which appear to be free of any other cellular contamination and which show a significant degree of morphological
intactness and metabolic activity (36, 66) enabled extensive biochemical studies on BBB functioning to be made.
Previous studies from our laboratory (20,66) established
the activity characteristics of a large number of enzymes
in the brain microvessels vs. the brain parenchyma. Because of the resulting better comprehension of the dynamics
of changes in these metabolic compartments, we considered
it worthwhile to study the behavior of some membrane enzymes
(ATPase, AChE) and of enzymes with higher activity in
microvessels (BuChE) in ECS.

Table I ENZYMATIC ACTIVITIES IN ISOLATED RAT FOREBRAIN
        MICROVESSELS AND PARENCHYMA, AND ACTIVITY RATIO:
        (BRAIN MICROVESSEL ACTIVITY)/(BRAIN PARENCHYMA
        ACTIVITY)

| Enzyme | Activity | | Ratio |
	in microvessels	in parenchyma	
AChE	318 ± 18	1302 ± 54	0.24
BuChE	72 ± 6	9 ± 0.3	8.00

The activities are given as nanomoles of substrate converted per hour per milligram of protein at 37°C.

     Figure 4 shows the proportionate changes in the AChe
and BuChE activity in the brain parenchyma and microvessels of the animals sacrificed immediately after ECS and
in different periods of recovery after ECS, in comparison
with findings in the control animals. In a single ECS
with completely developed electroconvulsive seizure, AChE
activity was significantly reduced in the brain parenchyma
compartment, while this enzyme activity was slightly
increased in microvessels. In the recovery periods, 30 min
and even more evidently 120 min, after ECS, drastic reduction of the enzyme activity of AChE was apparent in both
compartments examined. Complete normalization of this
enzyme appeared 240 min after ECS in the microvessels,
but in the brain parenchyma, despite marked tendency toward
normalization, enzyme activity values in this period not
only were still significantly lower than control values,
but also were lower than the values immediately after
seizure. The BuChE activity was significantly increased
immediately after ECS in both compartments examined. The
activity of this enzyme decreased rapidly in the recovery
period, and 30 min after ECS was approaching the control
values; 120 min after ECS, activity approached values lower
than the control values, especially in the microvessels.
After 240 min, this activity reached control levels in the
brain parenchyma, and in the microvessels it was again increased to the ECS values. The BuChE activity of the control animals was 8 times higher in the microvessels than
in the parenchyma, while the AChE activity was more than
4 times lower in microvessels than in parenchyma (Table I);
these values are thus in accordance with the findings of
Djuričić et al. (20).

Table II   ENZYMATIC ACTIVITIES IN ISOLATED RAT FOREBRAIN
            MICROVESSELS AND PARENCHYMA, AND ACTIVITY RATIO:
            (BRAIN MICROVESSEL ACTIVITY)/(BRAIN PARENCHYMA
            ACTIVITY)

| Enzyme | Activity | | Ratio |
	in microvessels	in parenchyma	
ATPase (total)	63 ± 8	100 ± 16	0.63
Na^+-K^+-ATPase	36 ± 3	40 ± 2	0.90
Mg^{2+}-ATPase	27 ± 4	60 ± 3	0.45

The activities are given as nanomoles of substrate converted per minute per milligram of protein at 37°C.

     As previously shown (20), ATPase activity in control
animals (Table II), although higher in the brain parenchyma than in the microvessels, was nevertheless relatively
high in the microvessels as well, particularly the activity
of Na^+-K^+-ATPase. This enzyme constitutes 52.6% of total
ATPase activity in the microvessels, compared with 47.4%
in the parenchyma. During a single ECS, ATPase activity
(total and selective) increased significantly in both compartments (Fig. 5). In the recovery periods, 30 min after
ECS, total ATPase activity increased further in the brain
parenchyma because of the Na^+-K^+-ATPase activity, the values of which were double the values immediately after ECS,
while the Mg^{2+}-ATPase activity decreased slightly. In the
brain microvessels, 30 min after ECS, total ATPase activity
decreased, mainly because of the Na^+-K^+-ATPase activity.
At 120 min after ECS, there was a drastic fall in total
ATPase activity in the brain parenchyma, the values being
slightly higher than control values; Mg^{2+}-ATPase values
were slightly lower than control values while Na^+-K^+-ATPase
activity was even higher than in the control animals.
There was a marked decline of total ATPase activity in
this period in the microvessels, where the values were 50%
of control values; Mg^{2+}-ATPase activity was almost unde-
tectable, while the Na^+-K^+-ATPase values were slightly
below control values. At 240 min after ECS, total ATPase
activity again increased, particularly in the parenchyma,
where it amounted to 248% of the control activity. In the
same period, this activity in the brain microvessels was
only 30% higher than control values. Such an increase of
total ATPASE activity in the parenchyma was due to the
simultaneous increase in activity of both enzymes, while

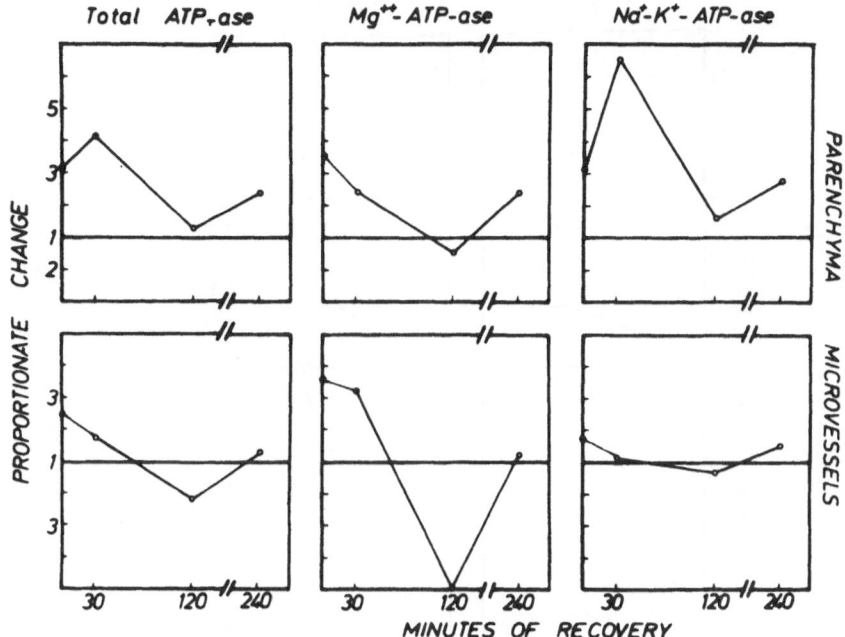

Fig. 5. ATPase activity in and after a single ECS

in the brain microvessels there was a pronounced increase in Na^+-K^+-ATPase activity.

Repeated ECSs (6 during a day, with an interval of 1 hour between ECSs) were followed by the characteristic variations of all the enzymes examined. AChE activity increased in the repeated ECSs, compared with the values after a single ECS, in both compartments; thus, the activity values of the brain parenchyma enzymes tended to approach control values and in the microvessels, the activity was twice as high as in the control animals. BuChE activity decreased in the parenchyma, but under such conditions significantly increased in the microvessels (Fig. 6). Total ATPase activity was decreased by the repeated ECSs, compared with single ECS values, with a tendency to reach the control values, particularly in the brain parenchyma (Fig. 7). The decrease in ATPase activity values was due to mainly the decrease in Mg^{2+}-ATPase activity, while the Na^+-K^+-ATPase activity, in such conditions, was either slightly decreased (brain parenchyma) or slightly increased (microvessels).

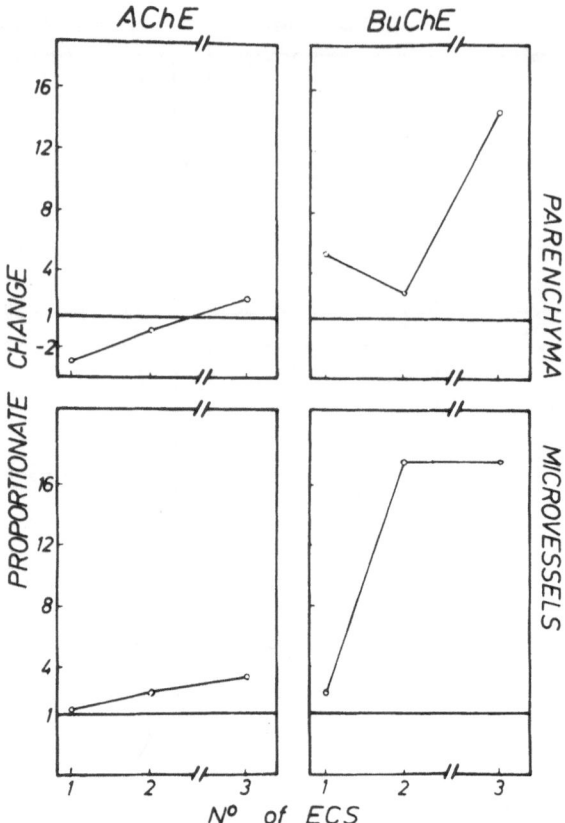

Fig. 6.   AChE and BuChE activity in single (1), 1-day
          repeated (2) and 5-days repeated (3) ECS.   For
          details, see Methods

          Multiple daily repetition of ECS (6 seizures daily,
with an interval of 1 hour between seizures for 5 days)
was accompanied by an accentuation of the AChE and BuChE
activity changes in both compartments.   The AChE activity
showed a further increase compared with the values after
6 repeated convulsions during one day, the values being
higher than control values in the parenchyma as well as
in the microvessels.   BuChE activity increased significantly in the parenchyma after several days ECS, compared
with the values obtained not only after 6 repeated convulsions during one day, but also after the single ECS, whereas in the microvessels, where it reached extremely high
values after repeated ECS during one day, the high activity
was only maintained under these conditions (Fig. 6).   The

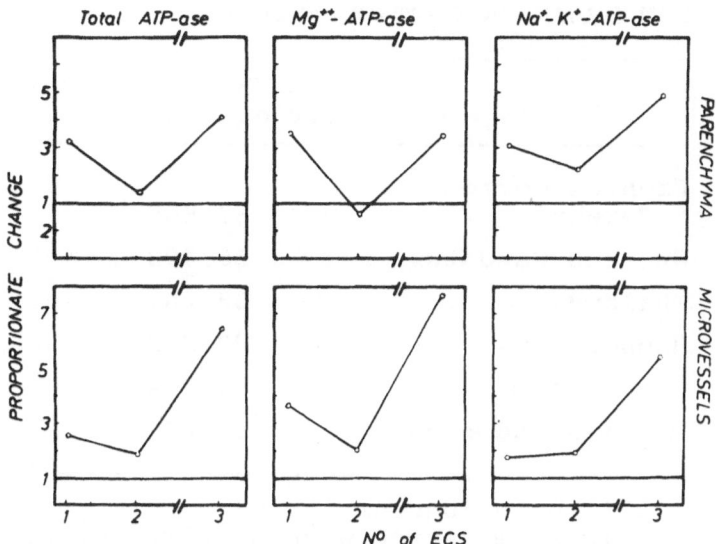

Fig. 7. ATPase activity in single (1), 1-day repeated
         (2) and 5-days repeated (3) ECS. For details,
         see Methods

ATPase activity showed a similar trend of changes resulting from repeated ECS applications during several days:
total ATPase activity increased together with the simultaneous increase in activity of both Mg^{2+}-ATPase and Na^+-
K^+-ATPase (Fig. 7).

B) Tubulin in ECS

Evidence that microtubules play a key role in axoplasmic transport has come from studies of colchicine and
vinblastine (5, 17, 43, 51).

These substances block cell division and bind to the
protein subunit of microtubules, tubulin. Binding to
tubulin, colchicine and vinblastine cause the microtubules
to disassemble, with the subsequent interruption of fast
axoplasmic transport (51). The microtubules undergo a
continual partial disassembly all along their length and
do not rapidly grow into fiber (79), which has a special
functional importance. It is worth stressing that electrical stimulation of nervous tissue does not induce an
increase in the number of filaments and tubules by de novo

Table III   TUBULIN CONTENT IN DIFFERENT REGIONS OF
             RAT BRAIN IN CONTROL ANIMALS

	cpm ^{3}H-colchicine per gram tissue*)
Frontal Cortex Cerebri	67.566
Nucleus Caudatus	56.492
Thalamus	83.487
Hippocampus	82.535
Cerebellum	37.964
Medulla and Pons	57.637

*)All the values are triplicate mean
values, measured with an error of ±1%.

protein synthesis, but triggers the association of preformed protein subunits (84), which represents a dynamic
and reversible phenomenon related to physiological conditions. The experiments of Rose et al. (81) suggest
that the role of tubulin production is related to functional stimulation. A particularly interesting finding
was that there was a considerable quantity of particulate
colchicine-binding activity in brain, which may be associated with the synaptosomal membrane (31). Because of
the known importance of tubulin in neural functions, it
seemed reasonable to investigate, by using some of its
physicochemical characteristics, the stability of this
structural and regulational protein in ECS.

During the study of tubulin content in particular
brain regions, characteristic and specific regional differences were noted. The highest content of this protein
was found in Hippo and Th, then in Cx, MP and N.Cd, while
significantly lower values were noted in Cer (Table III).

In single ECS with completely developed electroconvulsive seizure, selective changes at the level of tubulin
or tubulin activity to bind colchicine, compared to values
in control animals, appeared in the brain regions examined. ECS leads to a marked decrease in tubulin levels

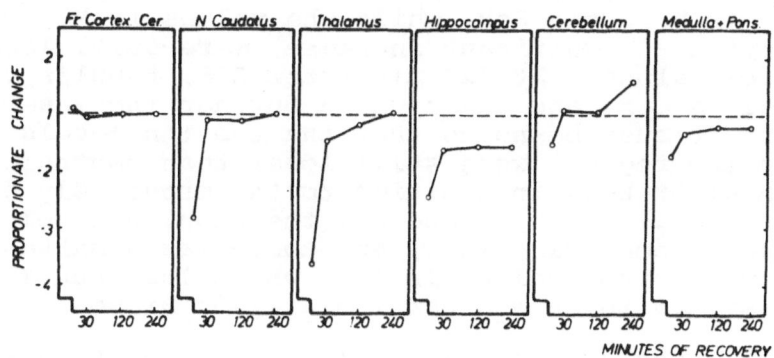

Fig. 8. Tubulin content in and after a single ECS

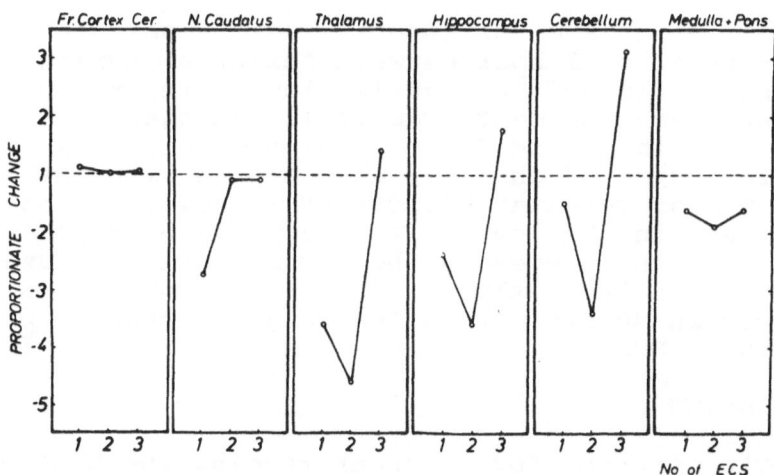

Fig. 9. Tubulin content in single (1), 1-day repeated (2) and 5-days repeated (3) ECS. For details, see Methods

in Th, N.Cd and Hippo and a less marked decrease in Cer
and MP, while in Cx no changes are observed. During the
recovery period, 30 min after ECS, a general tendency of
tubulin levels to reach control values was noted (Fig. 8).
Within this period, the tubulin level was completely
normalized in N.Cd and Cer, while the values in MP, Th and
Hippo, despite a significant increase, were still lower
than control values. At 120 min after ECS, tubulin values
were normal in Cer, and there was a further increase in Th
and MP, but it must be noted that the tubulin levels in
the latter two regions were still lower than control values.
At 240 min after ECS, in addition to the previously obtained normal values in Cs and Cd, the tubulin level in
the Th reached control values, and there was a marked increase above control values in Cer, while low tubulin
values persisted in MP and particularly in Hippo.

Repeated ECS (6 seizures during a day, with an interval of 1 hour between seizures) did not affect the tubulin
levels in Cs and N.Cd, but greatly decreased levels of
this protein in Th, Hippo and Cer, compared not only with
control values, but also with those after single ECS.
This drop was marked in MP (Fig. 9).

Repetition of ECS for several days (6 seizures daily,
with an interval of 1 hour between seizures, during 5 days)
also did not influence the tubulin levels in Cx and N.Cd.
Such a course of treatment drastically changed not only the
tubulin levels in Th and Hippo, and particularly in Cer,
but also the trend of changes, as compared with the trend
in single ECS or repeated ECS during a day, thus leading
to an increase in the levels of this protein in comparison
not only with the changes in the convulsive states examined,
but also with control values. Under such conditions, tubulin levels in MP are within the range of changes noted
after a single ECS.

DISCUSSION

Efforts directed toward understanding the biochemical
mechanisms related to the effects of ECS aid in better
understanding of the large scale of the accompanying behavioral manifestations related to convulsions, affective
behavior, learning and memory. ECS, being as it is a drastic agent, causes not only massive discharge over wide
areas of the brain, but also activation of the peripheral
autonomic nervous system, with the subsequent release of
secretions of many endocrine glands and powerful muscle
contractions throughout the body, thus changing the

chemical homeostastasis of the body and accentuating biochemical changes in the brain initially induced by the
electroshock itself. In this context, taking into consideration changes caused by the ECS, attention must be
paid to these secondary effects, which could play a significant role, particularly in the repeated convulsions,
the convulsions of longer duration, and the accompanying
recovery periods.

The results obtained in the comparative biochemical
studies of membrane enzyme activities (AChE, ATPase) and
the enzymes with higher activities in the brain microvessels (BuChE) vs. the brain parenchyma emphasize the
fact that ECS selectively and specifically changes the
homeostasis of the systems forming the blood-brain barrier.
With the use of the relatively simple technique of iso-
lating cerebral capillaries having a significant degree of
morphological intactness and metabolic activity (36, 67),
it was shown that the enzymatic activity of enzymes located in the microvessels, as compared with the activity of
enzymes located in the brain parenchyma, does not indicate
an identical trend of changes during ECS. The AChE enzyme
activity during ECS increased slightly in the microvessels,
but dropped significantly in the parenchyma, while the
BuChE activity increased significantly in both compartments (Fig. 4); this finding probably reflects the particular functional roles of these enzymes under such conditions. Adams et al. (1) have also found that AChE in
the rat brain increased immediately after ECS, reaching
normal levels after approximately 96 hours. Our results,
however, indicate that normalization lasted significantly
longer, and that the activity values of this enzyme fluc-
tuate in post treatment periods--decreasing further 120
min after ECS and falling short of control values in the
brain parenchyma even after 240 minutes. In agreement
with the above results are the findings of Essman (29)
which show a fall in acetylcholine (ACh) in the post convulsive period in rats in the first 10 min after ECS, with
a subsequent increase, and, 120 min after ECS, total and
free brain cholin-acetyl-transferase, the enzyme regulating
ACh synthesis, showed the lowest activity following a
single ECS treatment (28). The normalization of BuChE
in the brain parenchyma after ECS was relatively rapid.
After the increase in enzyme activity in the microvessels
occurring during ECS, a significant fall was recorded in
the recovery period of 120 min and 240 min after ECS there
was another increase in BuChe activity, the values reaching those in ECS. The marked BuChe activity was long
ago demonstrated histochemically in the capillary walls

of various organs, being particularly marked in the rat
brain (50) but absent from capillaries of brain areas
outside the blood-brain barrier (44, 45). In the past,
numerous studies have pointed out that ECS changes the
permeability of several vascular and vasculocerebral compartments to the selective transport of different molecules (7). An increased cerebrovascular permeability
after ECS as measured with trypan blue (7), trypan red
(3) and cocaine (2) has been described. ECS increases
the permeability of the barrier system to noradrenalin
(82). In rats given ECS, filtration through the capillary
epithelium was partially or completely inhibited during
1 hour following treatment (37). More recent electron-mi-
croscopic studies of cerebral ultrastructures following
seizures have renewed the controversy surrounding this
subject. De Robertis et al. (19) have reported swelling
of perivascular astroglia in rats following Metrazol
induced seizures. In contrast, Brennan et al. (11) found
no significant ultrastructural differences between control
and ECS brain tissue. Capillary and endothelial cells
showed no abnormality.

The results obtained in our studies, based on the
biochemical changes in the brain capillaries during ECS
conditions, offer the possibility of better comprehension
of the dynamics of changes in the BBB, in the context of
alterations in the rest of the brain tissue. The variations in BuChE enzyme activity in the brain parenchyma
and especially in the microvessels could be related to the
role of the enzyme in the regulation of vascular permeability, which has been previously mentioned (39, 40, 50),
despite the absence of ultrastructural changes under such
conditions (11). In this context, we would like particularly to point out the role of BuChE in the microvessels,
where its activity is enormously higher and the dynamics
of changes during ECS and recovery periods far more complex.
The alterations of this enzyme in the brain parenchyma were
recorded mainly in the initial period of ECS, and were
most probably due to the changes in satellite cells, where
a large quantity of BuChE was found (19, 33, 40). During
ECS, when the AChE activity was drastically reduced,
simultaneous activation of capillary and glial BuChE most
probably provided additional homeostatic mechanisms of
the acetylcholine system.

ECS produces massive central neuronal excitation and
a tremendous increase in the metabolic rate. These changes
are accompanied by a marked fall in ATP and phosphocreatine
(48, 65). Cerebral excitation under these conditions involves large transmembrane movements of sodium and potassium.

With the continuous recording of electrolytes of arterial
and venous blood with electrodes sensitive to sodium and
potassium, it was shown that during seizure sodium entered
the brain and potassium was lost (63). The behavior of
ATPase during ECS (Fig. 5), serves as a complement to
these findings. The highest ATPase activity, particularly
Na^+-K^+-ATPase, was recorded in the grey matter of the
brain (9). This membrane-bound enzyme is located at the
outer and inner cell walls and plays a role in active
sodium and potassium transport through the cell walls (83).
Na^+-K^+-ATPase activity is stimulated by extracellular (but
not intracellular) potassium and by intracellular (but not
extracellular) sodium; therefore, ATPase exhibits the
attributes expected of a Na^+-K^+ pump. Addition of either
Na^+ or K^+ alone does not affect the rate of enzymatic
activity. When both ions are present, ATP hydrolysis is
markedly accelerated. ATPase activity is $Mg^{2+}-Na^+-K^+$-de-
pendent and exhibits both phosphokinase and phosphatase
activity. The increase in total ATPase activity in ECS
(Fig. 5) which is attributable mainly to the increase in
Na^+-K^+-ATPase activity, could possibly be explained by the
phenomenon, already mentioned, of Na^+-K^+ transmembrane
movement (83) influencing activation of this enzyme and
the subsequent ATP hydrolysis. The high level of ATPase
activity in the recovery periods even 240 min after ECS
suggests the existence of long-lasting fluctuations in
the transmembrane movements of ions, which, in different
compartments (brain parenchyma, microvessels), could have
different levels and trends of changes. Accordingly many
metabolic relationships involved in total cerebral homogenate levels show no significant deviations from control
values in the recovery periods because they are compensated
by the differences between particular compartments. The
changes in ACh level, as well as in AChE and ATPase activity, should be considered in the same context. In the
recovery periods 30 min following ECS, an increase in
ATPase (Fig. 5) and a decrease in ACh were observed (28)
because the formation of ACh was reduced in the presence
of high ATPase activity (87), despite the low AChE values
(Fig. 4). It is possible that in these interactions both
systems, ACh-AChE and ATP-ATPase, have some common independent role in ion movements across cell membranes, useful for normal functioning of certain cells. There are
some suggestions that the ATPase enzyme-substrate complex
itself may be the carrier involved in the Na^+-K^+ transport
system (94). ACh and AChE have been implicated in the
control of both passive and active transport. The dependence on cholinesterase of the permeability of cerebral
tissue to barbital and chloralose was suggested many years
ago (40), based on the finding that physostigmine increased

Table IV   INCORPORATION OF ^{3}H-LEUCINE INTO THE
           NERVE-ENDING PROTEINS*)

Control	ECS
cpm/mg/protein	
3504	1641
2641	1227
2916	1381

*)All the values are triplicate mean
  values, measured with an error of
  ± 1%.

the rates of onset of narcosis and uptake of barbital by
the brain following intravenous injection. The previously
mentioned common role in the relationship during ECS would
be not only in the regulation of the ion-transport system
but also in the energy processes, because it is well known
that in the sequence of energy transformations, ACh hydrolysis precedes that of ATO (69).

     The changes in tubulin content in ECS and during the
recovery periods indicate that this protein adapts itself
specifically to the functional requirements. ECS did not
lead to changes in the tubulin level in Cx, but the subcortical regions were noted to have the highest decrease
in this protein in Th, N.Cd and Hippo, and less in Cer and
MP. Remarkably higher regional differentiation in tubulin
levels was recorded in the recovery periods. Numerous
past studies have pointed out that ECS inhibits protein
synthesis in the brain. Dunn et al. (26) have found a
significant reduction of leucine incorporation into the
brain protein immediately after ECS, this reduction persisting for only 15 minutes. The same authors have found
a significant reduction in polyribosomes during the same
brief interval following ECS. At intensities below the
brain-seizure threshold, protein synthesis was not significantly altered (15). Particularly significant inhibition of protein synthesis by a single ECS was demonstrated
in synaptosomes (Table IV), which agrees with the findings
of others (29).

Taking into consideration previously mentioned results (Fig. 8), we are not able to answer the question whether changes in tubulin levels in ECS reflect changes in synthesis, redistribution or stability. Our preliminary experiments showed that a single ECS changed the stability and redistribution of tubulin between soluble and membrane-bound tubulin more than the synthesis processes (Mileusnić and Rakić, unpublished observation). The above-mentioned tubulin characteristics of rapid assembly and disassembly in response to a wide variety of chemical agents and physical conditions (70) have particular significance in the context of other biochemical changes in ECS. GTP and ATP are necessary for assembly of tubulin (86). Calcium also has the ability to block tubulin assembly (69). Cyclic AMP plays a considerable role in the turnover of tubulin, and this effect might be mediated by calcium (35). Cyclic AMP, cyclic GMP and calcium keep the pool of soluble tubulin and polymerased microtubules in equilibrium (86). ATPase dynein interacts with axonal microtubules (34), and ATPase with an ion-activation effect similar to that of dynein is present in the brain (84). During ECS, cyclic AMP concentrations are decreased (59). Thus, in parallel with the ATP system discussed above, ATPase changes create the conditions for the occurrence of changes in the affinity of tubulin for colchicine or for disturbance of the equilibrium between soluble tubulin and microtubules.

Comparison of regional changes in tubulin levels during ECS and in the recovery periods with the regional changes in protein synthesis emphasizes certain differences among them, not only in the proportions but also in the trend of changes. The greatest inhibition of protein synthesis during ECS was recorded in Cx (29), while the same brain region showed no changes in tubulin levels under such conditions (Fig. 8). Thus, the different trend of changes in response to ECS in various cerebral regions in the recovery periods at the tubulin level shows that the changes in Hippo, Cer and MP in such convulsions are longer lasting, most probably being the expression of particularity in functioning of the reassembly system. With this in mind, one should consider ECS effects in the context of the previously mentioned changes occurring in the motility transport systems of the cells.

Repeated ECSs, especially polydiurnal repeated ECSs, are followed by characteristic changes differing in several biochemical parameters from those caused by a single ECS. There were also differences between the biochemical changes

caused by repeated daily ECSs and those caused by repeated
polydiurnal ECSs related to the different cerebral regions
and to the different cerebral compartments. These differences point out that repeated convulsions activate biochemical mechanisms on the cellular, subcellular and regional levels, as well as on the blood-brain level, in
order to reorganize function according to differing requirements. Previous studies particularly emphasize the
effects of repeated ECSs on the synapses and the changes
in the release of many diverse neurotransmitters, which
could be related to the corresponding clinical and behavioral effects (8, 15, 46, 62, 75). Some authors suggest
the possibility of tissue and especially membrane damage
in these conditions (6, 26). According to our experiments,
polydiurnal repeated convulsions significantly increase
the activity of membrane enzymes (AChE, ATPase) and of the
enzymes with higher activity in microvessels (BuChE) in
comparison not only with control values but also with the
findings after a single ECS and daily repeated convulsions
(Figs. 6 and 7). It is quite apparent that membrane occlusion of the enzymes bound to membrane fragments can
limit their manifestation of activity, and that factors
that disturb the membrane equilibrium, including ECS,
influence the enzymatic activity (1). The increase of
enzymatic activity in both compartments (brain parenchyma
and microvessels) of not just membrane enzymes but particularly of BuChE points out that the release of occluded
enzymes from the occluded pool (29), although significant,
is not the most important factor in the maintenance of a
high level of enzymatic activity.

Polydiurnal repeated ECSs create completely new relationships in the tubulin level with distinct regional
characteristics. While a single ECS and the repeated convulsions during one day decreased the tubulin level in Th,
Hippo and Cer, polydiurnal repeated ECSs caused an increase
in the level of this protein, this increase being most
probably the result of increased synthesis. Regulatory
mechanisms of protein synthesis in such conditions were
triggered by the numerous imbalances in the homeostatic
mechanisms produced by each successive ECS. The absence
of changes in Cx and N.Cd and the persisting low tubulin
level in MP in such conditions not only emphasize the
selective regional response to ECS, but also indicate
that specific cerebral regions are particularly important
in the regulation of seizure-dependent functional conditions (61, 75, 78, 79). Studies of the kind reported can
be the basis for further molecular-anatomical studies of
behavior.

REFERENCES

1. Adams, H.E., Hoblit, P.R. and Sutker, P.B. (1969): Electroconvulsive shock, brain acetylcholinesterase activity and memory. Physiol. Behav. 4: 113-116.

2. Aird, R.B. (1958): Clinical correlates of electroshock therapy. Arch. Neurol. Psychiatry 79: 633-639.

3. Aird, R.B. and Strait, L. (1944): Protective barriers in the central nervous system: An experimental study of trypan-red. Arch. Neurol. Psychiatry 51: 54-66.

4. Angel, C. and Roberts, A.J. (1966): Effect of electro-shock and anti-depressant drugs on cerebrovascular permeability to cocaine in the rat. J. Nerv. Ment. Dis. 142: 376-380.

5. Banks, P., Mayor, D., Mitchell, M. and Tomlinson, O. (1971): Studies on the translocation of noradrenaline-containing vesicles in post-ganglionic sympathetic neurons in vitro. Inhibition of movement by colchicine and vinblastine and evidence for involvement of axonal microtubules. J. Physiol. 216: 625-639.

6. Bazán, N.G. (1971): Changes in free fatty acids of brain by drug-induced convulsions. Electroshock and anaesthetic. J. Neurochem. 18: 1379-1385.

7. Bjerner, B., Broman, T. and Swenson, A. (1944): Tier-experimentelle Untersuchungen über Schädigungen der Gefässe mit Permeabilitätsstörungen und Blutengen im Gehirn bei Insulin, Cardiazol und Elektroschockbehandlung. Acta Psychiat. Neurol. Scand. 19: 431-452.

8. Blachly, P.H. and Gowing, D. (1966): Multiple electro-convulsive treatment. Comprehensive Psychiatry 7: 100-109.

9. Bonting, S.L., Simon, K.A. and Howkins, N.M. (1961): Studies on sodium-potassium-activated adenosine tri-phosphatase. I. Quantitative distribution in several tissues of the cat. Arch. Biochem. Biophys. 95: 417-423.

10. Brajan, J. and Wilson, L. (1971): Are cytoplasmic microtubules heteropolymers? Proc. Natl. Acad. Sci. U.S.A. 1762-1977.

11.   Brennan, R.W., Petito, K.C. and Porro, R.S. (1972):
      Single seizures cause no ultastructural changes in
      brain. Brain Res. 45: 574-579.

12.   Cavanagh, J.B. and Thomson, R.H.S. (1954): Demyelin-
      ation. Brit. Med. Bull. 10: 47-51.

13.   Clark, G. and Sarkaria, D.S. (1958): Acid fuchsin
      convulsions and electroshock in the mouse. J. Neuro-
      pathol. Exp. Neurol. 17: 612-619.

14.   Cohen, H.B., Duncan, R.F. and Dement, W.C. (1967):
      Sleep: The effect of electroconvulsive shock in cats
      deprived of REM sleep. Science 156: 1646-1648.

15.   Cotman, C.W., Banker, G., Zornetzer, S.F. and McGaugh,
      J. (1971): Electroshock effects on brain protein syn-
      thesis: Relation to brain seizures and retrograde am-
      nesia. Science 173: 454-456.

16.   Crone, C. (1965): Facilitated transfer of glucose
      from blood to brain. J. Physiol. (London) 181: 103-
      113.

17.   Dahlström, A. (1971): Axoplasmatic transport with
      particular respect to adrenergic neurons. Physiol.
      Trans. R. Soc. London (Biol.) 261: 325-358.

18.   Dawson, R.M.C. and Richter, D. (1950): Effect of
      stimulation of the phosphate esters of the brain.
      Am. J. Physiol. 160: 203-211.

19.   DeRobertis, E., Alberic, M. and Delores-Arnaiz, G.R.
      (1969): Astroglial swelling and phosphohydrolases in
      cerebral cortex of metrazol convulsant rat. Brain
      Res. 12: 461-466.

20.   Djuričić, B.M., Rogač, Lj., Spatz, M. Rakić, Lj. and
      Mršulja, B.B. (1977): Brain microvessels. I. Enzymic
      activities. In: Pathology of Cerebrospinal Microcir-
      culation, J. Cervos-Navarro, E. Betz, and R. Wüllen-
      weber (eds.), pp. 197-205, Raven Press, New York.

21.   Dorfmann, L.J. and Jarvik, M.E. (1968): A parametric
      study of electroshock-induced retrograde amnesia in
      mice. Neuropsychology 6: 373-380.

22.   Dornbush, R.L. (1973): Memory and induced ECT con-
      vulsions. Semin. Psychiatry 4: 47-54.

23. Duffy, T.E., Howse, D.C. and Plum, F. (1975): Cerebral
     energy metabolism during experimental status epilep-
     ticus. J. Neurochem. 24: 925-934.

24. Dunn, A. (1971): Brain protein synthesis after elec-
     troshock. Brain Res. 35: 254-259.

25. Dunn, A. (1973): The dependence of brain ATP content
     on cerebral electroshock current. Brain Res. 61: 442-
     445.

26. Dunn, A., Guiditta, A., Wilson, J.E. and Glassman, E.
     (1974): The effect of electroshock on brain RNA and
     protein synthesis and its possible relationship to
     behavioral effects. In: Psychobiology of Convulsive
     Therapy, M. Fink, S. Kety, J. McGaugh and T.A. Willams,
     (eds.), pp. 185-197, V.A. Winston and Sons, Washington,
     D.C.

27. Essman, W.B. (1972): Neurochemical changes in ECS and
     ECT. Semin. Psychiatry 4: 67-77.

28. Essman, W.B., Baker, L.A. and Keller, A. (1972):
     Alterations of cholinergic neurons with electrocon-
     vulsive shock. Fed. Proc. 31: 249.

29. Essman, W.B. (1973): Neurochemistry of Cerebral Elec-
     troshock. Spectrum Publ., Flushing, New York.

30. Essman, W.B. (1973): Neuromolecular modification of
     experimental-induced retrograde amnesia. Confina
     Neurol. 35: 1-22.

31. Feit, H. and Barondes, S.H. (1970): Colchicine-binding
     activity in particulate fractions of mouse brain.
     J. Neurochem. 17: 1355-1367.

32. Ferrendelli, J.A. and McDougal, D.B., Jr. (1971):
     The effect of electroshock on regional CNS energy
     reserves in mice. J. Neurochem. 18: 1197-1205.

33. Giacobini, E. (1964): Metabolic relationship between
     glia and neurons studied in single cells. In:
     Morphological and Biochemical Correlates of Neuronal
     Activity, M.M. Cohen and R.S. Snider (eds.), pp. 15-
     56, Harper and Row, New York.

34. Gibbons, I.R. (1966): Studies on the adenosine-triphos-
     phatase activity of 14S and 30s dynein from cilia of
     Tetrahymena. J. Biol. Chem. 241: 5590-5596.

35.   Gillespie, E. (1971): Colchicine binding in tissue
      slices: decrease by calcium and biphasic effect of
      A-3´,5´-monophosphate. J. Cell Biol. 50: 544-551.

36.   Goldstein, G.W., Wolinski, J.S., Csejey, J. and
      Diamond, I. (1975): Isolation of metabolically active
      capillaries from rat brain. J. Neurochem. 25: 715-717.

37.   Gozsy, B., Kato, L., Roy, P.B., Grog, V. and
      Lallonde, M. (1965): Investigations into the mechanism
      of electroconvulsive shock; part I. Int. J. Neuro-
      psychatry 1: 623-625.

38.   Greig, M.E. and Mayberry, T.C. (1951): The relation-
      ship between cholinesterase activity and brain perme-
      ability. J. Pharmacol. 102: 1-4.

39.   Greig, M.E. and Holland, W.C. (1949): Increased perme-
      ability of hemoencephalic barrier produced by physo-
      stigmine and acetylcholine. Science 110: 237-238.

40.   Greig, M.E., Holland, W.C. and Mayberry, T.C. (1951):
      The relationship between cholinesterase activity and
      brain permeability. J. Pharmacol. Exp. Therap. 102:
      1-4.

41.   Howse, D.C., Caronna, J.J., Diffy, T.E. and Plum, F.
      (1974): Cerebral energy metabolism, pH, and blood
      flow during seizure in the cat. Am. J. Physiol. 227:
      1444-1451.

42.   Iuvone, M.P., Boast, C.A., Gray, H.E. and Dunn, A.
      (1977): Pentylentetrazol: Inhibitory avoidance be-
      havior, brain seizure activity and ^{3}H-lysine incor-
      poration into brain proteins of different mouse
      strains. Behav. Biol. 21: 236-250.

43.   James, K.A.C., Bray, J.J., Morgan, I.G. and Austin, L.
      (1970): The effect of colchicine on the transport of
      axonal protein in the chicken. Biochem. J. 117: 767-
      771.

44.   Joó, F. and Csillik, B. (1966): Topographical corre-
      lation between the hematoencephalic barrier and the
      cholinesterase activity of brain capillaries. Exp.
      Brain Res. 1: 147-151.

45.   Joó, F. and Várkonyi, T. (1969): Correlation between
      the cholinesterase activity and capillaries and the
      blood-brain barrier in the rat. Acta Biol. Acad. Sci.
      Hung. 20: 359-372.

46. Kety, S.S., Javoy, F., Thierry, A.M., Julon, L. and Glowinski, J. (1967): A sustained effect of electro-convulsive shock turnover of norepinephrine in the central nervous system of the rat. Proc. Natl. Acad. Sci. U.S.A. 58: 1249-1254.

47. King. L.J., Lwory, O.H., Passonneau, J.V. and Venson, V.V. (1967): Effects of convulsants on energy reserves in the cerebral cortex. J. Neurochem. 14: 599-611.

48. King, L.J., Schoepffe, G.M., Passonneau, J.V. and Wilson, S. (1967): Effect of electrical stimulation on metabolites in brain of decapitated mice. J. Neurochem. 14: 613-618.

49. King, L.J., Welb, O.L. and Carl, J. (1970): Effects of duration of convulsions on energy reserves of the brain. J. Neurochem. 17: 13-18.

50. Koella, G. (1954): The histochemical localization of cholinesterase in the central nervous system of the rat. J. Comp. Neurol. 100: 211-228.

51. Kreutzberg, G.W. (1969): Neuronal dynamic and axonal flow. IV. Blockade of intra-axonal enzyme transport by colchicine. Proc. Natl. Acad. Sci. U.S.A. 62: 722-728.

52. Kriendler, A. (1965): Biochemical aspects of the seizure in convulsive disorder. In: Progress in Brain Research. Elsevier, Amsterdam, 19: 168-181.

53. Ladish, W., Steinhauff, N. and Matussek, N. (1969): Chronic administration of electroconvulsive shock and norepinephrine metabolism in the rat. Psycho-pharmacology 15: 296-304.

54. Laemmli, U.K. and Favre, M. (1973): Maturation of the head of bacteriophage T_4. I. DNA packing events. J. Mol. Biol. 80: 575-599.

55. Lee, J.C. and Olszewski, J. (1961): Increased cere-brovascular permeability after repeated electroshock. Am. J. Psychiatry 104: 765-770.

56. Lee-Teng, E. (1969): Retrograde amnesia in relation to subconvulsive and convulsive current in the chick. J. Comp. Physiol. Psychol. 67: 135-139.

57.   Lovell, R.A. (1971): Some neurochemical aspects of
      convulsion. In: Handbook of Neurochemistry. A. Lajtha
      (ed.), Plenum Press, New York, 6: 63-102.

58.   Lowry, O.H. and Passonneau, J.V. (1964): The relation-
      ship between substrate and enzyme of glycolysis in
      brain. J. Biol. Chem. 239: 39-42.

59.   Lust, W.D. and Passonneau, J.V. (1976): Cyclic nucle-
      otides in murine brain: Effect of hypothermia on
      adenosine 3´,5´-monophosphate, glycogen and phos-
      phorylase glycogen synthesis and metabolites following
      maximal electroshock or decapitation. J. Neurochem.
      26: 11-16.

60.   McDonnell, P.C. and Greengard, O. (1974): Enzymes in
      intracellular organelles of adult and developing rat
      brain. Arch. Biochem. Biophys. 163: 644-649.

61.   McGaugh, J.L. (1968): Electroconvulsive shock. Int.
      Encycl. Soc. Sc. 5: 21-25.

62.   McGaugh, J.M. (1974): Electroconvulsive shock: Effects
      on learning and memory in animals. In: Psychobiology
      of Convulsive Therapy, M. Fink, S. Kety, J. McGaugh
      and T.A. Williams (eds.), pp. 85-97, V.H. Winston and
      Sons, Washington, D.C.

63.   McIlwain, H. (1963): Chemical Exploration of the Brain.
      A Study of Cerebral Excitability and Ion Movement.
      Elsevier, Amsterdam.

64.   Milligan, W.L. (1946): Psychoneurosis treated with
      electrical convulsions. Lancet 215: 516-520.

65.   Minard, F.N. and Davis, R.V. (1962): The effects of
      electroshock on the acid-soluble phosphates of rat
      brain. J. Biol. Chem. 237: 1283-1289.

66.   Mršulja, B.B., Djuričić, B.J., Mršulja, B.J., Rogač,
      Lj., Spatz, M. and Klatzo, I. (1977): Brain micro-
      vessels II. The effect of ischemia and dehydroergo-
      toxine on the enzymic activity. In: Pathology of Cere-
      brospinal Microcirculation, J. Cerrvos-Navarro, E.
      Betz, and R. Wüllenweber (eds.), pp. 207-213, Raven
      Press, New York.

67.   Mršulja, B.B., Mršulja, B.J., Fujimoto, T. and Klatzo,
      I. (1976): Isolation of brain capillaries: a simplified
      technique. Brain Res. 110: 361-365.

68. Mussacchio, J.M., Juon, L., Kety, S.S. and Glowinski, J. (1969): Increase in rat brain tyrosine hydroxylase activity produced by electroconvulsive shock. Proc. Natl. Acad. Sci. U.S.A. 63: 1117-1119.

69. Nachmansson, D. (1959): Chemical and Molecular Basis of Nerve Activity. Academic Press, New York.

70. Nathanson, J.A. and Greengaard, P. (1976): Cyclic nucleotides and synaptic transmission. In: Basic Neurochemistry, G.J. Siegel, R.W. Albers, R. Katzman and B.W. Agranoff (eds.), pp. 246-262, Little, Brown, Boston.

71. Ocks, S. (1975): Axoplasmatic transport. In: The Nervous System, D.B. Tower (ed. in chief), Vol. 1, The Basic Neurosciences, R.O. Brady (ed.), pp. 137-146, Raven Press, New York.

72. Oldendorf, W.H. (1971): Brain uptake of radiolabeled amino-acids, amine and hexose after arterial injection. Am. J. Physiol. 221: 1629-1639.

73. Ottoson, J.O. (1968): Psychological or physiological theories of ECT. Int. J. Psychiatry 5: 170-174.

74. Plumm, F., Posner, J.B. and Troy, B. (1968): Cerebral metabolic and circulatory responses to induced convulsions in animals. Arch. Neruol. 18: 1-13.

75. Pogodaev, K.I. (1964): Biochimia Epilepticheskovo Pristupa. Medicina, Moscow.

76. Pryor, G.T. and Otis, L.S. (1969): Brain biochemical and behavioural effects of 1, 2, 4 or 8 weeks electroshock treatment. Life Sci. 8: 387-399.

77. Pryor, G.T. (1974): Effect of repeated ECS on brain weight and brain enzymes. In: Psychobiology of Convulsive Therapy, M. Fink, S. Kety, J. McGaugh and T.A. Williams (eds.), pp. 185-197, V.H. Winston and Sons, Washington, D.C.

78. Rakic, Lj.M., Buchwald, N.A. and Wyers, E.Y. (1962): Induction of seizures by stimulation of the caudate nucleus. EEG Clin. Neurophysiol. 14: 809.

79. Rakić, Lj.M. (1966): Cortical inhibition and subcortical inhibitory influences. In: Impact of Basic Science on Medicine, B. Shapiro and M. Prywes (eds.), pp. 298-305, Academic Press, New York, London.

80. Ray, O.S. and Barrett, R.J. (1969): Disruptive effects of electroconvulsive shock as a function of current levels and mode of delivery. J. Comp. Physiol. Psychol. 67: 110-116.

81. Rose, S.P.J., Sinha, A.K. and Jones-Lecointe, A. (1976): Synthesis of tubulin-enriched fraction in rat visual cortex is modulated by dark-rearing and light-exposure. FEBS Lett. 65.2: 135-139.

82. Rosenblatt, S., Chanley, J.D., Sobotka, H. and Kaufman, M.R. (1960): Interrelationship between electroshock, the blood-brain barrier and catecholamines. J. Neurochem. 5: 172-176.

83. Seiler, N. (1969): Enzymes. In: Handbook of Neurochemistry. Vol. I: Chemical Architecture of the Nervous System, A. Lajtha (ed.), pp. 325-468, Plenum Press, New York.

84. Seite, R., Noel, M. and Vuillet-Luciani, J. (1973): Effect of electrical stimulation on nuclear microfilaments and microtubules of sympathetic neurons submitted to cycloheximide. Brain Res. 50: 419-423.

85. Silver, A. (1974): The Biology of Cholinesterase. North Holland Publ. Co., Amsterdam.

86. Shelanski, M.L. (1973): Microtubules. In: Proteins of the Nervous System, D.J. Schneider (ed.), pp. 227-241, Raven Press, New York.

87. Smallman, B.N. and Pal, R. (1957): The activity of intra-cellular distribution of choline acetylase in insect nervous tissue. Bul. Entomol. Soc. Amer. 8: 25.

88. Sokoloff, L. (1969): Cerebral blood flow and energy metabolism. In: Basic Mechanisms of the Epilepsies, H.H. Jasper, A.A. Ward and A. Pope (eds.), pp. 639-646, Little, Brown, Boston.

89. Tower, D.B. (1960): Neurochemistry of Epilepsy. C.C. Thomas, Springfield, Illinois.

90. Van Buskirk, R. and McGaugh, J.L. (1974): Pentylene-tetrazol-induced retrograde amnesia and brain seizure in mice. Psychopharmacology 40: 77-90.

91. Vinitsky, I.M. and Abuladze, G.V. (1971): Retrograd-naya amneziya vyzyvemaya karazolom na fone destviya narkoza. Zh. Vyssh. Ner. Deyatel. Im. I.P. Pavlova 21: 572-575.

92. Weisenberg, R.C., Borrisy, G.G. and Taylor, E.W. (1968): The colchicine binding protein in mammalian brain and its relation to microtubules. Biochemistry 7: 4466-4479.

93. Weisenberg, R.C. and Timasheff, S.F. (1970: Aggrega-tion of microtubule subunit protein. Effect of diva-lent cations, colchicine and vinblastine. Biochemistry 9: 4110-4116.

94. White, A., Handler, P. and Smith, E.L. (1973): Prin-ciples of Biochemistry. McGraw-Hill, New York; Kogakusha, Ltd., Tokyo.

95. Wolfe, L.C. and Elliott, K.A.A. (1962): Chemical studies in relation to convulsive conditions. In: Neurochemistry, K.A.C. Elliott, J.H. Page and J.H. Quastel (eds.), pp. 694-727, C.C. Thomas, Springfield, Illinois.

INITIATION, PROPAGATION AND ARREST OF SEIZURES

D.M. Woodbury and J.W. Kemp

Department of Pharmacology
University of Utah College of Medicine
Salt Lake City, Utah 84132, U.S.A.

Ideally, the best model to study the mechanisms of seizures is man himself. However, other than EEG recordings, depth electrode studies and occasional biopsies of epileptogenic foci this is difficult to do. This is the case because detailed analysis of the basic neurophysiological and biochemical mechanisms of seizures involves work on the brain itself and sometimes removal of brain samples and this is not possible to do in humans. Consequently, experimental models of epilepsy must be used instead. The relevance of such studies is clearly related to the degree to which the experimental models approximate the disease in humans. The ideal model for studying the basic mechanisms of seizures is one that closely approximates human epilepsy and yet is readily available, inexpensive and easy to work with. No model at present meets all these criteria, but all models are potentially useful for studying at least one aspect of the mechanisms of seizures. However, because of the variety of clinical types of epilepsy with different types of onset and manifestations, it is essential to have models in animals that mimic the different forms in man, that is, focal and generalized seizure types. It is the purpose of this chapter to discuss some of the various experimental models and how study of them helps to elucidate the basic mechanisms of seizures. It is important to note that the basic properties of seizures, regardless of etiology are similar and any one model can be used to study these properties (2). Nevertheless, it is imperative to study the mechanisms by which the seizure in each model

Table I EXPERIMENTAL MODELS OF EPILEPSY USEFUL FOR
        ELUCIDATION OF MECHANISMS OF SEIZURE INITIATION,
        SPREAD AND ARREST

I MODELS FOR INDUCING LOCAL EPILEPTOGENIC ACTIVITY

   A. ACUTE MODELS
      Topical Convulsant Metals: Cobalt, Tungstic
                                         Acid Gel
      Topical Freezing
      Topical Convulsant Drugs: Penicillin, Ouabain,
                                Strychnine
                                Conjugated Estrogens,etc.
      Focal Electrical Stimulation

   B. CHRONIC MODELS
      Topical Convulsant Metals: Alumina Cream

II MODELS FOR INDUCING GENERALIZED EPILEPTOGENIC ACTIVITY

   A. SENSORY-PRECIPITATED MODELS (GENETIC)
      Photogenic Seizures-Papio Papio Audiogenic
                                       Seizures-Mice

   B. ELECTRICALLY INDUCED SEIZURES

   C. CHEMICALLY INDUCED SEIZURES
      Stimulate Excitatory Systems-Phenylenetetrazol
      Block Inhibitory Systems
         Block gycine receptors-strychnine
         Block GABA receptors-picrotoxin, penicillin,
                                          bicuculline
      Block Electrolyte Transport Systems in Neurons
                        and/or Glia and/or
      Elevate Extracellular Ion Concentrations
         Cation-Ouabain, Lithium, High Potassium
                      Perfusion
         Anion-Thiocyanate, Perchlorate, Low Chloride
                      Perfusion, High Bicarbonate

   D. WITHDRAWAL SEIZURE MODELS
      ACUTE   - CO_2 Withdrawal Seizures
      CHRONIC   - Barbiturate Withdrawal Seizures
                 - Alcohol Withdrawal Seizures
                 - Benzodiazepine Withdrawal Seizures
                 - Bromide Withdrawal Seizures

is initiated because the mechanisms may be different for
each one and give us information on the various ways
in which seizures start.

A simple classification of the various experimental models of epilepsy is presented in Table I. The models shown are those that have been most used for study of seizure mechanisms and, therefore, ones in which the majority of our current knowledge of the neuropathological, neurophysiological and biochemical properties of seizures has been obtained. Two classes of models are used: Those for inducing local epileptogenic activity acutely and chronically, and those for inducing generalized epileptogenic seizures. By definition, human epilepsies are recurrent self-sustained paroxysmal disorders of brain function characterized by excessive discharge of cerebral neurons. Among the many models for producing local epileptiform seizures in animals, the only one that fulfills the criterion of prolonged "spontaneous" recurrence is the topical or intracerebral application of alumina gel (56). It produces a long-lasting chronic epileptogenic lesion and it is, therefore, the method of choice for the reproduction of a lesion resembling very closely that seen in man. However, it is most useful in primates and requires elaborate secondary techniques to obtain good results. This is very costly; hence for biochemical studies requiring tissue samples, less elaborate and expensive acute models have been used and are satisfactory for studying seizure mechanisms. The agents of choice for such acute models, as shown, are local freezing, cobalt or penicillin. Local electrical stimulation with implanted electrodes in either cortical or subcortical sites is particularly useful for studying the neurophysiological mechanisms of seizures. Biochemical and neuropathological studies of generalized seizures are often best done in genetic models of epilepsy such as photomyoclonus in baboons (Papio papio) and audiogenic seizures in mice, and by use of chemically induced generalized seizures such as pentylenetetrazol, strychnine and picrotoxin, which usually act by different synaptic mechanisms involving excitatory and inhibitory systems. In this Chapter the emphasis will be on the correlation of neuroanatomical, neurophysiological and biochemical data obtained in various local and generalized models in a attempt to elucidate some of the basic mechanisms by which seizures occur.

Briefly, the pathophysiological events in epilepsy can be considered in three steps: I. Initiation: (A) the abnormalities that generate the discharge of an epileptogenic focus and (B) the local spread and the changes which precipitate the interictal activity into a seizure. II. The Propagation of discharge during a major seizure.III. The events that arrest or ameliorate the seizure.

I. INITIATION

A. ABNORMALITIES THAT GENERATE THE DISCHARGE OF
AN EPILEPTOGENIC FOCUS

1. Seizures with local onset

a. Pathological changes. The pathological changes
that occur in epileptogenic foci produced by various
experimental focal procedures are characterized by glial
proliferation, loss of nerve cells, and impairment in
local cerebral circulation (44). An impressive feature
of such lesions observed in epileptogenic foci in man and
in experimental animals by Westrum et al. (60) was, in
addition to the glial proliferation, the marked abnormalities in dendritic structure. The dendritic trees were
deformed and denuded of spines as compared to normal tissue. These changes were thought to result in a denervation
type of supersensitivity as a result of reduced synaptic
input (partial deafferentation) (56). There was an overall decrease in neuronal elements and the remaining neurons were of intermediate to small size, an observation
that was thought to be important, since small cells are easier to discharge (30). There was less dendritic branching,
and the course of the apical shaft and its branches was
frequently distorted in different planes. Dendrites were
marked by irregular varicose-like swelling on both the
large shafts and the finer branches. The dendritic surfaces were relatively smooth along most of their course
and were almost completely divested of their normal dendritic spines. The changes extended out from the focus a
short distance and then gradually disappeared. This geo-
graphical distribution of the anatomical changes correlated well with the physiological changes where abnormal
patterns of firing showed a similar transition to normal
activity as one moved radially from the focus. Similar
changes have been observed by Scheibel and Scheibel (47).
Also Brown (9) has reported that electron microscopic
studies in man of temporal lobe removed at operation show
evidence of ongoing degeneration of boutons, spines and
axons in these cortices. Histological studies in the chronic epileptic monkey by Harris (27) indicate that continuing clinical seizures are associated with continuing neuronal damage. Both axonal and dendritic degeneration are
continuously occurring. This continuing neuronal degeneration is related to the continuing seizures and their
frequency and severity rather than to other variables.
This is in line with the data of Westrum et al. described
above. Thus it appears that the occurrence of clinical
seizures is not innocuous. The prudent strategy, as stated

by Ward (57), "would appear to consist of aggressive medical therapy in the attempt to achieve complete control of clinical seizures as quickly as possible." If this does not result in control then surgical therapy is necessary to ablate the lesion in order to prevent "continuing neuronal damage in the vicinity of the focus [which] may add to the pathophysiologic alterations and augment the total epileptic process, thereby making future seizure control more difficult by any means."

Pathological changes in the secondary focus projected from the primary side are minimal as compared with the primary focus. Mild glial proliferation is present (53, 54, 59, 61), as well as some small changes in the dendritic spines. Both these changes appear to be secondary to transcallosal degeneration as a result of the lesion in the primary area (59, 61). Often no changes or only a mild glial proliferation are seen in secondary foci that have become independently active and it is evident from the morphological as well as the neurophysiological evidence discussed below that partial denervation is not a necessary condition for establishing a chronic epileptic focus. The role of glial cell proliferation in epileptogenic foci will be discussed below.

b. Neurophysiological events (See excellent summary by Merlis, 41). It is important to discuss the main neurophysiological changes that occur in epileptogenic foci of animals and man for both the primary and secondary foci. A number of different experimental models of the local epileptic process have been studied intensively. No matter what method is used, acute topical application of chemical agents such as strychnine, pentylenetetrazol or penicillin, or more chronic agents such as freezing or alumina cream, all are characterized by the sporadic occurrence of large potentials recorded from the cortical surface or scalp as spikes or sharp waves. Coincidental with these there is, in some cells, a paroxysmal depolarization shift (PDS), upon which there may appear a burst of axon spikes (Fig. 1). The PDS is a graded potential, frequently so large as to inactivate the spike-generating mechanism by excessive depolarization. It is frequently followed by a prominent hyperpolarization of IPSP, attributed to recurrent inhibition. Extending outward from the focus, hyperpolarization becomes more prominent and, at the periphery, many of the neurons generate IPSPs which are not preceded by PDS. This is called "surround inhibition" and it presumably serves to limit the spread of the discharge. Neurons generating PDSs appear to be otherwise normal. In the intervals between paroxysmal

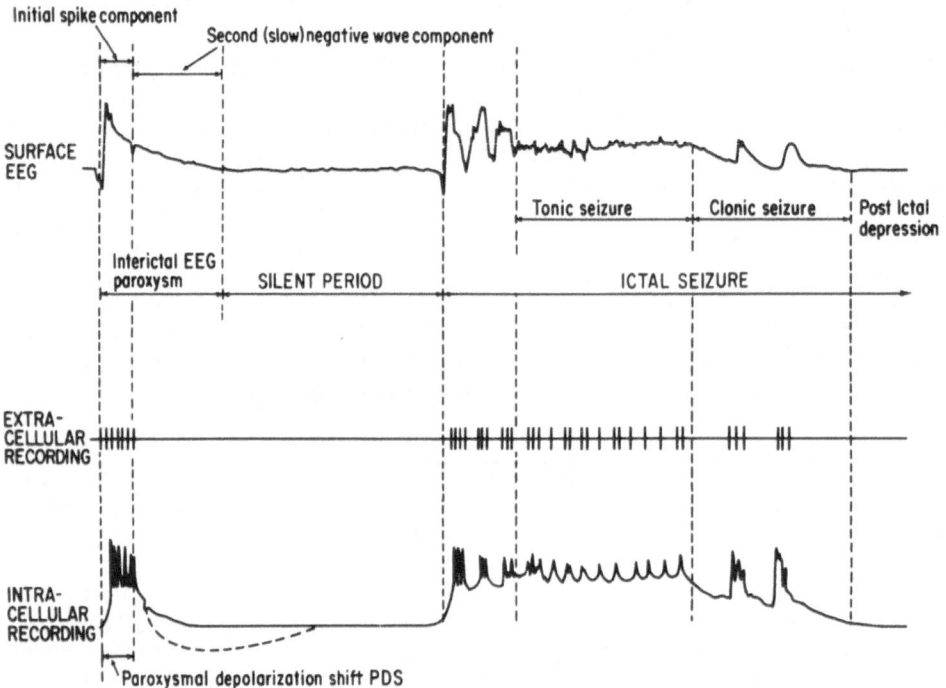

Fig. 1. Schematic diagram of relations between EEG discharge and microelectrode recordings from a neuron in a penicillin focus (from Ayala et al., 5). Used by permission of publisher.

discharges they may generate normal EPSPs and spike potentials. A PDS is thought to be a large EPSP as a result
of synaptic activation, by an as yet unknown mechanism.
Part of the effect may be due to the depression of recurrent inhibition. However, a PDS is not present in all epileptogenic discharges. This is especially the case in the
chronic focus produced by alumia cream or by freezing. In
freezing lesions PDSs were recorded by Goldensohn (24) is
only about 14 per cent of neurons and even less frequently
by Ward (56) in the alumina cream focus. The chronic focus
appears to be less active than the acute one, probably for
the reason, as discussed below, that more glial cell proliferation has occurred. The partial denervation sensitivity as a result of changes in dendritic structuresas described by Ward and colleagues, i.e., "epileptic neurons,"
and which result in generation of repetitive firing with
abnormal sites of spike generation can explain the sequence of events which occur in the epileptic focus. Final
proof of the hypothesis, however, is still lacking.

The interictal cortical spike may propagate via association tracts to the homologous region of the opposite
hemisphere and evoke a secondary spike. If this secondary
spiking becomes independently active, as it often does in
chronic preparations, it is called mirror focus and may
persist even after the primary focus becomes inactive or
is ablated (61). Mirror foci can also be established in limbic and other subcortical nuclei that are synaptically related to a primary subcortical focus. The
activity of neurons in the secondary focus, recorded by
microelectrodes, does not differ from that in the primary
focus. It is, therefore, evident that at least for the
mirror focus partial denervation is not a necessary condition for establishing a chronic epileptic focus. The
secondary focus seems to result in some manner from bombardment by neurons of the primary focus so that they become independently epileptic. This phenomenon is related to
kindling, as originally described by Goddard (22) and Goddard
et al. (23), which is the development of an epileptic response
to repeated minimal electrical stimulation of various regions
of the brain. The mechanism(s) of kindling is not known.
A schematic diagram of the relations between EEG discharge and microelectrode recordings from a neuron in a penicillin focus is shown in Fig. 1.

c. Biochemical changes. A summary of the biochemical
changes that have been observed in epileptogenic foci
(primary and secondary) taken from animals and man is
shown in Table II. Changes in the primary focus are those

Table II   SUMMARY OF BIOCHEMICAL CHANGES IN EPILEPTOGENIC FOCI OF ANIMAL AND MAN

	PRIMARY FOCUS	SECONDARY FOCUS
ELECTROLYTES		
Water	increased	very slightly increased
Na^+	increased	slightly increased
K^+	decreased	slightly decreased
Ca^{++}	increased	no change
Mg^{++}	increased	no change
Cl^-	--	--
$H^{14}CO_3^-$ space	decreased	decreased
Na, K^+-ATPase	increased (8 hr) normal (24 hr) or decreased	increased
CARBONIC ANHYDRASE	increased	increased
SYNAPTOSOME K^+ TRANSPORT	decreased	decreased
DNA	increased	slightly increased
RNA SYNTHESIS	decreased	decreased
PROTEINS	decreased	little change
AMINO ACIDS		
Glutamic Acid	decreased	decreased or no change
Glutamine	increased	no change
Aspartic Acid	decreased	decreased or no change
GABA	decreased	decreased or no change
Glycine	increased	no change
Taurine	decreased	
Alanine	no change	
Serine	no change	

		Area adjacent to lesion	
GLUTAMIC ACID DECARBOXYLASE	decreased		transient small decrease
CHOLINESTERASES Acetyl	increased		
CHOLINEACETYLASE	decreased		decreased – no change
CYCLIC AMP	increased		increased
CO$_2$ FIXATION	increased		increased
LACTIC DEHYDROGENASE	decreased		increased
OXIDATIVE METABOLISM	no change or decrease		no change
LIPIDS	decrease		
Total Lipid	decreased	no change	no change
Free Cholesterol	decreased	no change	no change
Total Phospholipid	decreased	no change	no change
Cholesterol Esters	increased	increased	no change
Triglycerids	increased	no change	no change
Phosphatidyl Ethanol-amine	decreased	decreased	no change
Oleic, Arachidonic & Nervonic acids	decreased	decreased	no change
Lignoceric acid	decreased	increased	slightly increased
Palmitoleic Acid	increased	decreased	decreased
Ganglioside Sialic Acid	no change		increased

either in the actual area of the lesion or in the immediately surrounding area, where maximal paroxysmal discharges are measured by microelectrode recording. The
changes in the area of the lesion are those to be expected from destruction of neurons and their replacement by
glial cells. Thus, there is a decrease in protein, usually but not always of oxidative metabolism, RNA synthesis,
K^+, total lipid and phospholipid, cholesterol and various
fatty acids due to loss of neurons and an increase in Na^+,
water, probably Cl^-, Ca^{++}, Mg^{++}, carbonic anhydrase, DNA,
CO_2 fixation, and some fats, changes that reflect the glial proliferation, since most of the changes in these substances involve glial cell metabolism. Thus, glial edema
occurs as reflected in the increase in Na^+, Cl^- and water.
The rise in DNA reflects the increase in glial cell number inasmuch as neurons do not divide in the adult. Also
the glial cell proliferation is reflected by the increase
in carbonic anhydrase, an enzyme found only in glial cells
(19) or in glial cell products (myelin). This enzyme is
also involved in CO_2 fixation, hence the reason for the
increase in this parameter. The changes in the amino acids,
cyclic AMP, acetylcholinesterase, and some specific fatty
acids in the primary focus appear to reflect changes in
synaptic function since all these are involved in neuro-
transmission or in synaptic membrane function. Of particular interest are the decrease in glutamic and aspartic
acids (excitatory neurotransmitters) and of GABA (an inhibitory transmitter) and the increase in glycine (an inhibitory neurotransmitter in the spinal cord) (4, 51, 52).
The amino acid changes are found both at the lesion site
and in the immediately adjacent active discharge area and
occur prior to the seizure. There was a clear correlation
between the severity of epilepsy and the extent by which
the concentration of these amino acids deviated from normal. These data were interpreted to mean that epileptogenic cortex suffers from impaired energy metabolism. Glutamic and aspartic acids are linked to the tricarboxylic
acid cycle by transamination and dehydrogenation, and
thus serve as secondary energy substrates of the CNS and
also are intermediate steps in the transfer of glucose
carbons into amino acids and protein. It is thought that
glucose oxidation and amino acid metabolism become uncoupled during chronic epileptogenic activity. The imbalance in taurine and glycine also indicates inhibition of
protein synthesis, as has been observed for such primary
focal lesions (Table II). It appears quite probable that
the later decrease in GABA content, which is secondary to
the low content of its precursor, glutamic acid, throughout the cortex as well as at the primary focus is an im-

portant factor in maintaining the hyperexcitable condition in the brain of animals in which chronic epileptogenic seizures have been produced. This later decrease in GABA concentration would release its strong inhibitory effect on the cortex and result in an increase in excitability, a change which can probably explain the diminished inhibitory control seen in a focus and which appears to contribute to the spread of epileptic discharge. The prior decrease in glutamic acid content in the lesion area appears to be due to increased liberation of this dicarbonic amino acid early in the sequence of events leading to a focal discharging lesion and is probably partly responsible for the development of a pool of hyperactive neurons (34).

The origin of the increase in glycine is not known but certain data suggest that it may result from both inhibition of protein synthesis and its enhanced catabolism that results from the neuronal hyperexcitability in the focus. The increase in glutamine concentration similarly seems to be the result of ongoing hyperexcitability and enhanced protein breakdown in the focal region.

In both the primary and secondary focus Escueta and Appel (16) have shown that synaptosomes isolated from these areas have decreased ability to transport K^+ activity. Thus these epileptic areas have impaired K^+ metabolism and this coupled with reports of decreased Na^+, K^+-ATPase activity in the focus indicated defects in the Na^+, K^+ active transport system in synaptic terminals. Such defects obviously can lead to enhanced excitability and seizures as is the case for local application of ouabain, a selective inhibitor of Na^+,K^+-ATPase. The mechanism of the failure of the Na^+, K^+ pump in the synaptosomal membranes from epileptogenic foci is not known, but appears to result from the pathological change in the neurons induced by the seizure provoking agent. It is likely that the changes in some of the lipid components (phospholipids, fatty acids, etc.) in the membrane (10) noted in Table II may be involved.

In the secondary focus the biochemical changes are generally similar to those observed in the primary focus but are quantitatively less marked. Again the changes are characteristic of glial proliferation as described for the primary focus. The significance of this in terms of the hyperexcitability of the secondary focus is described below. The lack of or only small changes in amino acids and other transmitter substances and their enzymes indi-

cates that the spiking activity in the secondary focus
is not related to transmitter levels in any simple way.
Studies on the release and synthesis of these amino acids
in the mirror focus are obviously necessary.

B. THE LOCAL SPREAD AND THE CHANGES WHICH PRECIPITATE THE INTERICTAL ACTIVITY INTO A SEIZURE

As described by Merlis (41), "Any focus discharging
sporadic spikes may develop a self-sustained discharge of
rhythmic potentials, i.e., a local seizure. In the development of seizure activity, cells generating PDSs may
demonstrate a progressive decrease in the hyperpolarization which follows PDS, with the development of afterdepolarizations and, finally, rhythmic depolarizations."
The paroxysmal discharges activate recurrent and surround
inhibitory receptors which tend to delimit the discharge.
However, by some as yet unknown mechanism these inhibitory
influences are dissipated, and the IPSP´s are converted
to EPSP´s, and an ictal episode occurs. Since this episode is highly synchronous a combination of recurrent excitatory and inhibitory systems appears to be involved
(50). Convulsant drugs applied locally probably act by
increasing potency of the excitatory system and reduce
inhibitory input, a response that produces a hypersynchronous discharge (41).

The experiments of Dichter et al. (13) provide some
evidence on the mechanism of transition from the interictal state to seizures. They recorded from glial cells
during penicillin-induced interictal discharges and seizures in the cat hippocampus. The changes in glial cell
membrane potentials were used as an index of changes in
local extracellular K^+ concentrations. They observed that
K^+ accumulated in the extracellular space during an interictal discharge and interpreted these results to indicate
that the K^+ accumulation is a result of the neuronal activity underlying the discharge. However, since it has been
shown that such accumulation can depolarize cells and
thereby lower their excitability threshold or increase
the effectiveness of synaptic transmission, Dichter et al.
(13) presented evidence that the interictal K^+ accumulation may play a causative role in the transition from
interictal spiking to seizure activity. Furthermore such
accumulation around neurons might enable many neurons to
become synchronized during paroxysm. After each isolated
interictal discharge, the excess K^+ in the extracellular
space returns to neurons via the $Na^+ - K^+$ pump, or enters
the glial cells also actively, or diffuses away into the

large volumes of extracellular fluid and the glial membrane potential returns to its value. However, when the excitability of the focus increases as interictal discharges occur more frequently, the hippocampal cortex is unable to clear the accumulated K^+ adequately. Thus, as each interictal discharge occurs more K^+ is released from neurons into the already loaded·extracellular space. The consequence of this is an even greater increase in excitability leading to a seizure. Thus they hypothesized that a seizure will tend to develop when subsequent interictal discharges occur before the cortex (presumably the glial cells) has had the chance to clear all the extra K^+ in the extracellular space from previous discharges.

The epileptogenic focus is composed of pathologically hyperactive neurons. The critical mass of hyperactive neurons necessary for spontaneous seizures is unknown but it is clearly more than one. In contrast to the firing patterns of normal neurons in cortex, the interictal firing patterns of neurons in the focus is characterized by recurrent, high-frequency bursts of action potentials. Firing rates in a burst may vary from 200 to 900/second; the burst usually starts at high frequency, and there is then the question how seizures are precipitated. Ward (57) has described the events that lead to seizure as follows:

"The current model of the epileptic process proposes that the epileptic focus is composed of autonomous epileptic cells (group 1) as well as much larger number of group 2 epileptic neurons whose activity , however,varies between normal activity and high-frequency burst activity. Under certain circumstances, appropriate synaptic input to the latter cells drives them to the upper limits of their range of pathologic activity and their activity now can be recorded." Other variables which appear to play a role in the precipitation of spontaneous epileptic seizures are those that affect brain excitability. These include such factors as endocrine changes, blood gas changes, water and electrolyte and acid-base changes and nutritional and temperature changes.

2. Seizures with generalized onset

Seizures may be generalized from the start, in the form of myoclonic jerks, tonic or tonic-clonic seizures. In many examples, there is widespread and diffuse neuronal involvement such as diffuse degenerative disease, metabolic disturbance or toxic state, in which there may be a diffuse increase in excitability of neurones, and explosively rapid involvement of cortex and subcortex.

But such diffuse disturbance is not essential, and generalized seizures have been elicited by focal electrical
stimulation of mesial thalamus or midbrain (see discussion by 41, and summary of the mechanism of action of
convulsants by 17).

a. Pentylenetetrazol. One of the drugs that produces
generalized seizures which mimic an absence-type seizure
in man, as validated by electrocorticographic recording
and clinical seizures, is pentylenetetrazol (PTZ). Drugs
that prevent PTZ seizures are those that are effective in
the treatment of absence epilepsy. Therefore, an understanding of its mechanisms of action should give a clue
to the processes by which generalized seizures, such as
absence, are triggered. From preceding discussion on mechanisms of focal epilepsy it is evident that initiation
of seizures involves effects on cell membranes and is related to active movement of ions across these membranes.
It is difficult in the intact brain to separate the effects of convulsant drugs on cell membranes with regard
to whether they influence active transport or membrane
permeability. Since there is evidence that PTZ may alter
both these parameters (11; 62), a summary of the
possible modes of action of this drug will be given.
PTZ has been shown in isolated systems to have a direct effect on neuronal membranes to decrease Cl conductance and thereby cause spontaneous firing in the
nerve cell. It has also been shown to block the enhancement of chloride conductance induced by GABA. Thus it
appears to act by increasing the excitability of membranes and to block the inhibitory effect of GABA. Also in
the nonexcitable membranes of isolated toad bladders,
PTZ has been shown to increase short-circuit current by
increasing K permeability of the serosal membrane, an
effect that is blocked by trimethadione, ethosuximide and
diazepam, selective antagonists of PTZ-induced seizures
in animals and of absence seizure in man (26). This effect on K^+ permeability of epithelial cells suggests that
PTZ may also act on glial cells, which are also epithelial
cells, to cause loss of K^+ from them, as it does in toad
bladder cells.

A PTZ -induced increase in K^+ permeability of the
glial cell membranes could cause rapid and massive release
of K^+ into the extracellular space where its accumulation would depolarize neurons to such an extent that
seizures might result, particularly if this occurred in
concert with enhanced excitatory transmitter release from
neuronal hyperpolarization and/or release of an ex-

citatory mediater such as glutamic or aspartic acids from glial cells.

b. Strychnine and picrotoxin and the inhibitory GABA and glycine receptors. Although there are a large number of convulsant drugs known, the mechanisms of action of only a few of these have been worked out. Strychnine causes convulsions by blockade of postsynaptic inhibitory pathways thus releasing excitatory pathways which fire excessively on sensory stimulation. The neurotransmitter for postsynaptic inhibition is glycine and it acts on the postsynaptic membrane via the glycine receptor. Strychnine blocks the glycine receptor. The effect of glycine on the glycine receptor is to increase Cl^- permeability of the postsynaptic membrane. The increase in Cl^- permeability results in an IPSP and hyperpolarization of the membrane. Strychnine blocks the increase in Cl^- permeability and the resulting IPSP induced by glycine.

Picrotoxin, like strychnine also causes seizures by release of inhibitory control, but in this case the effect is blockage of the GABA receptor. GABA is an inhibitory mediator of postsynaptic inhibition in higher cortical centers and in the cerebellum, and also in some subcortical areas, and of presynaptic inhibition in the brain stem and spinal cord.

GABA acts on its receptor to increase Cl^- permeability and produce an IPSP. Picrotoxin blocks these effects of GABA on the receptor. Bicuculline and penicillin also act by blocking the effects of GABA to increase Cl^- conductance (62).

c. Convulsants that affect various putative neurotransmitters

1. Anticholinesterases. These agents cause accumulation of acetylcholine, an excitatory transmitter, and thereby cause seizures.

2. Effects on biogenic amines. Many drugs inhibit synthesis, release, or the effects of biogenic amines at synapses and since these agents are generally inhibitory in nature they can result in seizures, as is the case with blockade of the inhibitory transmitters glycine and GABA. However since inhibitory synapses due to biogenic amines constitute 5% or less of total synapses, inhibition of these synapses does not generally result in seizures.

3. Cyclic GMP has been implicated as an excitatory agent. Whether this is a direct or an indirect effect has not been established.

d. Covulsants that are thought to act by effects on environment of neurons

1. Effects on glial cells.

a. Pentylenetetrazol. As already discussed, the possibility exists that this agent may cause seizures, at least in part, by an action on glial cells to increase K^+ permeability and cause a massive release of K^+ and/or excitatory amino acid transmitters (glutamic acid, aspartic acid) into the extracellular fluid.

b. Covulsant monovalent anions - thiocyanate and perchlorate. These can cause seizures by inhibiting the transport of anions (HCO_3^-, Cl^-, I^-) into glial cells, an active porcess that requires carbonic anhydrase and HCO_3^--activated ATPase and also is K^+-dependent. Some experiments concerned with the role of this anion transport system in epileptogenesis will now be discussed.

The evidence to be presented suggests that glial cells, in addition to regulating the cation content of the extracellular space of the brain, particularly K^+, also regulate the anion content of this fluid. This is accomplished by actively transporting monovalent anions utilizing carbonic anhydrase and HCO_3^-- activated Mg-ATPase. Inhibition of this system, as is the case for inhibition of the cation system in glial cells (e.g., with ouabain), has dire consequences·on neuronal excitability and can result in seizures. This monovalent anion transport system is selective for glial cells and has the ability , like the choroid plexus, to transport actively chloride, iodide, bromide, thiocyanate, perchlorate and bicarbonate into glial cells (6, 7, 20, 35, 65, 66). The system is K^+-dependent and is carrier-mediated since the transported monovalent anions compete for the carrier system. In Table III is shown the distribution of radioactive iodide, thiocyanate, perchlorate and bromide with and without added carrier (5 mEq/kg) in brain and CSF of adult rats. The values given are the ratios of the concentration of the specific anion in the brain cells or CBF as compared to its concentration in the extracellular space in the case of the brain or in the plasma in the case of CSF. In the brain of the control rats the ratio is greater than 1.0 for iodide and thiocyanate and in the case of perchlorate also greater than

Table III   DISTRIBUTION OF RADIOACTIVE IODIDE, THIOCYA-
NATE, PERCHLORATE AND BROMIDE WITH AND WITH-
OUT CARRIER (5 m Eq/kg) IN BRAIN AND CSF OF
ADULT RATS

$$\text{RATIO:} \frac{\text{BRAIN CELL ANION}}{\text{EXTRACELLULAR ANION}} \qquad \text{RATIO:} \frac{\text{CSF ANION}}{\text{PLASMA ANION}}$$

RADIOACTIVE ANION	CONTROL	CARRIER LOADED	CONTROL	CARRIER LOADED
IODIDE	1.30	0.22	0.02	0.42
THIOCYANATE	1.33	0.19	0.06	0.74
PERCHLORATE	0.71	0.15	0.10	0.80
BROMIDE	0.15	0.25	0.78	1.00

the predicted ratio for passive distribution of monovalent
anion (approximately 0.25). Thus the three anions are
transported into cells against an electrical gradient.
The bromide ratio is less than the predicted passive ratio; hence it appears to be transported out of the cells.
A large load of carrier blocks this transport and the
anions are then distributed passively across the cell membrane in accordance with the membrane potential. All these
anions including bromide are also transported out of the
CSF across the choroid plexus and the system is blocked by
carrier loading.

The distribution of radioactive perchlorate ion with
and without loading in the brain and CSF of rats during
maturation has also been tested. In 3-day-old rats, an
age when no functional glial cells are present, radioactive perchlorate is passively distributed in the brain and
carrier loading has little effect on its distribution, nor
does it influence brain excitability. The active anion
transport system appears after 10 days of age when glial
cells develop, and at this time carrier loading blocks
this transport. It is also at this time that perchlorate
and thiocyanate begin to produce prominent excitatory
effects on the nervous system as described in a moment.
This anion transport system is in the glia and these an-

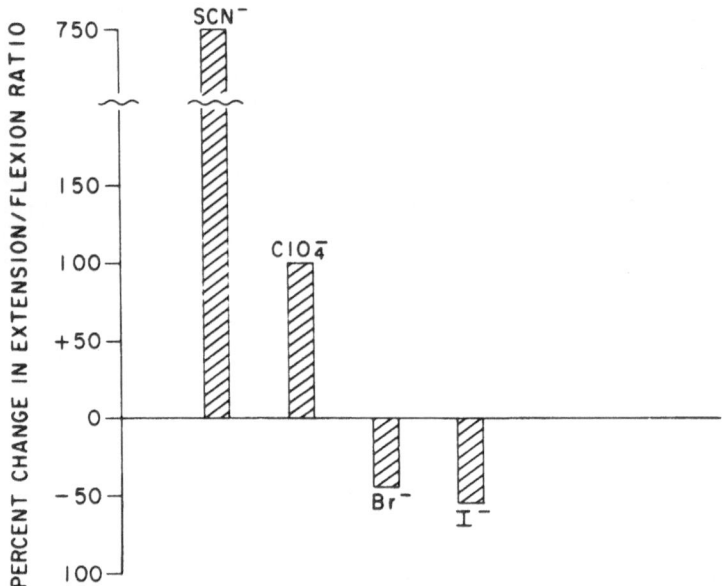

Fig. 2. Effect of various monovalent anions (5 mEq/kg)
        on the extensor/flexor (E/F) ratio of rats.
        The ordinate is the percent change in E/F rates
        from the pretreatment control value. See text
        for discussion.

ions seem to be passively distributed across neurons as
is evident from data in 3-day old rats in which no functional glial cells are present. Also resolution of radioactive perchlorate uptake curves, allows separation of
the glial and neuronal uptake components and calculation
of the glial and neuronal cell concentrations. The results
from such calculations show active transport into glial
cells and passive distribution across neurons. Carrier
loading blocks the glial transport but does not alter the
distribution across neurons.

     The evidence for active transport of radioactive
chloride into glial cells has been presented by Bourke,
Tower and colleagues (6, 7, 18). Its transport into glial
cells is K^+-dependent but that into neurons is not (18).
As is the case for the transport of perchlorate, iodide
and thiocyanate, chloride transport into glial cells is
blocked by large loads of these same anions as has been

demonstrated by Gill et al. (20) for cultured astrocytes. In addition, the system is inhibited by acetazolamide, showing the involvement of carbonic anhydrase, and 2,4-dinitrophenol and fluoride, evidence that the process is energy-dependent.

The effect of inhibition of the glial anion system by large loads (5 mEq/kg) of the various monovalent anions on brain excitability as assessed by the ratio of the tonic extensor to the tonic flexor component of the maximal electrshock seizure (E/F ratio) is shown in Fig. 2. The E/F ratio was increased 750% by thiocyanate, 100% by perchlorate, and decreased about 45% by bromide and 50% by iodide. Thus thiocyanate and perchlorate exert marked excitatory effects on the CNS whereas bromide and iodide exert anticonvulsant effects. We have also determined the values for the convulsive dose 50s (CD50) and lethal dose 50s (LD50) of perchlorate, alone and in combination with acetazolamide in rats. The CD50 for perchlorate is 9 mEq/kg and the lethal dose 50 (LD50) is 12 mEq/kg. When acetazolamide (10 mg/kg) was given in combination with perchlorate the CD50 decreased to 7.5 mEq/kg and the LD50 to 10 mEq/kg. Thus inhibition of carbonic anhydrase enhanced both the convulsive and the lethal effects of perchlorate. It is evident therefore, that inhibition of glial cell anion transport results in seizures in adults, but not in neonatal animals, in which no functional glial cells are present. The observation of Ransom (45) that seizures can be produced in rats when the cortical surface is irrigated with low-chloride concentration solutions also demonstrates the importance of anions in the regulation of brain excitability. In addition they showed that irrigation of the cortex with solutions of high chloride concentration prevented the seizures induced by elevation of K_o^+. High-bicarbonate-concentration solutions also cause seizures. How inhibition of anion transport into glial cells causes seizures is not clear, but may be due either to elevation of K_o^+, since anion transport is K^+-dependent, or to HCO_3-accumulation in the extracellular space, since HCO_3^- appears to be transported into glial cells by this same system.

Kimmelberg and Bourke (36) have characterized a HCO_3^--activated Mg-ATPase (HCO_3-ATPase) in brain and have found that it is located in the mitochondria of both glial and neuronal cells. It is postulated to play some role in anion transport. That this is the case is suggested by the fact that it is inhibited by thiocyanate, perchlorate and other monovalent anions. The effects of these anions on HCO₃-ATPase have been assessed in our laboratory (65) and it was found that thiocyanate completely inhibited this enzyme at a concentration of 12 mM/l, and perchlorate at a concentration of 16 mM/l. Bromide and iodide were much less effective and inhibited HCO₃-ATPase only about 15% at a concentration of 10 mM/l. The inhibition of the enzyme by perchlorate was found to be noncompetitive, an indication of a multienzyme system, which probably involves inhibition of both HCO₃-ATPase and Mg-ATPase.

A comparison (65) of the concentrations of perchlorate ion in various body fluids 15 minutes after injection of 5 or 10 mM/kg with the concentration that inhibits HCO₃-ATPase in vitro shows that at 5 mM/kg body weight, the fluid concentrations of perchlorate in plasma, CSF and brain cell are 14, 3 and 0.6 mM/l, respectively. At these concentrations, the HCO₃-ATPase in brain would be 93% inhibited at the concentration of perchlorate in plasma, 30% inhibited at the concentration in CSF, and 5% inhibited at the concentration observed in brain cells. At 10 mM/kg, the concentration of perchlorate in these fluids is calculated to be 28 mM/l for plasma, 13 mM/l for CSF, and 2 mM/l for brain cells. The HCO₃-ATPase in brain would be inhibited 100% at the concentration in plasma, 92% at the concentration in CSF and 20% at the concentration in brain cells. At 5 mM/kg perchlorate produces an increase in the E/F ratio but no seizures, whereas, at 10 mM/kg perchlorate produces seizures in 100% of animals and 17% of the animals died. The results suggest that the effect of perchlorate to inhibit HCO₃-ATPase and to increase excitability and produce seizures correlates best with its concentrations in CSF, which is also the brain extracellular concentration. This is also the concentration that inhibits anion transport into glial cells. The ability of perchlorate and thiocyanate to elicit seizures appears to be related to their capacity to inhibit HCO₃-ATPase an enzyme, together with carbonic anhydrase, involved in glial cell anion transport processes. The ability of acetazolamide to enhance both the seizure-evoking effects and the inhibition of anion transport by perchlorate appears to be due to its inhibition of carbonic anhydrase. Hence it is evident that inhibition of both enzymes results in a greater effect than blockade of either enzyme alone.

2. In addition to effects on glia, convulsants may also affect the blood-brain barrier (BBB) and the choroid plexus, both epithelial-cell-secreting tissues, similar

to toad bladder and glia. Thus PTZ and electroshock have
been shown to increase permeability of the endothelial
cells of the cerebral capillaries, the site of the BBB
(39). This allows leakage of electrolytes and proteins
into the interstitial space of the brain and can cause
swelling, thus compromising circulation and can lead to
hypoxia and seizures. Similar effects on choroid plexus
can alter CSF secretion and affect the electrolyte content of the CSF, which is in equilibrium with the interstitial fluid. Changes in extracellular electrolytes,
particularly K^+, can alter brain excitability.

II. THE PROPAGATION OF DISCHARGE DURING A MAJOR SEIZURE

Once a critical mass of neurons is firing in high-
frequency bursts at the focus, their propagating burst
activity can now recruit normal cells to fire in bursting
mode-and this regenerative process leads to propagation
of the seizure widely throughout the central nervous
system. The circuits of spread obviously depend on the
cells of origin. With a focus in motor cortex, the axonal
propagation of the seizure will obviously be rapidly manifested by appropriate motor levels. From foci in temporal lobe, there are preferential pathways of spread
within the limbic system, and, in fact, the discharge may
commonly remain restricted to limbic circuits.

The seizure discharge may propagate (spread) to other
areas of the brain by either a rapid or a slow process.
The rapid spread of discharge involves all synaptically
related neurons available to the epileptogenic focus. However, not all routes are of equal importance in the spread
of the epileptic discharge. The propagation of discharge
involves local spread, presumably utilizing short inter-
nuntial cortical neurons, spread to the contralateral homologous region (mirror focus) via transcallosal and other
interhemispheric or subcortical pathways, and spread via
corticofugal pathways to the mesencephalon, where presumably by involvement of the reticular substance the spread
is projected diffusely to the entire telencephalon, as
well as to the spinal cord. However, the generalization
of seizures also involves a number of other subcortical
centers in addition to the reticular activating system
(see 1, 33, 41 for summaries).

Spread of the discharge to other regions usually requires a high frequency drive from the primary focus which
appears to involve post-tetanic potentiation. This excessive bombardment causes the other areas to become inde-

pendently active and this serves to initiate discharges
in still other areas, with diffuse spread of seizure activity. Anticonvulsant drugs act by either raising the
threshold for firing of the focus or by preventing the
spread of seizure activity. Their ability to prevent spread
is different for different drugs. Thus, phenytoin prevents spread in the cortex whereas trimethadione prevents
spread subcortically and phenobarbital has some effects
in both places, but is excellent at raising threshold.

A seizure discharge in a cortical epileptic focus
may spread to neighboring cortex at a <u>slow rate</u>. This is
the case for the Jacksonian motor seizure where the wave
creeps across the cortex at a rate of only a few mili-
meters per minute. This slow spread of activity appears
to be related to the spreading depression of Leao (38)
and involves accumulation of K^+ in the extracellular fluid
and depolarization of both glia and neurons. Grafstein
(25) presented evidence to suggest that the K^+ released
during excessive activation of clusters of neurons leads
to depolarization of the same and adjacent neurons, thus
leading to further release of K^+ and the recruitment of
more and more neurons into the process. This leads to
diffuse spread of the seizure discharge and this is followed by profound postictal depression as a result of the
spreading depression. As expected from the role of glial
cells in K^+ homeostasis, the accumulated K^+ in spreading
depression would depolarize glia as well as neurons and
cause stimulation of the K^+ transport system in the glia.
This would then reduce Ko^+ and lead to recovery from
spreading depression. Measurements with potassium selective electrodes have shown that during spreading depression Ko^+ levels in the cortex rise to very high values
(see 49 for summary).

III. THE EVENTS THAT ARREST OR AMELIORATE THE SEIZURE

Although the seizure focus is always present, ictal
episodes are only occasional and short-lived events in the
behavior of an epileptogenic focus. It is essential therefore, to consider the various processes that delimit focal
firing and that terminate a seizure once it has been initiated. The most important methods for carrying out these
processes appear to be inhibitory mechanisms, accumulation of metabolic products that inhibit seizure activity
and glial uptake of Ko^+, anions and released neurotransmitters. The intrinsic cortical inhibitory mechanism, i.e.,
recurrent and surround inhibition, have been discussed
above and appear to be most important processes to res-

train a focus from firing and to arrest an ongoing seizure discharge. Extracortical inhibitory systems also exert some measure of control as well. For example, stimulation of portions of the cerebellum suppresses interictal spiking activity in the penicillin- or cobalt-induced cortical focus, and cerebellar ablations facilitate the spike activity (14, 32). Also, stimulation of the caudate suppresses the acute cortical penicillin focus (see 41 for summary). Whether the same or similar inhibitory systems are operative to terminate seizures in parts of the cortex other than the sensory motor areas is not known. During severe tonic-clonic seizures, respiratory arrest occurs briefly and there is a transient hypoxia and hypercapnia. Also since intense muscular activity takes place during the seizure, lactic acid accumulates in the plasma. This results in a mixed systemic metabolic and respiratory acidosis. In experimental animals, possibly in man, CO_2 accumulates in brain to a level sufficiently high to block seizure discharges (58, 64). The seizure arrest may be partly due to accumulation of CO_2. The hypoxia and increased CO_2 accumulation also cause increased cerebral blood flow which in man may be sufficient to prevent CO_2 accumulation in brain, but that this is the case has not been proved. A number of metabolic and other factors affect duration of seizure activity, susceptibility to seizures, and recovery from seizures. These factors are important in the control of epileptic seizures and will be discussed briefly. The rate of recovery from seizures depends on a number of factors, among the most important of which is an adequate supply of its energy source, glucose. Since very little glucose is stored in the brain as glycogen, the main source of glucose for the brain is from blood and from glycogenolysis from liver. Because of its dependence on glucose, we tested the effects of alterations in this parameter in blood on the ability of rats to recover from maximal electroshock seizures (MES). The recovery time was measured as the time for 50% of the animals to recover sufficiently to be able to have another MES. This is called the RT50. In Fig. 3 the relation between the Rt 50 and the blood sugar is plotted. The points represent various procedures or treatments in rats that change the blood sugar by altering the endocrine status, such as pancreatectomy, alloxan treatment, cortisol treatment and injections of insulin and glucagen. It is evident from Fig. 3 that there is a high inverse correlation between the Rt50 and the blood sugar level. Thus recovery is enhanced by a high blood sugar and slowed by a low level. Thus, providing an adequate source of energy hastens the ability of the brain to recover normal function.

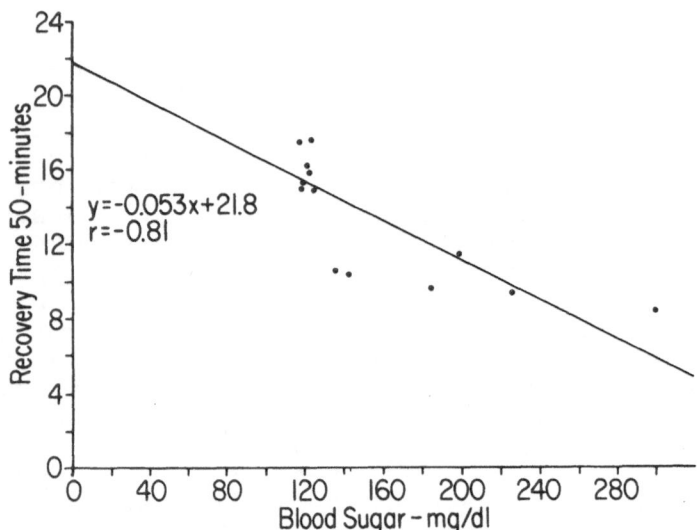

Fig. 3. Relation between recovery time 50 in minutes
         (ordinate) and blood sugar concentration (mg/dl)
         (abscissa) in rats subjected to maximal electro-
         shock seizures (MES). Recovery time 50 is time
         in minutes for 50% of rats to recover the ability
         to have a second MES following the first MES.
         See text for discussion.

        Evidence by many workers has shown that the glial
cells play an important role in modulating K_o^+ and acid-
base balance and also in ameliorating chronic neuronal
hyperexcitability. Thus this K^+-modulating system appears
to play an important role in delimiting focal spread an
in arresting ongoing seizure activity. In addition the
ability of glial cells actively to take up neurotransmitters released as a result of neuronal activity may be important in terminating seizures (28, 48). It is important
therefore, to discuss the role of glial cells in modulating brain excitability and ameliorating seizures.

ROLE OF GLIAL CELLS IN EPILEPTOGENIC FOCI

1. Buffering of potassium. Since glial proliferation is one of the major events in the development of the focus it is important to consider the role of glial cells in the seizure process. That glial proliferation is present is indicated by the anatomical studies already discussed and by the increase in DNA, and carbonic anhydrase, a glial marker. What is the effect of the glial proliferation? Many workers (see review by Somjen, 49) have demonstrated an important role of glial cells in K^+ homeostasis in the brain. During excessive nervous activity or in seizures, a marked increase in extracellular potassium (Ko^+) occurs (18, 31, 42, 45). However, this increase is generally ameliorated by K^+-stimulated active uptake of K^+ by glial cells (29); hence the change is only transient. Nevertheless, as a result of this rise in Ko^+ glial depolarization takes place. The increase in Ko^+ activates Na^+, K^+-ATPase in the membranes of glial cells and promotes K^+ active uptake into the cells.

During the large increase in spiking activity that occurs in an epileptic focus, Ko^+ is lost from the hyperactive neurons and leaks into the narrow interstitial channels between the neurons and glia. As shown by many workers hyperactivity of neurons is a stimulus for glial proliferation (3, 12, 15, 37, 56) presumably as a mechanism to increase the capacity of these cells to modulate the marked elevation of Ko^+ and possibly other electrolytes. Therefore, the lower epileptogenicity of chronic focal lesions as compared to acute lesions might well be due to glial proliferation and as a result, enhanced ability of glial cells to "buffer" Ko^+. The experiments of Glötzner (21) provide further evidence that glial cells modulate the increase in Ko^+ that occurs in a chronic aluminum hydroxide epileptogenic focus. He concluded from experiments to determine the membrane properties of glial cells in the scar as compared with normal glial cells that the neuroglial buffering system in the focus is able to transport even more K^+ away from sites of K^+ release than it does under normal conditions. If K^+ were allowed to accumulate in the interstitial channels to a sufficiently high level, the resulting depolarization of neurons and glia would lower the excitability thresholds of neurons and/or increase the effectiveness of synaptic transmission, and thereby cause seizure discharges. Such an accumulation of Ko^+ does not normally occur because it is prevented by the glial cell modulation system which is sensitive to Ko^+. The data of Glötzner suggest an adap-

tive response of this system during chronic epileptogenesis to ameliorate the enhanced Ko^+ accumulation caused by
hyperactivity of the neurons. This adaptive response would
limit the spread of the seizure discharge which is at
least partly due to Ko^+ accumulation as was discussed
above (13).

    2. Role of carbonic anhydrase. In addition to buffer-
ing Ko^+, glial cells appear to play an important role in
the modulation of acid-base changes in the extracellular
fluid. This process utilizes the glial enzyme carbonic
anhydrase and probably a Mg-ATPase that is activated by
HCO_3^- ion. This enzyme is part of the system involved in
the transport of monovalent anions (Cl^-, I^-, HCO_3^-, etc.)
across glial cell membranes. The role of carbonic anhydrase in epileptogenic foci can be elucidated in animals
implanted with cobalt in the frontal cortex (40). When
this is done spiking activity is generated in the region
of the focus, in the parietal cortex on the ipsilateral
side, and in the secondary focus on the contralateral side.
Such activity begins 5 days after implantation and reaches
a peak at 12 days. By 19 days the spiking has decreased
and by 30-40 days it has almost disappeared. Accompanying
the increased spiking activity is a gliosis characterized
by a marked increase in carbonic anhydrase, the time
course of which follows pari-passu with spiking activity
in the electrocorticogram. The changes with time in days
of carbonic anhydrase activity in the secondary focus located in the left frontal cortex as compared with the
right occipital cortex in which no spiking activity occurs are shown in Fig. 4 taken from observations of Mc-
Queen and Woodbury (40). There is a marked and significant
increase in carbonic anhydrase activity at 5 and 12 days
after cobalt implantation and a slight increase at 19
days. These values are significantly increased as compared
to the same tissue obtained from control animals at the
same time. These authors found a direct correlation between spiking activity and carbonic anhydrase activity in
the various regions of the brain (right and left parietal
cortex) and at various times after implantation, particularly in the secondary focus where few pathological changes are produced.

    The effect of cobalt implantation in the right fron-
tal cortex of the rat on the total free $^{14}CO_2$ space, $^{14}CO_2$
fixation, and carbonic anhydrase activity in the projected
(secondary) focus of the left frontal cortex as compared
with the same tissue from control animals not implanted
with cobalt was also measured. The samples were removed

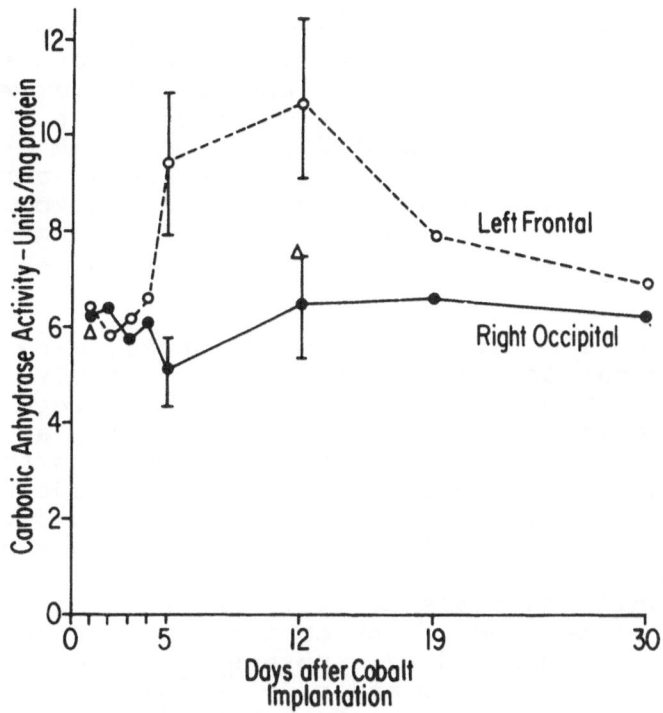

Fig. 4. Effect of implantation of cobalt in the right
frontal cortex of rats on carbonic anhydrase
activity in the left frontal (projected focus)
cortex and right occipital cortex at various
days after implantation. See text for discussion.
(From McQueen, J.K. and Woodbury, D.M., 40).

12 days after cobalt implantation at the peak of the
spiking activity in the electrocorticogram. Compared to
the same area in control animals, the secondary focus in
the frontal cortex had a 23% lower total free $^{14}CO_2$ space,
a 44% greater rate of $^{14}CO_2$ fixation, and a 114% increase
in carbonic anhydrase activity. These data suggest that
carbonic anhydrase regulates the total free CO_2 concentration in glial cells and also the incorporation of CO_2 into
various metabolic products (CO_2 fixation), such as tri-

carbocyclic acid cycle intermediates, glutamic and aspartic acids, pyrimidines and purines and fatty acids. Enhanced CO_2 fixation would, therefore, increase synthesis of proteins and nucleic acids and result in glial growth and proliferation. The decrease in free CO_2 space suggests that carbonic anhydrase also regulates CO_2 movement into and out of glial cells and that increased amounts accelerate its removal from the brain and thereby prevent its accumulation. Such accumulation could occur in the focus where metabolic production of CO_2 is increased as a result of the enhanced neuronal activity if carbonic anhydrase were not increased to handle the greater metabolic load of CO_2. Accumulation of CO_2 results in blocking of synaptic transmission and, in high concentration, can produce seizures (63, 64, 67).

The effects of acetazolamide (Diamox), a selective inhibitor of carbonic anhydrase, on the activity of this enzyme in the primary and secondary foci induced by cobalt as related to its effects on spiking activity add further evidence to a modulating role of glial cell carbonic anhydrase on neuronal excitability. Acetazolamide given acutely or chronically inhibited carbonic anhydrase to normal or below normal levels and enhanced seizure activity. On withdrawal of acetazolamide after chronic administration , carbonic anhydrase activity returned to levels higher than those produced by cobalt implantation alone and this was accompanied by suppression of both spiking in the electrocorticorgam and of clinical seizure activity. Thus chronic administration of a carbonic anhydrase inhibitor induces increased activity of carbonic anhydrase in glial cells. This can explain the previously observed development of tolerance to the anticonvulsant effect of acetazolamide. Acetazolamide also inhibits CO_2 fixation in brain (unpublished observation of D.M. Woodbury). These data suggest a role of glial cells in the modulation of increased neuronal activity no matter how induced. This regulatory system involves control of HCO_3^- - H_2CO_3-CO_2 metabolism and distribution in the brain by glial cell carbonic anhydrase. Inhibiting the enzyme by acetazolamide or other carbonic anhydrase inhibitors compromises the modulating system and increases neuronal excitability. In the chronic cobalt lesion the increased neuronal activity enhances CO_2 production which appears to cause an adaptive increase in carbonic anhydrase activity. This increase in activity, however, is not only not prevented by chronic inhibition of the carbonic anhydrase, but is actually enhanced, and this results in modulation of the excessive neuronal activity present in the cobalt brain

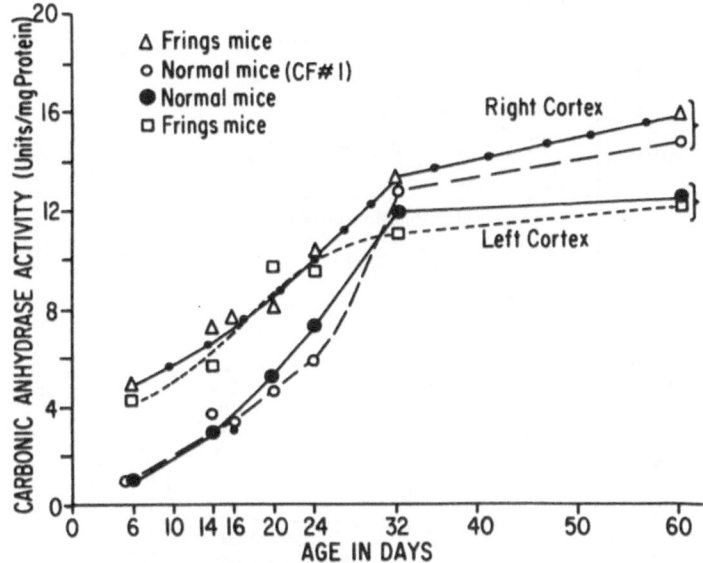

Fig. 5. Carbonic anhydrase activity (units/mg protein) in the left and right cerebral cortex of normal and audiogenic-seizure-susceptible mice (Fring´s strain) (ordinate) as a function of postnatal age in days (abscissa). See text for discussion.

and causes tolerance to develop to the inhibitor. It is evident from these observations that glial cells have a significant role as modulators of activity by regulating the concentrations of K^+ and of the CO^2-HCO_3^--H_2CO_3 system in the interstitial ionic milieu.

Other factors affecting excitability are shown in Figures 5 and 6. In Fig. 5 is shown the changes in cerebral cortical carbonic anhydrase activity with age in the cortex of mice normal and susceptible to audiogenic seizure. It is seen that the susceptible mice have considerably more carbonic anhydrase in their brain than do the non-susceptible mice. Thus genetic factors are important in determining the numbers of glial cells in the brain

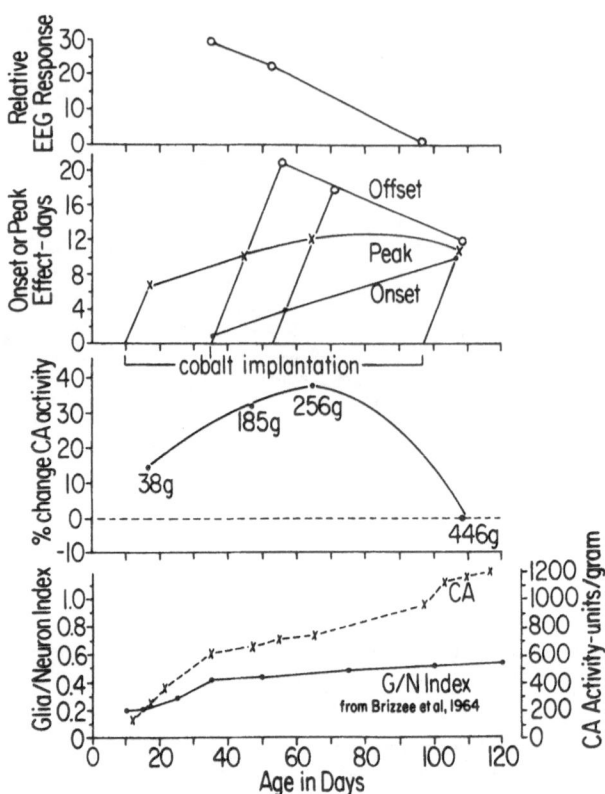

Fig. 6. Effect of aging on various brain responses of
rats to implantation of cobalt at different age
periods in the right frontal cortex. Abscissa
is postnatal age in days. The top ordinate is
the relative electroencephalographic response
to cobalt implantation as measured by polyspike
activity. The ordinate second from top is the
days after cobalt implantation for onset, peak
effect and offset of the polyspike activity in
the EEG of rats. The ordinate third from top is
the percent change in carbonic anhydrase activity in the left cerebral cortex in response to
implantation of cobalt in the right frontal
cortex at different ages. The bottom ordinate
on the left side is the change in the glia/neuron
index in rats with age as plotted from the data
of Brizzee et al. (8) and that on the right side
the changes in carbonic anhydrase activity
(units/gram) with age in rats. Both indices of
glial growth increase with age. See text for
discussion.

and their change with maturation. How this change in the number of glial cells is related to the susceptibility of the mice to sound-induced seizures awaits further experimentation. However, as shown in Fig. 6 (Chow, F. and D.M. Woodbury, unpublished observations) aging factors are important in determining susceptibility to seizures. In man, seizure susceptibility decreases with advancing age. The reason for this is unknown. The data shown in Fig. 6 suggest a role of glial cells in decreasing seizure susceptibility with age. Rats of different ages were implanted with cobalt in the right frontal cortex and their EEG´s monitored daily. At various intervals the rats were killed and the brain sampled for carbonic anhydrase analysis. The abscissa is the age of rats in days. In the top graph is shown the relative responsiveness of the EEG to the implanted cobalt with age. The spiking response to cobalt is highest in the 35-day-old animals and decreased to practically zero responsiveness in the 97-day-old rats (446 grams). In the second graph from the top are plotted the time of onset, the peak time, and the offset of the seizures induced by cobalt with age. Animals implanted with cobalt at 10 days had the peak effect of the seizure 7 days later. Animals with cobalt at 35 days had the onset of the seizure at one day, the peak effect at 10 days, and the offset of the seizure at 21 days after implantation. Thus the seizure activity in the EEG lasted for 20 days. The spiking activity in the EEG was marked. This represented the maximal effect of the cobalt. Animals implanted with cobalt at 53 days of age had the onset of their seizures (spiking activity in EEG) at 4 days after implantation; the peak time was at 12 days and the offset of the activity was at 18 days. The seizure activity thus lasted 14 days. In the old animals, the cobalt was implanted at 97 days, the onset of the maximal spiking activity that occurred was at 10 days, the peak activity was at 11 days and the offset was at 12 days. The maximal spiking activity lasted a total of only 2 days. Thus the EEG response to cobalt epilepsy decreases dramatically with age. In the graph third from the top the changes in carbonic anhydrase activity in the secondary focus with aging are shown. The curve is similar to that of the peak spiking activity curve in response to cobalt. The 35 and 53-day old rats showed the greatest carbonic anhydrase response to cobalt, whereas the old rats did not show any carbonic anhydrase response to cobalt. The immature 17-day old rats showed only a slight response. The lower curve shows the changes in brain carbonic anhydrase and the glia/neuron index with age (8). Both of these parameters increase with age, an indication

that glial cells increase continously with age. We have
interpreted these results to suggest that the decrease
in seizure responsiveness to cobalt with age is a result
of the increase in glial cells which via their anion and
cation transport systems ameliorate the excitatory effects
of the cobalt epileptogenic focus. The responsiveness of
the glial transport systems in the younger animals to
cobalt-induced hyperexcitability is high and results in
induction of carbonic anhydrase, which acts also to limit
the spread of the seizure activity as already described.
As glial cells increase with age the adaptive response to
cobalt decreases, because they have already proliferated
maximally and thereby prevented the cobalt from inducing
seizure activity in the EEG.

SUMMARY

The study of seizure mechanisms is best carried out
in experimental models that resemble as closely as possible human epilepsy. Both local and generalized models are
used. Local foci induced by implantation of alumina cream,
cobalt or tungstic acid or by local freezing or application of penicillin are characterized by neuronal loss,
glial proliferation and abnormalities in dendritic structure, which include loss of dendritic spines, and results
in deafferentation and supersensitivity. Spikes and large
slow waves are observed in the surface EEG and intracellular recordings show a paroxysmal depolarizing shift with
bursts of axon spikes. This is frequently followed by a
hyperpolarization particularly in the area outward from
the focus. Biochemically the lesion and its surrounding
areas and the secondary (mirror) focus are characterized
by glial edema, decreased protein synthesis and oxidative
metabolism, decreased K^+ transport and Na-K-ATPase, defects in neurotransmitters, and increase in carbonic anhydrase, a glial enzyme. These changes are compatible
with the pathological alterations and the increased neuronal activity. Glial proliferation represents an adaptive
response to the increased neuronal discharge such that
the spread of the discharge is limited. Glial cells in a
focus have a greater ability to modulate changes in extracellular K^+, Cl^-, HCO_3 and CO_2 and thereby regulate the
ionic composition of the neuronal environment. In addition, glial cells actively transport the monovalent anions
chloride, bicarbonate, iodide, thiocyanate, perchlorate
and bromide by a system that involves HCO_3^--activated
ATPase and carbonic anhydrase. Inhibition of this anion
transport system by thiocyanate and perchlorate produces
seizures which are enhanced by acetazolamide. The increase

in excitability produced by increasing doses of thiocyanate and perchlorate correlates with the degree to which
they inhibit HCO_3^--ATPase.

Generalized seizures are usually induced by drugs
that block the effect of the inhibitory mediater (GABA
and glycine) to increase chloride permeability in the
postsynaptic membrane or have a direct effect on the
membrane to decrease chloride permeability. GABA blockers
are picrotoxin, penicillin, bicuculline and pentylenetetrazol all of which also have direct effects on the
membrane as well. Glycine blockers are strychnine and related congeners. Other drugs cause seizures by interference with the synthesis or release of GABA in the presynaptic endings (pyridoxal phosphate antagonists, ally-
glycine, 3-mercaptopropionic acid). Drugs that block
electrolyte transport systems in neurons and/or glia, such
as ouabain and thiocyanate also cause seizures. Arrest of
seizures is usually due to recruitment of inhibitory mechanisms, depletion of substrate, accumulation of end
products, and glial modulating mechanisms.

ACKNOWLEDGMENTS

Unpublished data presented in this paper were supported by Grant No. NS 12812 from the National Institute
of Neurological Diseases and Stroke. One of us (DMW) is
a Research Career Awardee (5-K6-NS-13, 388) of the National Institute of Neurological Diseases and Stroke.

REFERENCES

1. Aird, R.D., and Woodbury, D.M. (1974): The Management
   of Epilepsy. Charles C. Thomas, Springfield, Ill.

2. Ajmone-Marsan, C. (1969): Acute effects of topical epileptogenic agents. In: Basic Mechanisms of the Epilepsies, edited by H.H. Jasper, A.A. Ward, Jr., and
   A. Pope. Little, Brown and Co., Boston, 299-319.

3. Altman, J., and Das, G.D. (1964): Autoradiographic
   examination of the effects of enriched environment
   on the rate of glial multiplication in the adult brain.
   Nature, 204: 1161-1163.

4. Ashcroft, G.W., Dow, R.C., Emson, P.C., Harris, P.,
   Ingleby, J., Jospeh, M.H., and McQueen, J.K. (1974):
   A collaborative study of cobalt lesions in the rat as
   a model for epilepsy. In: Epilepsy. Proceedings of
   the Hans Berger Centenary Symposium, ed. P. Harris and
   C. Mawdsley, Churchill Livingstone, Edinburgh, 115-124

5. Ayala, G.F., Matsumoto, H., and Gumnit, R.J. (1970):
    Excitability changes and inhibitory mechanisms in neo-
    cortical neurons during seizures. J. Neurophysiol.
    33: 73-85.

6. Bourke, R.S. (1969): Evidence for mediated transport
    of chloride in cat cortex in vitro. Exp. Brain Res.
    8:219-231.

7. Bourke, R.S., and Nelson, K.M. (1972): Further stu-
    dies on the K^+-dependent swelling of primate cerebral
    cortex in vivo: The enzymatic basis of the K^+-depen-
    dent transport of chloride. J. Neurochem. 19:663-685.

8. Brizzee, K.R., Vogt, J. and Kharetchko, X. (1964):
    Postnatal changes in glia/neuron index with a compari-
    son of methods of cell enumeration in white rat.
    Progr. Brain Res. 4: 136-149.

9. Brown, W.J. (1973): Structural substrates of seizure
    foci in the human temporal lobe. In: Epilepsy: Its
    Phenomena in Man, edited by M.A.B. Brazier. Acad.
    Press, New York.

10. Cendella, R.J. and Craig, C.R. (1973): Changes in ce-
    rebral cortical lipids in cobalt-induced epilepsy.
    J. Neurochem. 21: 743-748.

11. DeRobertis, E., Rodriquez de Lores, Arnaiz, G., and
    Alberici, M. (1969): Ultrastructural neurochemistry.
    In: Basic Mechanisms of the Epilepsies, edited by
    H.H. Jasper, A.A. Ward, Jr., and A. Pope, 137-158.
    Little Brown and Co., Boston.

12. Diamond, M.C., Law, F., Rhodes, H., Lindner, B.,
    Rosenzweig, M.R., Krech, D., and Bennett, E.L. (1966):
    Increases in cortical depth and glial numbers in
    rats subjected to enriched environment. J. Comp.
    Neurol. 128: 117-126.

13. Dichter, M.A., Herman, C.J. and Selzer, M. (1972):
    silent cells during interictal discharges and seizures
    in hippocampal penicillin foci. Evidence for the role
    of extracellular K^+ in the transition from the inter-
    ictal state to seizures. Brain Res. 48: 173-183.

14. Dow, R.C. (1965): Extrinsic regulatory mechanisms of
    seizure activity. Epilepsia 6: 122-140.

15. Dropp, J.J. and Sodetz, F.J. (1971): Autoradiographic study of neurons and neuroglia in autonomic ganglia of behaviorally stressed rats. Brain Res. 33: 419-430.

16. Escueta, A.V., and Appel, S.H. (1972): Brain synapses. An in vitro model for the study of seizures. Arch. Intern. Med. 129: 333-344.

17. Esplin, D.W., and Zablocka-Esplin, B. (1969): Mechanisms of action of convulsants. In: Basic Mechanisms of the Epilepsies, edited by H.H. Jasper, A.A. Ward, Jr., and A. Pope, Little,Brown and Co., Boston, 167-183.

18. Fertiziger, A.P., and Ranck, J.B. (1970): Potassium accumulation in interictal space during epileptiform seizures. Exp. Neurol. 26: 571-585.

19. Giacobini, E. (1962): A cytochemical study of the localization of carbonic anhydrase in the nervous system. J. Neurochem. 9: 169-177.

20. Gill, T.H., Young, O.M., and Tower, D.B. (1974): The uptake of ^{36}Cl into astrocytes in tissue culture by potassium-dependent, saturable process. J. Neurochem. 23: 1011-1018.

21. Glötzner, F.L. (1973): Membrane properties of neuroglia in epileptogenic gliosis. Brain Res., 55: 159-171.

22. Goddard, G.V. (1967): Development of epileptic seizures through brain stimulation at low intensity. Nature 214: 1020-1021.

23. Goddard, G.V., McIntyre, D.C., and Leech, C.K. (1969): A permanent change in brain function resulting from daily electrical stimulation. Exp. Neurol. 25: 295-330.

24. Goldensohn, E. (1969): Discussion. Experimental seizure mechanisms. In: Basic Mechanisms of the Epilepsies, edited by H.H. Jasper,A.A. Ward, Jr., and A. Pope, Little, Brown and Co., Boston, 289-298.

25. Grafstein, B. (1956): Mechanism of spreading cortical depression. J. Neurophysiol. 19: 154-171.

26. Gross, G.J., and Woodbury, D.M. (1972): Effects of pentylenetetrazol on ion transport in the isolated toad bladder. J. Pharmacol. Exp. Ther. 181: 257-272.

27. Harris, A.B. (1972): Degeneration in experimental epileptic foci. Arch. Neurol. 26: 434-449.

28. Henn, F.A., and Hamberger, A. (1971): Glial function: uptake of trasmitter substances. Proc. Natl. Acad. Sci. U.S.A. 68: 2686-2690.

29. Henn, F.A., Haljamäe, H. and Hamberger, A. (1972): Glial cell function: active control of extracellular K^+ concentration. Brain Res., 43: 437-443.

30. Henneman, E., Somjen, G., and Carpenter, D.O. (1965): Functional significance of cell size in spinal moto-neurons. J. Neurophysiol. 28: 560-580.

31. Hotson, J.R., Sypert, G.W., and Ward, A.A. Jr. (1973): Extracellular potassium concentration changes during propagated seizures. Exp. Neurol. 38: 20-26.

32. Hutton, J.R., Frost, J.D., and Foster, J. (1972): The influence of the cerebellum in cat penicillin epilepsy. Epilepsia, 13: 401-408.

33. Jasper, H.H. (1969): Mechanisms of propagation: extra-cellular studies. In: Basic Mechanisms of the Epilepsies, edited by H.H. Jasper, A.A. Ward, Jr., and A. Pope, Little, Brown and Co., Boston, 421-438.

34. Jasper, H.H. (1972): Application of experimental models to human epilepsy. In: Experimental Models of Epilepsy, edited by D.P. Purpura, J.K. Penry, D.B. Tower, D.M. Woodbury, and R.D. Walter, 585-601. Raven Press, New York

35. Kemp, J.W., and Woodbury, D.M. (1975): The effect of perchlorate and acetazolamide on brain excitability. Abstracts of Sixth International Congress of Pharmacology, Helsinki, Finland. 617.

36. Kimmelberg, H.K., and Bourke, R.S. (1973): Properties and localization of bicarbonate-stimulated ATPase activity in rat brain. J. Neurochem. 20: 347-359.

37. Kulenkampff, H. (1952): Das Verhalten der Neuroglia in den Vorderhörner des Rückenmarks der weissen Maus

unter dem Reiz physiologischer Tätigkeit. Z. Anat. Entwicklungs-gesch. 116: 304-312.

38. Leao, A.A.P. (1944): Spreading depression of activity in the cerebral cortex. J. Neurophysiol. 7:359-390.

39. Lorenzo, A.V., Hedley-Whyte, E.T., Eisenberg, H.M. and Hsu, D.W. (1975): Increased penetration of horse-radish peroxidase across the blood-brain barrier induced by Metrazol seizures. Brain Res. 88: 136-140.

40. McQueen, J.K., and Woodbury, D.M.: Carbonic anhydrase activity and cyclic AMP levels during the development of cobalt-induced epilepsy in the rat. Submitted for publication.

41. Merlis, J. (1974): Neurophysiological aspects of epilepsy. In: Epilepsy. Proceedings of the Hans Berger Centenary Symposium, edited by P. Harris and C. Mawdsley, Churchill Livingstone, Edinburgh, 5-19.

42. Moody, W.J., Futamachi, K.J. and Prince, D.A. (1974): Extracellular potassium activity during epileptogenesis. Exp. Neurol. 42: 248-262.

43. Murray, M. (1968): Effects of dehydration on the rate of proliferation of hypothalamic neuroglia cells. Exp. Neurol. 20: 460-468.

44. Pope, A. (1969): Perspectives in neuropathology. In: Basic Mechanisms of the Epilepsies, edited by H.H. Jasper, A.A. Ward, Jr., and A. Pope, Little, Brown and Co., Boston, 773-781.

45. Prince, D.A., Lux, H.D., and Neher, E. (1973): Measurement of extracellular potassium activity in cat cortex. Brain Res. 50: 489-495.

46. Ransom, B. (1974): The behavior of presumed glial cells during seizure discharge in cat cerebral cortex. Brain Res. 69: 83-99.

47. Scheibel, M.E., and Scheibel, A.B. (1968): On the nature of dendritic spines-report of a workshop. Commun. Behav. Biol. 1: 231-265.

48. Schrier, B.K., and Thomson, E.J. (1974): On the role of glial cells in the mammalian nervous system. Uptake, excretion and metabolism of putative neurotransmitters by cultured glial tumor cells. J. Biol.Chem.249:1769.

49. Somjen, G.G. (1975): Electrophysiology of neuroglia.
    Annu. Rev. Physiol. 37: 163-190.

50. Spencer, W.A., and Kandel, E.R. (1969): Synaptic
    inhibition in seizures. In: Basic Mechanisms of the
    Epilepsies, edited by H.H. Jasper, A.A. Ward, Jr.,
    and A. Pope, Little, Brown and Co., Boston, 575-603.

51. Van Gelder, N.M., and Courtois, A. (1972): Close cor-
    relation between changing content of specific amino
    acids in epileptogenic cortex of cats and severity
    of epilepsy. Brain Res. 43: 477-484.

52. Van Gelder, N.M., Sherwin, A.C. and Rasmussen, T.
    (1972): Amino acid content of epileptogenic human
    brain: focal versus surrounding regions. Brain Res.
    40: 385-393.

53. Velasco, M., Velasco, F., Estrada-Villanueva, F.,and
    Olivera, A. (1973): Alumina cream-induced focal motor
    epilepsy in cats. Part 1. Lesion size and temporal
    course. Epilepsia. 14: 3-14.

54. Velasco, M., Velasco, F., Lozoya, X., Feria, A. and
    Gonzales-Licea, A. (1973): Alumina cream-induced focal
    motor epilepsy in cats. Part 2. Thickness and cellula-
    rity of cerebral cortex adjacent to epileptogenic
    lesions. Epilepsia, 14: 15-27.

55. Vernadakis, A., Valcana, R. Curry, J.J., Maletta, G.J.
    Hudson, D., and Timiras, P.S. (1967): Alterations in
    growth of brain and other organs after electroshock
    in rats. Exp. Neurol. 17: 505-516.

56. Ward, A.A., Jr.(1969): The epileptic neuron: chronic
    foci in animals and man. In:Basic Mechanisms of the
    Epilepsies, edited by H.H. Jasper, A.A. Ward, Jr.,
    and A. Pope, Little, Brown and Co. Boston, 263-288.

57. Ward, A.A. (1975): Theoretical basis for surgical
    therapy of epilepsy. In: Advances in Neurology, Vol.
    8, edited by D.P. Purpura, J.K. Penry, and R.D.Walter.
    Raven Press, New York, 23-35.

58. Ward, J.R., and Call, L.S. (1949): Changes in blood
    chemistry in rats following electrically-induced sei-
    zures. Proc. Soc. Exp. Biol. Med. 70: 381-382.

59. Westmoreland, B.F., Hanna, G.R. and Bass, N.H. (1972):
    Cortical alterations in zones of secondary epilepto-
    genesis: a neurophysiologic, morphologic and micro-
    chemical correlation study in the albino rat. Brain
    Res. 43: 485-599.

60. Westrum, L.E., White, L.E., and Ward, A.A., Jr.(1964):
    Morphology of the experimental epileptic focus. J.
    Neurosurg. 21: 1033-1046.

61. Wilder, B.J. (1972): Projection phenomena and secon-
    dary epileptogenesis mirror foci. In: Experimental
    Models of Epilepsy, edited by D.P. Purpura, J.K.,
    Penry, D.B. Tower, D.M. Woodbury, and R.D. Walter,
    Raven Press, New York, 85-111.

62. Woodbury, D.M. (1978): Mechanisms of Action of Con-
    vulsants. In: Mechanisms of Action of Antiepileptic
    Drugs, edited by G. Glaser, D.M. Woodbury, and J.K.
    Penry, Raven Press, New York, in press.

63. Woodbury, D.M., and Esplin, D.W. (1959): Neuropharma-
    cology and neurochemistry of anticonvulsant drugs.
    Proc. Assoc. Res. Nerv. Ment. Dis. 37: 24-56.

64. Woodbury, D.M., and Karler, R. (1960): Role of carbon
    dioxide in the nervous system. Anesthesiology, 21:
    686-703.

65. Woodbury, D.M. and Kemp, J.W. (1977): Basic mechanisms
    of seizures: neurophysiological and biochemical etio-
    logy. In: Psychpathology and Brain Dysfunction, edi-
    ted by C. Shagass, S. Gershon, and A.J. Friedhof,
    Raven Press, New York, 149-182.

66. Woodbury, D.M., Johanson, C.E., and Brondsted, H.(1974):
    Maturation of the blood-brain and blood-cerebrospin-
    al fluid transport systems. In: Narcotics and the
    Hypothalamus, edited by E. Zimmerman and R. George,
    Raven Press, New York, 225-250.

67. Woodbury, D.M., Rollins, L.T., Gardner, M.D., Hirschi,
    W.C., Hogan, J.R., Rallison, M.L., Tanner, G.S., and
    Brodie, D.A. (1958): Effects of carbon dioxide on
    brain excitability and electrolytes. Am. J. Physiol.
    192: 79-90.

SOME MECHANISMS OF EPILEPTIC ACTIVITY GENERALIZATION

V. M. Okujava

Institute of Clinical and Experimental Neurology
Tbilisi, U.S.S.R.

Epileptic activity induced in a definite cerebral
focus may remain localized for some long periods of time.
On the other hand, epileptic activity evoked in one point
may easily spread to other brain regions thus leading to
extensive generalization. But even in the first case,
when focal epileptic activity is observed, the volleys
originating from the focus naturally exert their influence upon other anatomically more or less closely connected brain regions.

Accordingly, if epileptic activity is induced in a
definite point of the brain, paroxysmal activity can be
discovered in a number of brain regions, i.e. secondary
foci of paroxysmal activity can be recorded. Appearance
of the "mirror" focus in the symmetrical point of the
contralateral cerebral cortex in a case of circumscribed
cortical epileptogenic focus is an example of such induced activity.

Obviously, in such cases, an activity induced under
the influence of volleys from the primary focus must be
considered as evoked potentials. Here, consequently, one
has to deal with projected activity and not with the epileptic activity proper. While such electrical phenomena
represent evoked potentials they disappear immediately
after the termination of the primary focus (14, 16).

While under the influence of longlasting and strong
volleys arriving from the primary focus, the secondary

one can be transformed into an independent focus of self-
sustained epileptic activity. Presumably this phenomenon
must be very important in generalization of epileptic activity. Such an independent secondary focus starts to exert its influence upon other brain regions, thus leading
to further spread of epileptic activity (14, 16).

From this point of view the study of the role of
callosal and subcortically relayed extracallosal pathways
in the interhemispheric spread of paroxysmal discharges
and in formation of secondary epileptogenic foci must be
of great importance for understanding the mechanisms of
epileptic activity generalization.

The conclusions of a great number of investigations
dedicated to this problem markedly differ from each other,
evidently due to differences in experimental conditions.
Some authors (2, 7, 8, 11, 12) ascribe interhemispheric
spread of epileptic discharges exclusively to callosal
pathways, while the others ascribe a preferential role
to subcortical structures (4, 5, 6, 17) or a combination
of callosal and extracallosal influences (10, 14, 15).

In the present communication an attempt was made to
elucidate the role of these structures in the interhemispheric spread of seizure discharges and in formation of
secondary foci of epileptic activity. Gross- and microelectrode studies were performed on adult unanesthetized,
immobilized cats and rabbits. Preliminary operations
had been performed under ether anesthesia. Experiments
were carried out on "intact" brains and brains with various sections of interrupted central pathways, namely
either following callosotomy or callosotomy combined with
one of the following transsections: complete sectioning
of the diencephalon, precollicular or intercollicular
sectioning of the latter. In some experiments undercut-
ting of cortico-subcortical connections sparing the callosal pathways was undertaken (3, 10, 14). A primary
epileptogenic focus was produced by topical applicaton
of penicillin to the cortical surface or by its repetitive electrical stimulation.

In the intact brain the appearance of epileptic activity in a definite point of the cerebral cortex as a
rule is accompanied by inducation of a secondary focus
in a symmetrical point of the opposite hemisphere. But
the latter is not the focus of independent self-sustained
epileptic activity. It is a projected focus and its activity is induced under the influence of volleys arriving from

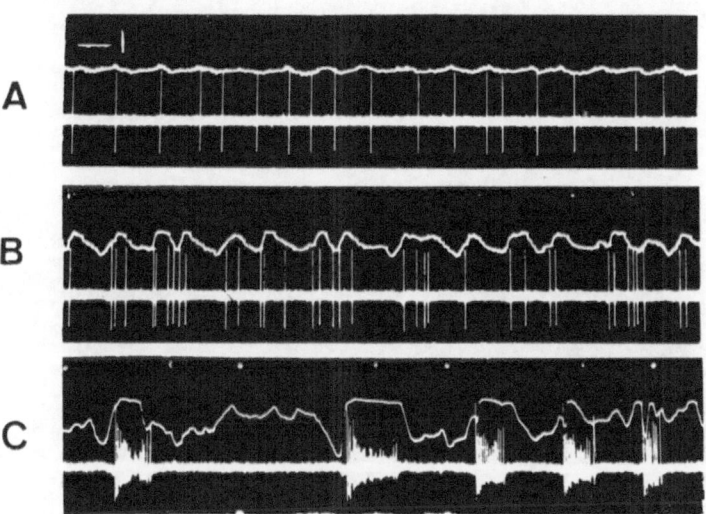

Fig. 1. Epileptization of a neuron in the left sensorimo-
        tor cortex of a cat during electrically induced
        epileptic activity in the contralateral cortex.
        Upper trace - electrocorticogram (ECoG), lower
        trace - extracellular recording.
        A - before stimulation; B - projected activity
        immediately following stimulation; C - several
        minutes later.
        Calibrations - horizontal bar - 100 msec, verti-
        cal bar - 400 microvolts for ECoG.

the primary one. If the activity of the primary focus is
artificially eliminated, or when it stops spontaneously,
paroxysmal discharges of the secondary focus disappear.

        Neuronal activity of secondary foci is very typical.
As illustrated by Fig. 1B, projected paroxysmal activity
at a neuronal level in extracellular records is characterized by single or low frequency discharge of spike potentials in correlation with paroxysmal electrocorticographic waves. These findings are in good accordance with
the results of Ajmone Marsan (1).

        However, such projected activity is not the only
pattern that can be observed in secondary foci. In cases
with prolonged and intensive epileptic activity, the number and the frequency of neuronal discharges in secondary

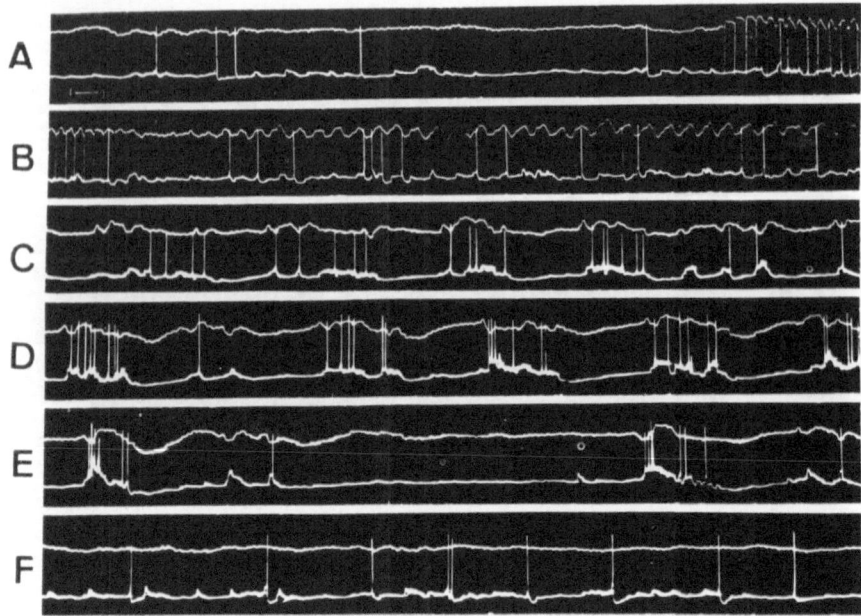

Fig. 2. Intracellular recording of a neuron in the senso-
        rimotor cortex of a cat following recurrent epi-
        leptogenic electrical stimulation of the contra-
        lateral cortex.
        Note gradual formation of PDS's.
        A - background activity; B - immediately following
        electrical stimulation.
        Between B, C, D, E - intervals of several seconds.
        Calibrations: horizontal bar - 50 msec, vertical
        bar - 15 milivolts for microelectrode and 200
        microvolts of gross electrode.

foci increase and finally high frequency discharges with
marked variability of spike amplitudes appear (Fig. 1C).

    Thus the situation appears to be essentially similar
to that in primary foci, i.e. neurons acquire "epileptic"
features. Activity in such foci becomes self-sustained and
does not depend any more on the volleys arriving from the
primary one.

    Such neuronal spike-potential behaviour must be
evidently conditioned by paroxysmal depolarization shifts

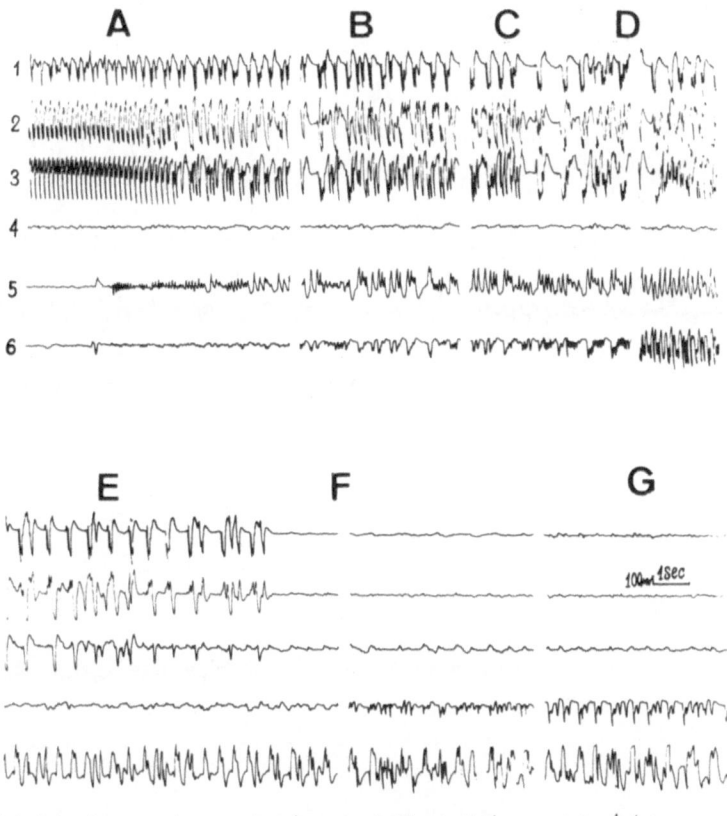

Fig. 3. ECoG of a rabbit. Appearance of seizure discharges in the contralateral cortex after sectioning
the corpus callosum.
Bipolar recording: 1 - motor, 2 - somatosensory,
3 - parietal, 4 - contralateral motor, 5 - contralateral somatosensory, 6 - contralateral parietal cortices.

(PDS's) of membrane potential, a phenomenon typical for
epileptic neurons (9, 13, 14).

    Actually, in secondary foci intracellular records
during long-lasting and recurring epileptic activity,
gradual formation of PDS's was observed (Fig. 2).

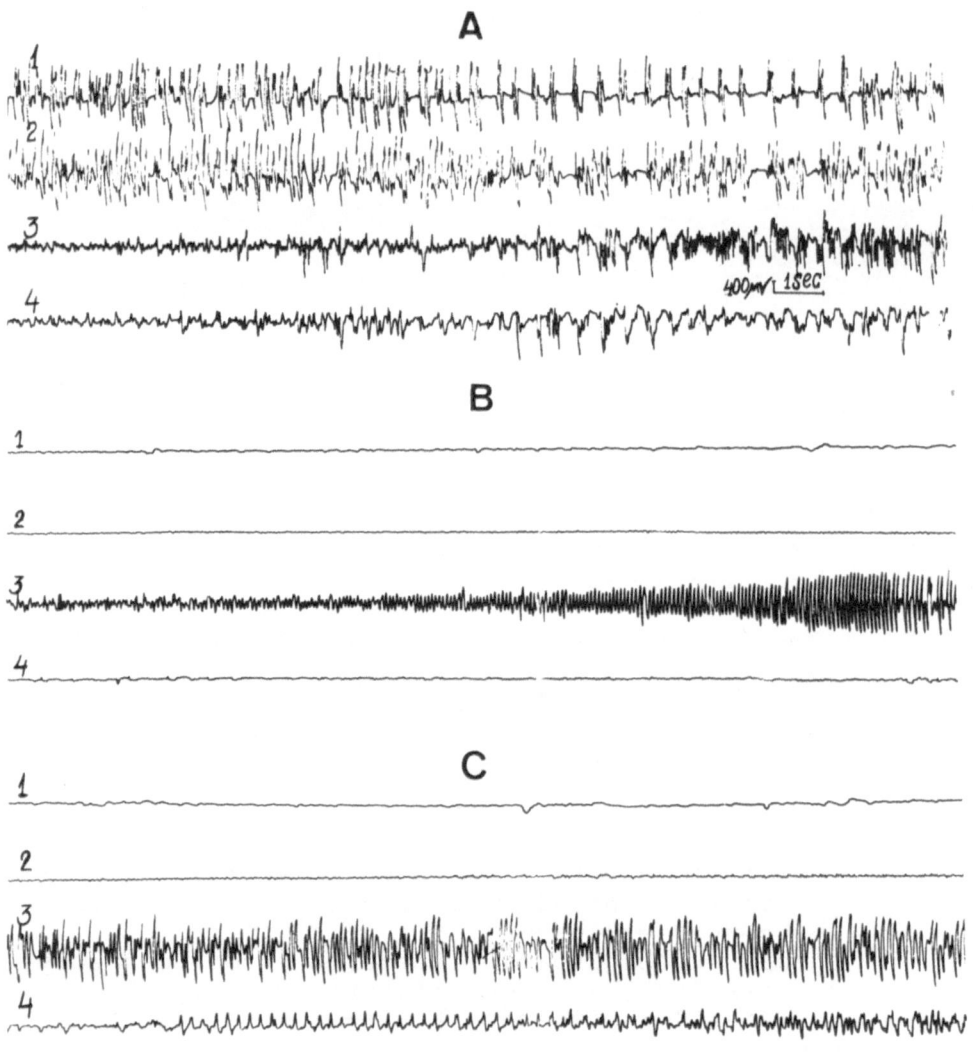

Fig. 4. ECoG of a rabbit.
         A - before and B, C - following depression of
         the epileptogenic focus under the influence of
         KCl.
         Bipolar recording of: 1 - sensorimotor, 2 - visu-
         al (focus), 3 - contralateral visual, 4 - contra-
         lateral sensorimotor cortices.

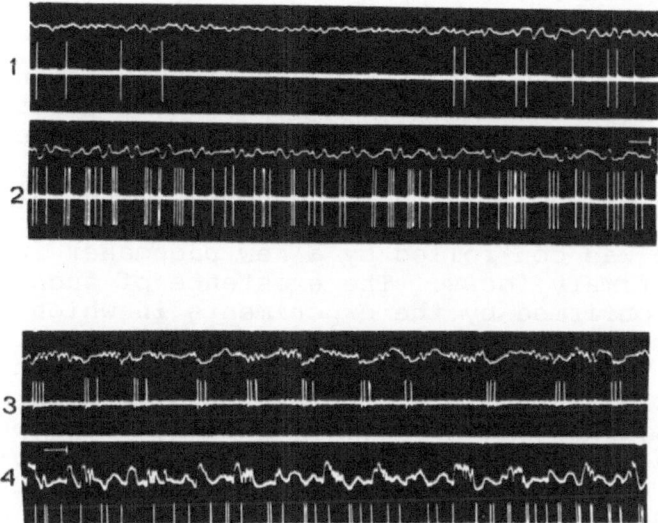

Fig. 5. Extracellular recording of neuronal discharges
in a secondary cortical focus appearing following
the callosotomy in a cat (1, 2) and in a rabbit
(3, 4). During the recording of (3) paroxysmal
activity was observed only in the secondary focus.
Calibrations: 1,2 - horizontal bar - 100 msec,
vertical bar - 300 microvolts for grosselectrode
recording, 3,4 - horizontal bar - 100 msec, vertical bar - 400 microvolts for ECoG.

In contradistinction to the intact brain in preparations with intact callosal connections and transected
cortico-subcortical pathways formation, independent epileptic activity in secondary foci occurs very seldom and
epileptic neuronal discharges are an extremely rare finding.

Thus combined action of both callosal and extracallosal volleys appears to be a favorable condition for the
independent secondary foci formation.

After sectioning, the corpus callosum paroxysmal
activity remains restricted to one hemisphere for a long
time. But in cases of long-lasting and recurrent epileptic activity it can spread to the opposite hemisphere.
Fig. 3 illustrates such findings. This record was preceded by a very long period of seizure activity on the

side of primary focus. At a definite moment paroxysmal discharges started to appear in the symmetrical region of the opposite hemisphere (Fig. 3a).

In a number of cases seizure discharges in the opposite hemisphere remained even after their cessation in the primary focus (Fig. 3e, f, g). Moreover, epileptic activity recorded from the opposite hemisphere could differ in its morphology and frequency from the discharges of the primary focus (Fig. 3). Evidently the opposite hemisphere was controlled by a new pacemaker independent from the primary focus. The existence of such a pacemaker was also confirmed by the experiments in which the primary focus was switched off by means of topical application of potassium chloride (Fig. 4). Epileptic activity in the opposite hemisphere persisted after depression of the primary focus (Fig. 4B, C). These electrophysiological findings of contralateral independent self-sustained epileptic activity could be the result of either a new subcortical independent secondary focus (a new pacemaker) or it could be formed under the influence of corticofugal volleys from the primary focus.

Microelectrode studies were undertaken in order to reveal the nature of paroxysmal discharges recorded in such cases in the contralateral cortex. As shown in Fig. 5 neuronal activity in such cases corresponded to the notion of projected discharges indicating thus the dependent nature of the epileptogenic focus. Fig. 5 (3) illustrates a case when epileptic activity in the primary focus was not recorded. In this case, too, neuronal activity in the symmetrical point where paroxysmal ECoG was observed, was characterized by low frequency discharge of spike potentials correlated with electrocorticographic waves, i.e. typical neuronal activity of a dependent secondary focus was observed (14). These findings confirm the existence of some independent subcortical focus of epileptic activity projecting to the symmetrical area of the contralateral hemisphere. Thus the results of these experiments indicate that in addition to callosal connections cortico-subcortical pathways play a definite role in interhemispheric spread of epileptiform discharges (14). Following the sectioning of the corpus callosum, projected seizure activity may be induced in the opposite hemisphere as a result of long-lasting action of extracallosal volleys. On the other hand association of callosal and extracallosal epileptic volleys appears easily to lead to the formation of true secondary cortical foci of epileptic activity (14).

All results, cited above, indicated the probable participation of subcortical structures (both diencephalic and brain stem) in the interhemispheric spread of epileptic activity. Therefore, the experiments with transverse brain stem section at different levels were performed in order to specify the role of different structures in the spread.

In callosotomized animals a precollicular brain stem section prevented the spread of paroxysmal discharges to the contralateral cortex, while the spread was still observed after cutting the brain stem at an intercollicular level. Thus, presumably, some critical amount of the mesencephalic reticular formation is crucial in such spread.

For further verification of this assumption, in a number of preparations callosotomy was combined with sagittal cutting of the midbrain. The interhemispheric spread was not observed in the single case.

REFERENCES

1. Ajmone-Marsan, C. (1963): Unitary analysis of projected epileptiform discharges. Electroen. Clin. Neurophysiol. 15: 197-208.

2. Erickson, T.C. (1940): Spread of the epileptic discharge. An experimental study of the afterdischarge induced by electrical stimulation of the cerebral cortex. Archs. Neurol. Psychiatr., Chicago, 43: 429-452.

3. Grafstein, B. (1959): Organization of callosal connections in suprasylvian gyrus of cat. J. Neurophysiol., 22: 504-515.

4. Guerrero-Figuerra, R., Barros, A., Heath, R.G., Gonsales, G. (1964): Experimental subcortical epileptiform focus. Epilepsia, 5: 112-139.

5. Holubar, J. (1965): Mode of activation of the mirror focus initiated by local penicillin application to the contralateral cerebral cortex in rats. Physiol. bohemoslovaca, 14: 509-511.

6. Holubar, J. (1966): Different mode of activation of foci. The penicillin and mirror cortical foci in rats. In: Z. Servit (ed.). Comparative and Cellular Pathophysiology of Epilepsy. Amsterdam.

7. Isaakson, R.L., Schwartz, H., Persoff, E., Pinson, L. (1971): The role of the corpus callosum in the establishment of areas of secondary epileptiform activity. Epilepsia, 12: 133-146.

8. Kopeloff, N., Kennard, M.A., Pacella, B.L., Kopeloff, L.M., Chusid, S.G. (1950): Section of corpus callosum in experimental epilepsy in the monkey. Arch. Neurol. Psychiatr. 63: 719-727.

9. Matsumoto, H., Ajmone Marsan, C. (1964): Cortical cellular phenomena in experimental epilepsy: interictal manifestations. Expl. Neurol. 9: 286-326.

10. Morrell, F. (1960): Secondary epileptogenic lesions. Epilepsia. 1: 538-560.

11. Mosidze, V.A., Rishinashvili, R.S., Kevanischvili, Z. Sh. (1967): On the pathways of transmission of strychnine discharges between cerebral hemispheres. Bull. Acad. Sci. of the G.S.S.R., 48: 201-206 (in Russian).

12. Mosidze, V.M., Rishinashvili, R.S., Kevanishvili, Z. Sh., Akbardia, K.K. (1972); The Split Brain. "Metsniereba", Tbilisi (in Russian).

13. Okujava, V.M. (1967): Characteristic of epileptic activity of cortical neurons. In: Problems of Modern Neurology. Tbilisi (in Russian).

14. Okujava, V.M. (1969): Basic Neurophysiological Mechanisms of Epileptic Activity. "Ganatleba", Tbilisi (in Russian).

15. Okujava, V.M., Chipashvili, S.A. (1973): On the role of callosal and extracellular connections in interhemispheric spread of epileptic activity. The Korsakov J. Neurol. Psychiatr. 11: 1679-1684 (in Russian).

16. Penfield, W., Jasper, H. (1954): Epilepsy and functional anatomy of the human brain. Little Brown & Co. Boston.

17. Shelikhov, V.N. (1960): Electrophysiological study of the role of subcortical structures in generalization of excitation in the cerebral cortex. The Korsakov J. Neurol. Psychiat. 2: 145-149 (in Russian).

THE INFLUENCE OF SEROTONIN ON SEIZURE SUSCEPTIBILITY

IN THE GERBIL

K.M.A. Welch, Janes C.H. Chan, Eva Chabi and
Tseng-Pu F. Wang

Laboratory of Clinical and Experimental
  Cerebral Metabolism
Baylor-Methodist Center for
  Cerebrovascular Research

Department of Neurology
Baylor College of Medicine
Houston, Texas 77030, U.S.A.

SUMMARY

5-hydroxytryptophan (5-HTP) (50 mg/kg, I.P.) increased central serotonin (5-HT) levels and delayed the onset of seizure during reflow in gerbils subjected to transient bilateral blockade with dimethyl tryptamine (DMT) (10 mg/kg I.P.) caused early onset of seizure. Results support an influence of 5-HT on seizure threshold in the central nervous system of the gerbil.

Previous studies have shown depletion of monoamines in ischemic brain of the gerbil (9,11). When we subsequently investigated the differences in brain monoamine levels between those animals with cerebral ischemia alone and animals with ischemia that developed seizures, we found that catecholamines were only depleted in the seizure group thus raising the possibility that this was a secondary effect of seizure activity (10). Serotonin (5-HT),however, was decreased in both groups of animals. Since pharmacological depletion or elevation of brain 5-HT respectively facilitates or decreases seizure threshold in other epilepsy models (8), we have tested the hypothesis that

depletion of brain 5-HT is a factor predisposing to the
development of seizures in gerbils subjected to ischemia.

METHODS

Adult male and female Mongolian gerbils (Meriones
unguiculatus) weighing 50-80 gm were studied. Under light
ether anesthesia, both common carotid arteries were dissected free of accompanying vagus nerves and veins and
occluded by the application of a miniaturized Heifetz
aneurysm clip to each artery. After 30 minutes of occlusion the clips were removed (reflow).

Animals were observed during occlusion and for up to
5 hours after reflow for evidence of abnormal motor activity suggestive of seizure. The behavioral criteria de-
manded for inclusion in the study were the observation in
each animal of solitary clonic jerks of a limb or whole
body, wild running, rolling fits and tonic-clonic convulsions (7).

The following experiments were performed:

1. Gerbils were treated at the end of the occlusive
period with 5-hydroxytryptophan (5-HTP) (50 mg/kg I.P.),
dissolved in 0.5 ml saline. Controls were injected with
0.5 ml saline at a similar time interval.

2. Gerbils were injected at the end of the occlusive
period with dimethyl tryptamine (DMT) (10 mg/kg, I.P.),
dissolved in 0.5 ml saline. Controls were injected with
0.5 ml saline at a similar time interval.

In both sets of experiments record was kept of the
seizure incidence, time after reflow of first seizure,
and mortality rate. All animals that survived up to 5
hours after reflow were sacrificed by whole-body immersion
in liquid nitrogen. Animals that died during seizure were
immediately frozen in liquid nitrogen.

Samples of brain cortex were chiseled out under liquid nitrogen. Serotonin (5-HT), 5-hydroxyindoleacetic acid
(5-HIAA) and 5-HTP were isolated and analyzed fluorometri-
cally according to the methods described by Atack and Lindqvist (1).

Statistical analysis was performed using Student's
t- and Chi-square tests.

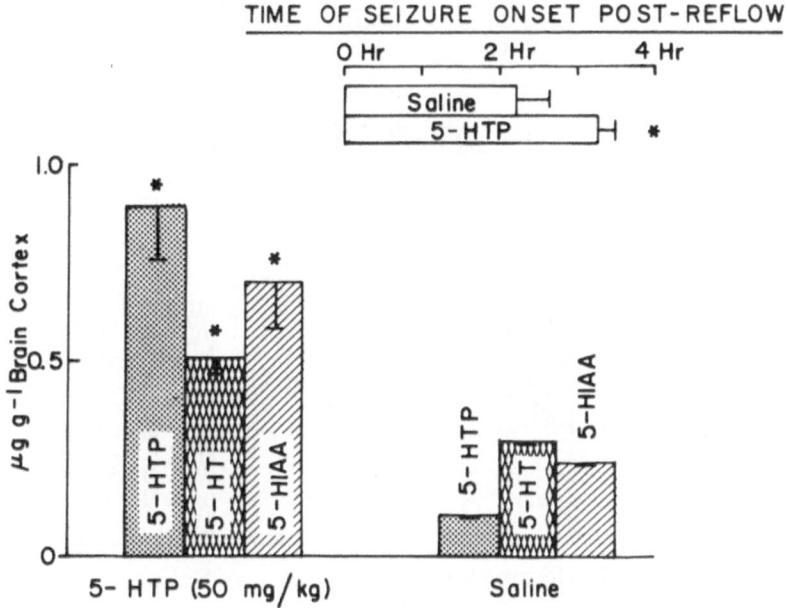

Fig. 1. The upper portion of the figure shows time
of seizure onset during reflow in animals
treated with 5-hydroxytryptophan (5-HTP)
and those given similar amounts of saline
at the end of 30 minutes of bilateral common
carotid artery ligation. In the lower portion,
bar graphs illustrate the levels of 5-HTP,
serotonin (5-HT) and 5-hydroxyindoleacetic
acid (5-HIAA) measured in brain cortex obtained from these animals.

RESULTS

Abnormal motor behavior suggestive of seizure activity was observed in all animals during the occlusive
period. In the 5-HTP-treated group (N = 16), 81% re-developed seizure after reflow and 50% died. Of the saline-
treated controls for this experiment (N = 10), 90% developed seizure and 70% died. In the DMT-treated group
(N = 10), 80% re-developed seizure after reflow and all
animals died. Of the control animals (N = 8), 75% developed seizure but 25% died. This difference in mortality
was, however, not significant.

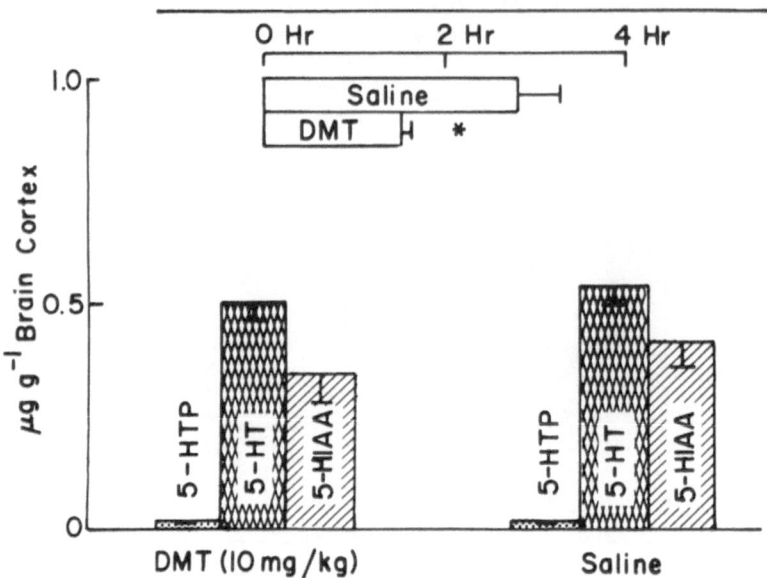

Fig. 2. The upper portion of the figure shows time
        of seizure onset during reflow in animals
        treated with dimethyl tryptamine (DMT) and
        those given similar amounts of saline at
        the end of 30 minutes of bilateral common
        carotid artery ligation. In the lower por-
        tion, bar graphs illustrate the levels
        of 5-HTP, 5-HT and 5-HIAA measured in brain
        cortex obtained from these animals.

Figure 1 shows that the onset of seizure activity
was delayed in 5-HTP-treated animals compared to saline-
treated controls. Measurement of 5-HTP, 5-HT and 5-HIAA
in these animals confirmed increase of cortical 5-HT
levels and of 5-HT turnover. Figure 2 shows that DMT
caused an earlier onset of seizure activity compared to
controls. There was no change in cortical 5-HT levels or
5-HT turnover compared to controls.

When death occured in either of the two experimental
or control groups it was always during or at the end of
a tonic-clonic convulsion.

DISCUSSION

Gerbils may spontaneously develop motor seizures, best classified as a form of reflex epilepsy (7), to which, from a behavioral viewpoint, seizures produced by ischemia in the gerbil are identical. It is unknown, however, if initiation of seizure activity is related to an ischemic neuronal focus or whether ischemia secondarily produces a shift in neurochemical balance whereby spontaneous seizures can be more easily provoked.

We had observed in untreated gerbils that seizure activity caused by bilateral common carotid artery occlusion ceased immediately after occlusion was removed and resumed approximately one hour later. This suggested to us some dynamic cerebral metabolic change in the reflow period that also influenced seizure threshold. Previous studies in the gerbil had shown that depletion of brain 5-HT occurred during ischemia and continued throughout the post-ischemic phase (4). Pharmacological depletion of brain 5-HT has been shown in other seizure models to increase seizure susceptibility (8). We therefore treated gerbils with 5-HTP in order to examine if normalization of brain 5-HT levels influenced post-ischemic seizure activity. The results, which showed delay in seizure onset without influence on the incidence or severity of seizure, seem in keeping with a modification of seizure threshold by the increased levels of 5-HT in the central nervous system.

Other investigators have reported an anti-convulsant action of 5-HTP (5) although this has not always been a consistent finding (3). Nevertheless, some recent studies of cerebrospinal fluid 5-HT metabolites in the epileptic patient have a part related the anti-convulsant action of diphenylhydantoin to an increase in central 5-HT turnover (2). Conversely, by depleting brain 5-HT, an increased seizure susceptibility can be achieved (6), probably by reducing delivery of 5-HT at central receptors. The early treatment with the predominantly 5-HT receptor antagonist DMT in the present study seems in accord with this observation. The results of both experiments reported here, therefore, support the theory that 5-HT modifies seizure threshold in the central nervous system of the gerbil.

ACKNOWLEDGMENTS

Reprint requests to: K.M.A. Welch, M.B.Ch.B.,
M.R.C.P. (UK), Associate Professor, Department of
Neurology, Baylor College of Medicine, Neurosurgery
Center, 6501 Fannin, NB 302, Houston, Texas 77030,U.S.A.

This work was supported by Grant NS 09287 from the
National Institute of Neurological and Communicative
Disorders and Stroke, National Institutes of Health,
Bethesda, Maryland 20014.

REFERENCES

1. Atack, C., Lindqvist, M. (1973): Conjoint native and
   orthophtaldialdehyde-condensate assays for the fluorimetric determination of 5-hydroxyindoles in brain.
   Naunyn Schmiedebergs Arch. Pharmacol. 279: 267-284.

2. Chadwick, D., Jenner, P., Reynolds, E.H. (1977): Serotonin metabolism in human epilepsy: The influence of
   anticovulsant drugs. Ann. Neurol. 1:218-224.

3. Chen, G., Ensor, C.R., Bohner, B. (1968): Studies of
   drug effects on electrically induced extensor seizures
   and clinical implications. Arch. Int. Pharmacodyn.
   172: 183-218.

4. Gaudet, R., Welch, K.M.A., Chabi, E., Wang, T.-P.F.
   (1977): Effect of transient ischemia on monoamine
   levels in the cerebral cortex of gerbils. J. Neurochem.
   (in press).

5. Kilian, M., Frey, H.H. (1973): Central monoamines and
   convulsive thresholds in mice and rats. Neuropharma-
   cology. 12: 681-692.

6. Koe, K.B., Weissman, A. (1966): p-chlorophenylalanine:
   a specific depletor of brain serotonin. J. Pharmacol.
   Exp. Ther. 154: 499-516.

7. Loskota, W.J., Lomax, P., Rich, S.T. (1974): The gerbil as a model for the study of the epilepsies. Epilepsia. 15: 109-119.

8. Maynert, E.W., Marczynski, T.J., Browning, R.A. (1975):
   The role of the neurotransmitters in the epilepsies.
   Advances in Neurology, W.J. Friedlander (ed.), Raven
   Press, New York (Volume 13), 79-147.

9. Welch, K.M.A., Chabi, F., Buckingham, J., Bergin, B.,
    Achar, V.S., Meyer, J.S. (1977): Catecholamine and
    5-hydroxytryptamine levels in ischemic brain. Influ-
    ence of p-chlorophenylalanine. Stroke 8:341-346.

10. Welch, K.M.A., Wang, T.-P.F., Chabi, E. (1977): Ische-
    mia-induced seizures and cortical monoamine levels.
    Ann. Neurol., (in press).

11. Zervas, N.T., Hori, H., Negora, M., Wurtman, R.J.,
    Larin, F., Lavyne, M.H. (1974): Reduction in brain
    dopamine following experimental cerebral ischemia.
    Nature, 247: 283-284.

ANTIEPILEPTIC DRUGS: PHARMACOLOGY AND MECHANISMS OF ACTION

D.M. Woodbury

Department of Pharmacology, University of Utah
College of Medicine, Salt Lake City, Utah, USA

The rational therapeutic use of antiepileptic drugs requires a complete knowledge of their pharmacology. This includes understanding their absorption, distribution, biotransformation and excretion in the body (ADBE), their pharmacological and toxic effects, and their mechanisms of action.

In order for a drug to act it must reach its site of action, which in the case of antiepileptic drugs is ultimately the central nervous system (CNS). The sequence of events between injection of these drugs and their action on the CNS receptors is shown in Fig. 1. Also shown are the dose-effect relation of drugs in man and the various factors that influence it.

Of the ADBE factors shown in Fig. 1, the completeness of absorption is one of the most important in determining the plasma level. Absorption depends on a host of physical and chemical factors such as solubility, disintegration and dissolution rates of solid drugs, particle size, pH of gastrointestinal fluids, pKa of the drug, Ca/PO_4 ratio, bile salt concentration, amount of food present, etc. In the case of phenytoin and other antiepileptics with very low solubility the amount absorbed is erratic. This results in considerable variation in plasma levels and inconstant therapeutic effects. Another important factor is the apparent volume of distribution of drugs, which depends on such factors as the size and extent of penetration by the drug of the extracellular and intracellular compartments, whether active transport

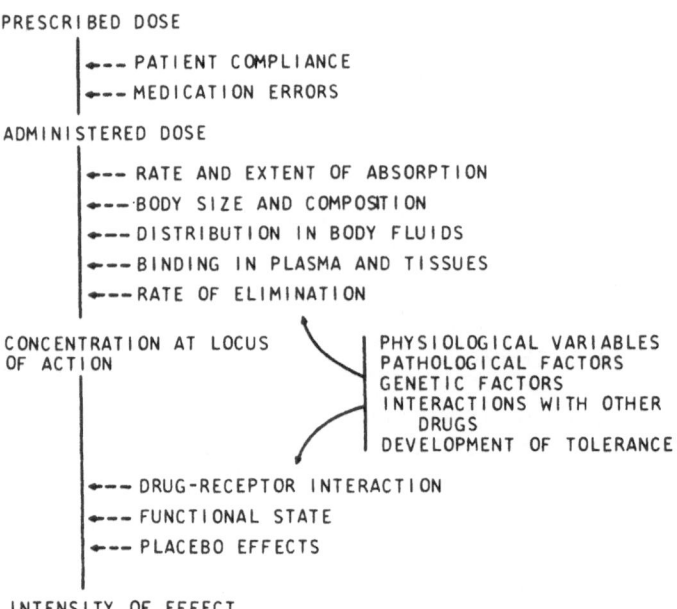

Fig. 1. Factors that determine the relationship between
        prescribed drug dosage and drug effect. (Modi-
        fied from Koch-Weser, 8).

is involved and whether binding to inactive sites occurs.
For example, the volume of distribution of phenytoin is
about 10 times the body weight because it is bound to
plasma and tissue constituents to the extent of about 90%.
On the other hand, trimethadione (TMO) is distributed in
total body water whereas bromide is distributed mainly in
extracellular fluid.

     The rate of elimination of anticonvulsant drugs is
probably the most important factor in regulating the plasma level. Elimination occurs either by biotransformation
of drugs from the active to the inactive form or by excretion in the urine or feces or by both processes.

     The intensity of the effect of the drug on the recep-
tor is altered by such factors as the functional state of
the body, pathologic disturbances, development of tolerance, and the presence of other drugs (described below).

     It is apparent, therefore, that the enormous indivi-
dual differences in the relation between the dosage of a

Table I PHARMACOKINETIC PARAMETERS OF ANTI-
EPILEPTIC DRUGS AND THEIR ACTIVE METABOLITES

| | PARENT | | METABOLITES | | | |
| | | | ACTIVE | | | INACTIVE |
NAME	PLASMA CONCENTRATION FOR EFFICACY µg/ml	PLASMA HALF-TIME HOURS	NAME	PLASMA CONCENTRATION µg/ml	PLASMA HALF-TIME HOURS	INACTIVE
PHENYTOIN	10-20	24	--			P-HPPH GLUCURONIDE
PHENOBARBITAL (PB)	10-30	96-144	--			pHYDROXYPHENOBARBITAL
CARBAMAZEPINE	3-7	12-30+	--			EPOXIDES, HYDROXYLATED DERIVATIVES
ETHOSUXIMIDE	40-100	36	--			HYDROXYLATED DERIVATIVES
SODIUM VALPROATE	66-117	9.5 (3.9-13.5)	--			VPA-GLUCURONIDE; 2-N-PROPYL-5-HYDROXY PENTANOIC ACID; 2-N-PROPYLGLUTARIC ACID
MEPHENYTOIN	2-3.2	<NIRVANOL	NIRVANOL 5-PHENYL-HYDANTOIN	23-37	LONG	HYDROXYNIRVANOL
ETHOTOIN						HYDROXYLATED DERIVATIVE 5-PHENYLHYDANTOIC ACID
MEPHOBARBITAL	5-14	<PB	PHENOBARBITAL	10-30	96-144	HYDROXYLATED DERIVATIVES
PRIMIDONE		3-12+	PHENYLETHYL-MALONAMIDE PHENOBARBITAL	4-17 / 10-30	24-48 / 96-144	HYDROXYLATED DERIVATIVES
DIMETHOXY-METHYL-PHENO-BARBITAL (DMMP, ETEROBARB)	NOT DETECTABLE (<50 PICOGRAMS/ml)	VERY FAST	MONOMETHOXYMETHYL-PHENOBARBITAL	<1.0 ?	1.5	HYDROXYLATED DERIVATIVES
			PHENOBARBITAL	10-30	96-144	HYDROXY-PHENOBARBITAL
TRIMETHADIONE	21 (6-41)	12-24	DIMETHADIONE	700 (415-1200)	144-312	
PARAMETHADIONE		<METABOLITE	N-DEMETHYLATED DERIVATIVE		SIMILAR TO DMO	
METHSUXIMIDE	0.04	1.4 (1.0-2.2)	2-METHYL-2-PHENYL-SUCCINIMIDE	20-40	38	HYDROXYLATED DERIVATIVES
PHENSUXIMIDE	5.7 (3.9-7.9)	7.8 (4.5-12)	2-PHENYL-SUCCINIMIDE	1.7 (1.5-2.1)	~8	HYDROXYLATED DERIVATIVES OTHERS
DIAZEPAM	0.16-0.70	8-10	N-DESMETHYLDIAZEPAM HYDROXIDIAZEPAM OXAZEPAM	1.2 (1.0-2.0)	48-192	GLUCURONIDES
CHLORAZEPATE	NOT DETECTABLE	VERY FAST	N-DESMETHYLDIAZEPAM		48-192	GLUCURONIDES

drug and the intensity of its pharmacologic action are
a result of the many factors shown in Figure 1 that
influence the steps between dosage and action. It is
important to realize that the variability of the dose-
effect relation among patients is primarly due to individual differences in the plasma concentration achieved
with a given dosage schedule rather than to a different
intensity of action associated with the same plasma concentration. Thus the intensity of the pharmacologic effect correlates remarkably well with the concentration
of the free drug in the plasma and unpredictably with
the amount administered. Hence a more accurate prediction
of the therapeutic response can be made from the plasma
levels than from the dosage, and dosage adjustments can
often be better guided by knowledge of plasma levels.

     Pharmacokinetics constitutes that aspect of pharma-
cology which provides concise descriptions of the absorption, distribution and elimination of drugs. The concepts
derived from pharmacokinetics form a useful framework for
the quantitative analysis of pharmaceutical dosage forms.
They also provide a semiquantitative guide for the interpretation of measured plasma concentrations of drugs and
for the choice and adjustment of dosage schedules for
therapy.

     Pharmacokinetic principles are useful only insofar
as the data used to formulate the dosage schedules are
reliable. The most critical need concerns accurate bio-
logic half-life data for drugs and the metabolites. Also
important are data on plasma and tissue binding, absorption half-life and fractional absorption. A summary of
the plasma levels of both parent and the daughter metabolites of the antiepileptic drugs that produce therapeutic effects, and also their half-lives is shown in
Table 1. These data are useful for setting up dosage schedules for administration of antiepileptic drugs to patients.
The table also gives information about the active form of
these drugs in man.

     It is important to emphasize that in the case of
some of these drugs the plasma half-life is a function
of the dosage. For example, as the dosage of phenytoin
is increased the half-time is prolonged. The dose-depen-
dency is best explained by saturation of a rate-limiting
enzyme reaction in the metabolism of the drug. The question that arises is when should plasma levels be determined. The plasma levels of patients receiving antiepi-

leptic drug therapy should be measured under the following circumstances: when there are symptoms and/or signs of INTOXICATION; when there is failure of response to therapeutic doses; and to determine the baseline plasma-dose ration when first initiating therapy.

In many cases toxicity due to drug overdosage cannot be distinguished from toxicity due to pathological processes. Any epileptic patient on antiepileptic therapy with toxicity should have plasma levels of the drug in question measured.

If a patient does not respond to a drug when given in the usual therapeutic doses blood levels should be measured for several reasons. In many cases the patient is not taking the drug. Examples of noncompliance after phenytoin and ethosuximide have been described in previous publications (4, 7, 11).

It is evident from these investigations that outpatients do not reliably ingest their medication and that counseling the patient on the importance of taking their medication is of considerable value.

Other reasons for failure of response to therapeutic doses are defects in absorption, genetic factors, diseases of the gastrointestinal tract, etc. and the presence of other drugs.

DRUG INTERACTIONS

The administration of other drugs simultaneously with antiepileptic drugs may alter the clinical response to the usual therapeutic doses in many ways. They may increase the biotransformation of the drug by inducing the enzyme system in the endoplasmic reticulum of the liver concerned with metabolism of drugs (See Mannering, 10, for summary). This would reduce the plasma level of the drug and decrease its half-life. Usual doses, would, therefore, not give the same response. Many drugs can induce this system, and lower antiepileptic drug levels. Among these are the barbiturates, particularly phenobarbital, ethanol, phenytoin, DDT, glutethimide, and griseofulvin. Induction of liver drug-metabolizing enzymes does not always decrease the therapeutic response since many antiepileptic drugs are biofransformed into products that are also active antiepileptics. For example, TMO is converted by the liver to its N-demethylated derivate, dimethadione (DMO), which has anticonvulsant activity slightly higher than that of TMO.

Table II   SUMMARY OF CLINICALLY IMPORTANT ADVERSE

ANTICONVULSANT	INTERACTING DRUGS
PHENYTOIN	ALCOHOL
	AMINOGLYCOSIDE ANTIBIOTICS
	CHLORAMPHENICOL
	ISONIAZID
	ANTICOAGULANTS, ORAL
BARBITURATES	ALCOHOL
	TETRACYCLINES
	ANTICOAGULANTS, ORAL
	TRICYCLIC ANTIDEPRESSANTS
	CORTICOSTEROIDS
	TETRACYCLINES
	ALCOHOL

INTERACTIONS OF ANTICONVULSANTS

ADVERSE EFFECT	PROBABLE MECHANISM
ENHANCED ANTICONVULSANT EFFECT WITH ACUTE INTOXICATION	REDUCED METABOLISM
DIMINISHED ANTICONVULSANT EFFECT WITH CHRONIC ALCOHOL ABUSE	ENHANCED METABOLISM
INCREASED TOXICITY OF PHENYTOIN	INHIBITION OF MICROSOMAL ENZYMES
INCREASED TOXICITY OF PHENYTOIN	INHIBITION OF MICROSOMAL ENZYMES
INCREASED TOXICITY OF PHENYTOIN	INHIBITION OF MICROSOMAL ENZYMES
INCREASED TOXICITY OF PHENYTOIN WITH DICUMAROL	INHIBITION OF MICROSOMAL ENZYMES
ENHANCED CNS DEPRESSION WITH ACUTE INTOXICATION	ADDITIVE: REDUCED METABOLISM
DIMINISHED SEDATIVE EFFECT WITH CHRONIC ALCOHOL ABUSE	ENHANCED METABOLISM
DECREASED DOXYCYCLINE EFFECT	INDUCTION OF MICROSOMAL ENZYMES
DECREASED ANTICOAGULANT EFFECT	INDUCTION OF MICROSOMAL ENZYMES
DECREASED ANTIDEPRESSANT EFFECT	INDUCTION OF MICROSOMAL ENZYMES
DECREASED STEROID EFFECT	INDUCTION OF MICROSOMAL ENZYMES
DECREASED DOXYCYCLINE EFFECT	INDUCTION OF MICROSOMAL ENZYMES
ENHANCED CNS DEPRESSION	ADDITIVE

Table III   CONCENTRATIONS OF ANTIEPILEPTIC DRUGS
            IN TISSUES RELATIVE TO PLASMA CONCENTRATION
            (PLASMA = 1.00)

DRUG	V_D Liters/kg	BRAIN	MUSCLE	LIVER	CSF
PHENYTOIN	0.6	1.6	1.5	3.2	0.10-0.2
PHENOBARBITAL	1	0.66	0.69	1.1	0.45
CARBAMAZEPINE	4-8.8	2.45	---	2.89	0.22
ETHOSUXIMIDE	0.9	1.04	0.94	1.10	0.90
TRIMETHADIONE	0.82	0.82	0.94	0.95	1.0
DIMETHADIONE	0.52	0.35	0.30	0.45	0.70
DIAZEPAM	1.75	0.70	1.1	14.6	
DIPROPYLACETATE	0.2	0.36	0.32	0.66	0.16

Enzyme induction by another drug might actually enhance
the effect of TMO by increasing the amount of DMO, although the plasma level of TMO would be decreased. Other
examples are primidone which is converted by the liver to
phenobarbital and phenylethylmalonamide (PEMA) both of
which are active anticonvulsants (Gallagher and Baumel,
3), and diazepam which is biotransformed to oxazepam and
other metabolites which possess anticonvulsant activity.

     Other drugs may cause inhibition of the drug-metabo-
lizing system in the liver. In this case the plasma level
and half-lives increase and toxicity often occurs. Examples of this are the effects of sulthiame, disulfiram,
phenyramidol and isoniazid to inhibit the biotransformation of phenytoin.

     Alteration of the therapeutic response to an antiepi-
leptic agent resulting from drug interactions can also result from competition for plasma binding sites. Thus phenytoin is about 90% bound to plasma proteins. The free
level of phenytoin in plasma can be increased by administration of salicylates, thyroxine, phenylbutazone, and
others that complete for the binding sites on the protein.
The unbound fraction of phenytoin increases more than

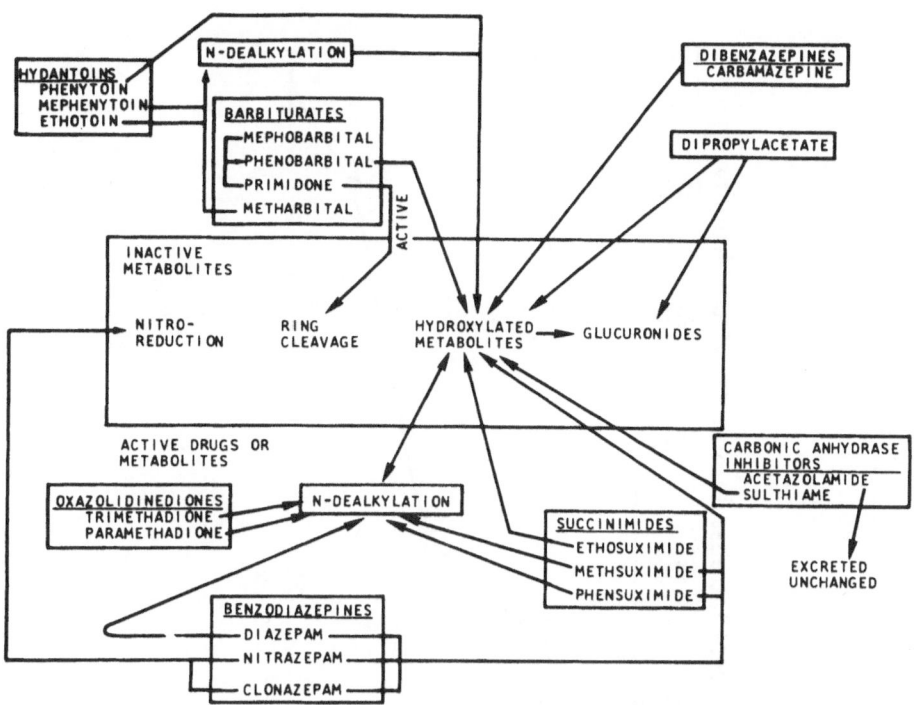

Fig. 2. Pathways for biotransformation of the antiepilep-
         tic drugs. The outer boxes contain drugs that are
         active or metabolized to active drugs. The inner
         box contains the metabolites of the active drugs
         that are inactive and usually represent the final
         product of the biotransformation process.

twofold in the presence of salicylic acid. Interactions
also can occur at the receptor level and thereby alter
the therapeutic response to the drug.

    A summary of the clinically important adverse inter-
actions of the anticonvulsants is presented in Table II.
Most of these involve enzyme induction of liver drug-metabolizing enzymes.

    A final reason for measuring plasma levels of anti-
epileptics is to determine the base line plasma-dose ratio.
This is essential when beginning therapy with a new drug
in order to determine the variability of the dosage-plasma
level response. It is evident from an examination of the

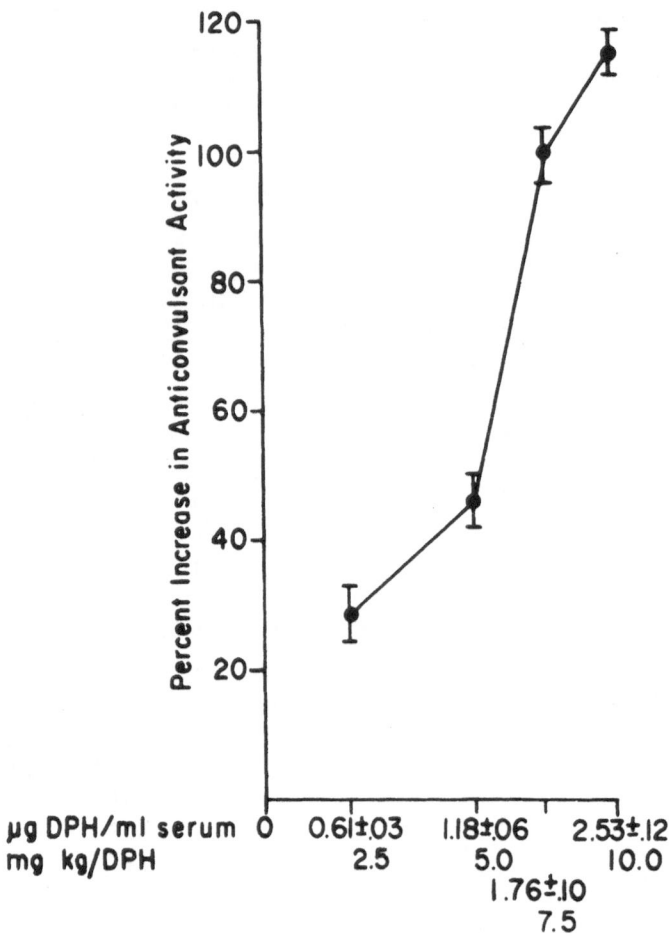

Fig. 3. Relationship between plasma phenytoin concentra-
        tion, or administered dose, and anticonvulsant
        activity as assessed by the changes in the exten-
        sion to flexion ratio observed in rats given maxi-
        mal electroshock seizures. From DeVore and
        Woodbury (1). Used by permission.

available data that although there is a relation between
the two the variation is so large that the plasma level
cannot reliably be predicted from the dosage although the
response can be predicted accurately from the plasma levels.

Distribution. Once the drug leaves the plasma it is distributed to the various tissues of the body where it may bind to non-specific receptors or to the drug receptor where it exerts its action. Measurement of the volume of distribution of these drugs (V_D) gives some clues about the mechanisms of distribution. If the V_D is greater than body water then binding to tissue constituents, active transport into cells, or accumulation in fat or bone, or other storage areas, is present. If V_D is less than total body water but more than extracellular fluid volume, then a cellular action is suggested. If V_D is extracellular then an extracellular action offers the most likely possibility. In Table III is shown a summary of the distribution of some of the clinically useful antiepileptic drugs in the whole body (V_D) and in various tissues and cerebrospinal fluid (CSF). The levels in the CSF generally reflect the free level in the plasma and, therefore, are a measure of plasma binding. However, active transport out of the CSF can occur with weak electrolytes and this compounds the problem of interpreting the meaning of the CSF levels. If large loads of the anticonvulsant drug in question are given this will saturate any transport system and raise the CSF level. If no change in CSF level results from a carrier dose then its level probably represents the free level in the plasma and would be a measure of plasma binding. In any event the Brain/CSF ratio is a measure of the ability of the brain to accumulate the drug since the CSF and not the plasma represents the extracellular fluid of the brain. Most accumulation in brain seems to be by binding to tissue constituents (phenytoin, phenobarbital, carbamazepine, diazepam). The accumulation of dipropylacetate appears to be active transport (see below), and the distribution of dimethadione is by non-ionic diffusion to pH. Trimethadione is distributed in total brain water.

Biotransformation. A summary of the biotransformation pathways of the various anticonvulsant drugs is shown in Figure 2. These reactions involve mainly the liver microsomal drug metabolizing system. The principal pathways are side-chain aromatic ring oxidations, N-dealkylation, deaminations and glucuronide conjugations. The outer box indicates active drugs or active metabolites of the parent drug. The inner box represents metabolites that are inactive and usually are the final product of the biotransformation process. The final fate of these metabolites is excretion from the body, mainly in the urine, but also to a minor extent in the feces.

Table IV   RELATIONSHIP BETWEEN MEAN PLASMA CONCENTRATION
            OF DIPHENYLHYDANTOIN AND NUMBER OF GRAND MAL
            SEIZURES*   (FROM LUND, 9)

PLASMA CONCENTRATION OF DIPHENYLHYDANTOIN, μg/ml					NO. OF GENERALIZED SEIZURES		
YEAR	MEAN	SD	N[+]	TOTAL	MEAN	RANGE	t TEST
1	6.1	2.9	99	186	5.8	0-43	$p < .001$
2	11.7	3.3	203	132	4.1	0-22	
3	15.0	2.5	222	52	1.6	0-9	$p < .001$

* Thirty-two outpatients observed prospectively for
  three years.

[+] N = No. of plasma samples each year.

RELATION OF PLASMA CONCENTRATION TO ANTICONVULSANT
EFFECT

It is important to assess whether the plasma levels
reflect the ability of the drugs to protect against seizures. Experimentally this can be readily verified. In
Fig. 3 is depicted the relation in rats between plasma
concentrations of phenytoin (abscissa) and its effect on
maximal electroshock seizures as a measure of anticonvulsant effect (ordinate) (1). It is evident that there is
an excellent dose-effect relationship. In addition,
Guberman et al. (6) have shown an excellent correlation
between ethosuximide plasma concentrations and percentage
of remaining epileptic activity in the EEG of cats, compared with the control values. The epileptic activity was
induced by penicillin. This correlation is demonstrated
in Fig. 4. There is also a correlation between anticonvulsant activity and brain concentrations of drugs. This has
been demonstrated by Gallagher and Baumel (3) in rats for
primidone and phenobarbital. In man Lund (9) has shown an
excellent correlation between plasma levels of patients
with generalized tonic-clonic seizures (grand mal) and
control of these sizures, as shown in Table IV. The patients were followed for a three-year period. Each year the
plasma levels were elevated and the control of seizures
improved.

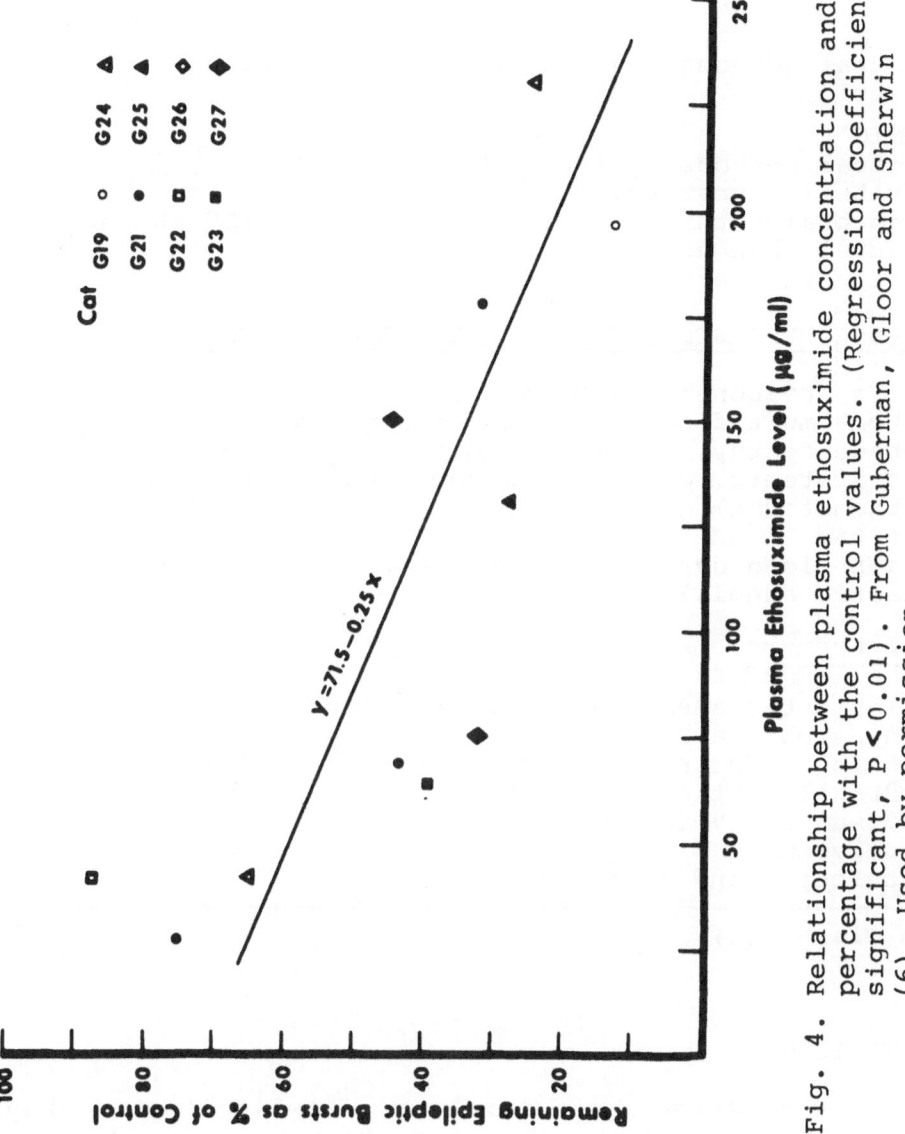

Fig. 4. Relationship between plasma ethosuximide concentration and percentage with the control values. (Regression coefficient significant, $P < 0.01$). From Guberman, Gloor and Sherwin (6). Used by permission.

Table V   DPH TOXICITY

OVERDOSAGE EFFECTS

ACUTE
  Central nervous system (cerebellar-vestibular,delirium)
  Gastric

CHRONIC
  Central nervous system
  Cerebellar-vestibular
  "Encephalopathy": increased seizures, EEG changes,
   mental changes
  Peripheral nervous system

PROBABLE HYPERSENSITIVITY, "ALLERGIC" EFFFCTS

Febrile responses with various dermatoses
Erythema multiforme, Stevens-Johnson syndrome
Lymphadenopathy, pseudolymphoma (? lymphoma)
Acute systemic collagen disorder reactions (lupus
 erythematosus)
Hepatitis
Certain blood dyscrasias (neutropenia, agranulocytosis,
 aplastic anemia)

OTHER EFFECTS

Megaloblastic anemia (folate deficiency)
Gum hyperplasia
Endocrine effects (role in toxicity unclear)
   Adrenal-cortical, pituitary
   Hirsutism (young females)
Hyperglycemia
Hypocalcemia and osteomalacia

From Glaser (5)

      TOXICITY

    The side effects and toxicity of antiepileptic drugs
are many. Those for phenytoin have been well documented by
Glaser (5) and are summarized in Table V (taken from his
paper). Two general classes of toxicity are noted: those
that are dose-dependent and those that are hypersensitivity reactions.

Table VI   DOSE AND PLASMA LEVEL OF DIPHENYLHYDANTOIN
              AND THE CLINICAL SYMPTOMS IN SIX PATIENTS
              WITH ACUTE SIDE EFFECTS*   (FROM LUND, 9)

PATIENT	DOSE OF DIPHENYL-HYDANTOIN, mg/kg	PLASMA LEVEL OF DIPHENYL-HYDANTOIN, ug/ml	FA-TIUGE	NYSTAG-MUS	ATA-XIA	SOM-NO-LENCE
1	9.7	34.9	+	+	+	-
2	10.0	34.1	+	+	+	+
3	7.3	30.9	+	+	-	-
4	6.5	30.9	+	+	-	-
5	9.0	28.9	+	+	-	-
6	7.0	24.8	+	+	-	-

*+Indicates symptoms present; -, absent.

There is also a relationship between plasma concentration of a drug and those aspects of toxicity that are dose-dependent. For example, in Table VI taken from the work of Lund (9) is seen the relation between dose, plasma level and the clinical symptoms in six patients with acute side effects to phenytoin. The effects are roughly dose-dependent. Others have also demonstrated this relationship.

PRINCIPLES OF PHARMACOTHERAPY OF SEIZURES

A summary of the recommended therapy for treatment of the various types of seizures in epileptic patients is shown in Table VII taken from the excellent analysis of Porter and Penry (12). The choice and efficacy of an antiepileptic agent depends first and foremost on seizure type. Hence correct diagnosis is a cardinal rule for therapy of seizure disorders. The responsiveness of the patient, then, depends on choosing the right drug for the identified seizure type and administering the drug according to sound pharmacological principles.

There seems little doubt that many patients can be controlled with only one drug, and there is little justification for starting a new patient on more than one medi-

Table VII   RECOMMENDED THERAPY FOR TREATMENT OF VARIOUS
            SEIZURE DISORDERS IN EPILEPTIC PATIENTS.
            (FROM PORTER AND PENRY, 12)

Seizure Type	Drugs That are Most Efficacious With the Least Toxicity	Drugs That Have Some Efficacy and Toxicity
Simple Partial Seizures	Phenytoin Carbamazepine Valproate	Primidone Phenobarbital Metharbital Mephenytoin
Complex Partial Seizures	Carbamazepine Phenytoin Valproate (?)	Primidone Phenobarbital Mephenytoin Clonazepam (?)
Generalized Tonic-Clonic Seizures	Phenytoin Carbamazepine Valproate (?)	Primidone Mephenytoin Phenobarbital Clonazepam (?)
Absence Seizures	Ethosuximide Valproate	Clonazepam Trimethadione Methsuximide Paramethadione(?) Phensuximide (?)
Infantile Spasms	ACTH or Corticosteroids (for Short-Term Therapy) Prolonged Therapy ?	Nitrazepam Clonazepam Valproate (?)
Atonic/Akinetic Seizures	Valproate	Clonazepam Nitrazepam

cation. The drug of choice should be administered and the
dose increased until mild toxicity is noted and/or therapeutic drug concentrations reached. If this does not control the patient then a second drug may be added to the
first. If seizure control is obtained by the two drugs,
an attempt gradually to remove the first medication is in

Table VIII   SUMMARY OF POSSIBLE MECHANISMS OF ACTION OF ANTIEPILEPTIC DRUGS

DRUG	NEUROPHYSIOLOGICAL EFFECTS	BIOCHEMICAL AND BIOPHYSICAL EFFECTS	PROPOSED MECHANISMS
PHENYTOIN	↓Spread of seizure discharge ↑Cerebellar discharge ↓PTP ↓Threshold in neonatal animals	↑Na transport ↓Na permeability ↑Na permeability ↓Protein synthesis ↑Liver cell EM ↑K efflux ↓Ca^{++} influx	Stimulate Na transport ↑Na permeability ↓Na permeability ↓Ca influx into synaptic endings ↓Synthesis of excitatory transmitter
PHENOBARBITAL	↓Spread of seizure discharge ↑Convulsive threshold for electrical stimulation of brain	↓Q_{O_2} of brain ↓ACh production ↑K permeability of Aplysia neurons	↑K permeability of neurons and thereby cause hyperpolarization
TRIMETHADIONE	↓Repetitive discharge ↑Convulsive threshold for electrical and pentylenetrazol (PTZ) stimulation ↓SCC↑ by PTZ	↓Permeability to K of epithelial cell membrane Alters metabolism of GABA	↓Permeability of glial cells to K ↑GAD activity and reverse inhibition by isoniazid
PRIMIDONE	↓Spread of seizure discharge ↑Convulsive threshold	?	?
ETHOSUXIMIDE	↑PTZ convulsive threshold ↓SCC↑ by PTZ	↓Permeability to K Alters GABA metabolism	↓Permeability to K ↑GAD activity and reverse inhibition by isoniazid
DIAZEPAM	↑Convulsive threshold to PTZ ↓SCC↑ by PTZ	↓K permeability Alters GABA metabolism	↓K permeability ↑GAD activity and reverse inhibition by isoniazid
ACETAZOLAMIDE	↓Spread of seizure discharge ↓CO_2-induced seizures and withdrawal seizures ↓Monosynaptic action potential in spinal cord	↓Carbonic anhydrase activity in brain and choroid plexus ↑Total CO_2 in brain ↓Na uptake in brain ↓CSF flow	↓Carbonic anhydrase activity of glial and choroid plexus cells ↑CO_2 in brain and ↓CSF flow
Na VALPROATE	↑Convulsive threshold for electrical, PTZ picrotoxin and strychnine seizures	Inhibits GABA-T activity in brain. ↑K^+ permeability in Aplysia neurons ↑GAD activity	↑Brain GABA concentration hyperpolarizes Aplysia neurons by increase in K^+ permeability

order, leaving the second drug, which may also alone
control the attacks. However, both drugs may be necessary
to obtain maximal control. Occasionally even a third drug
may be necessary, but this increases the chance for drug
interactions and should be tried only after vigorous
attempts, including serial determination of plasma concentrations of all the drugs used, to control the patients
with fewer drugs have failed. Some patients may require
multiple medications because of multiple seizure types
that have different mechanisms, e.g., patients with generalized tonic-clonic seizures and absence attacks. Even in
this case, single drug therapy is possible by use of sodium valproate.

MECHANISM OF ACTION OF ANTIEPILEPTIC DRUGS

In Table VIII is summarized briefly the neurophysiological and biochemical effects of the commonly used antiepileptic drugs and their proposed mechanisms of action,
where known. It is apparent from this table that there are
large gaps in our knowledge of these drugs and that much
more research is essential. It also seems evident from the
results shown in this table that many of the effects of
these drugs are exerted on cell membranes and related to
the active or passive movement of ions across these membranes.

In summary, an attempt has been made to provide an
insight into the pharmacology of anticonvulsant drugs and
to summarize the various sites at which anticonvulsant
drugs might act. Although much information has been collected and many insights into possible mechanisms of action obtained, the precise mechanism by which any anticonvulsant drug acts is still not known, even for a simple
ion such as bromide, a drug whose anticonvulsant effect
has been known since 1857.

REFERENCES

1.   DeVore, G.R. and Woodbury, D.M. (1977): Phenytoin:
      an evaluation of several potential teratogenic mecha-
      nisms. Epilepsia. 18: 387-396.

2.   Fingl, E. (1972): In: Antiepileptic Drugs, edited by
      D.M. Woodbury, J.K. Penry and R.P. Schmidt. Raven
      Press, New York. 7-21.

3. Gallagher, B.B. and Baumel, I.P. (1972): Primidone: biotransformation. In: Antiepileptic Drugs, edited by D.M. Woodbury, J.K. Penry and R.P. Schmidt. Raven Press, New York. 361-366.

4. Gibberd, F.B., Dunne, J.F., Handley, A.J. and Hazleman, B.L. (1970): Supervision of epileptic patients taking phenytoin. British Med. J. i: 147-149.

5. Glaser, G.H. (1972): Diphenylhydantoin. Toxicity. In: Antiepileptic Drugs, edited by D.M. Woodbury, J.K. Penry and R.P. Schmidt. Raven Press, New York, 219-226.

6. Guberman, A., Gloor, P. and Sherwin, A.L. (1975): Response of generalized penicillin epilepsy in the cat to ethosuximide and diphenylhydantoin. Neurology 25: 758-764.

7. Haerer, A.F., Buchanan, R.A. and Wiggul, F. (1970): Ethosuximide blood levels of epileptics. J. Clin. Pharmac. 10: 370-374.

8. Koch-Weser, J, (1972): Serum drug concentrations as therapeutic guides. New Engl. J. Med. 287: 227-231.

9. Lund, L. (1974): Anticonvulsant effect of diphenylhydantoin relative to plasma levels. A prospective three-year study in ambulant patients with generalized epileptic seizures. Arch. Neurol. 31: 289-294.

10. Mannering, G.J. (1972): Biotransformation. In: Antiepileptic Drugs, edited by D.M. Woodbury, J.K. Penry and R.P. Schmidt. Raven Press, New York. 23-43.

11. Penry, J.K., Porter, R.J. and Dreifuss, F.E. (1972): Relation of plasma levels to clinical control. In: Antiepileptic Drugs, edited by D.M. Woodbury, J.K. Penry and R.P. Schmidt. Raven Press, New York. 431-441.

12. Porter, R.J. and Penry, J.K. (1978): Efficacy and choice of antiepileptic drugs. Proceedings of the Congress and Symposium on Epilepsy. Amsterdam, Sept. 1977.

MECHANISM OF ACTION OF PHENYTOIN: EVIDENCE FOR A CEREBELLAR LOCUS

D.W. McCandless, W.D. Lust, G.K. Feussner and
J.V. Passonneau

Laboratory of Neurochemistry
National Institute of Neurological and
   Communicative Disorders and Stroke
National Institutes of Health
Bethesda, Maryland 20014, U.S.A.

The metabolic events in the brain associated with experimental seizures have been the subject of recent investigations in this laboratory. Maximal electroconvulsive shock (MES) has been used to induce seizures in mice (8, 9, 10). While the stimulus to the brain is undoubtedly excessive and does not exactly mimic the true seizure state, the nature and characteristics of the response are highly reproducible. Using this model, the effect of MES on cerebral and cerebellar energy metabolites and cyclic nucleotides has been investigated. The anticonvulsant, phenytoin, was demonstrated to have a locus of action in the whole cerebellum. Consequently, further studies were made on the effect of MES in the absence and presence of phenytoin on the layers of frozen-dried sections of the cerebellum in an attempt to localize more accurately the effects of convulsions and drug actions (4, 11).

MATERIALS AND METHODS

The experimental animals used were male, NIH general purpose mice weighing 25-27 g and fed ad lib. Maximal electroshock was administered by corneal electrodes at an intensity of 50 mA for 0.2 sec. At the appropriate time intervals, the whole animals were frozen in liquid N_2 with rapid stirring. When phenytoin was administered,

a dose of 25 mg/kg IP was given 25 min prior to electroshock; the animals were subsequently treated like the
animals without drug. Animals injected with saline and
then frozen served as controls.

The animals were stored at $-60^{\circ}C$, and dissected in
a cryostat at $-20^{\circ}C$. A wedge of cerebellum and a 1-2 mm
layer of superficial temporal cortex were removed and
extracted in one ml of 0.3 N perchloric acid containing
1 mM EGTA. The homogenate was centrifuged, the supernatant fluid removed and neutralized with one-tenth volume
of 3.0 M $KHCO_3$. The remaining pellet was dissolved in
one ml of 1 N NaOH for protein analysis. Protein was
measured according to Lowry et al. (7), and lactate, P-
creatine and ATP according to Lowry and Passonneau (6).
Cyclic nucleotides were measured using the radioimmuno-
assay of Steiner et al. (20).

For the analysis of cerebellar layers, areas 6 and
7 (according to Larsell, 5) of the cerebellar vermis were
dissected at a thickness of 20 μm at -20°. The sections
were dried at -40°, and 10-15 μg sections dissected from
the molecular, granular and white areas using the techniques of Lowry and Passonneau (6). The cyclic nucleotides were measured in these samples, using a micro-modi-
fication of the method of Steiner et al. (12). Statistical
significance was determined using the student's t-test (14).

RESULTS

Maximal electroshock applied to mice resulted in a
reproducible behavioral response; there was a 1-2 sec
latency period and tonic flexion; a 13 sec tonic-exten-
sion, and a 7.5-sec intermittent terminal clonus followed
by a period of postical depression. When phenytoin was
administered, the tonic extension was prevented and was
replaced by a pronounced bilateral clonic seizure.

The effects of MES on cortical and cerebellar P-creatine, ATP, and lactate are shown in Fig. 1. P-creatine
decreased markedly at 10 sec after MES to 25% of control.
Recovery was rapid and at 60 sec P-creatine was restored
to normal values. At 10 sec after MES, ATP decreased to
30 and 60% of normal concentrations in the cerebral cortex and cerebellum respectively, and were near normal
values at 60 sec. The concentration of lactate increased
4-fold in the cortex and 3-fold in the cerebellum, and
remained elevated at 4 min.

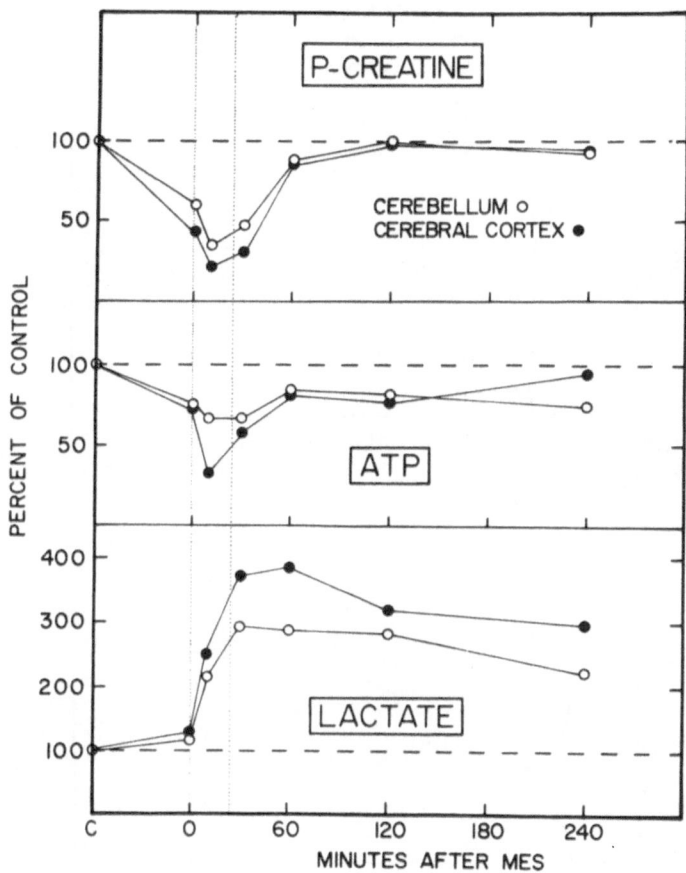

Fig. 1. The concentrations of P-creatine, ATP and lactate
        in the cerebellum and cerebral cortex of mouse
        brain after MES. The electroshock was adminis-
        tered as described in Materials and Methods. The
        vertical dotted lines indicate the excitable
        period of convulsion (tonic flexion, tonic exten-
        sion and terminal clonus). The horizontal dashed
        lines indicate the control values for the meta-
        bolites. A total of 28 mice were used, with at
        least 4 in each group.

    The effect of MES on lactate concentrations with and
without pretreatment with phenytoin is shown in Fig. 2.

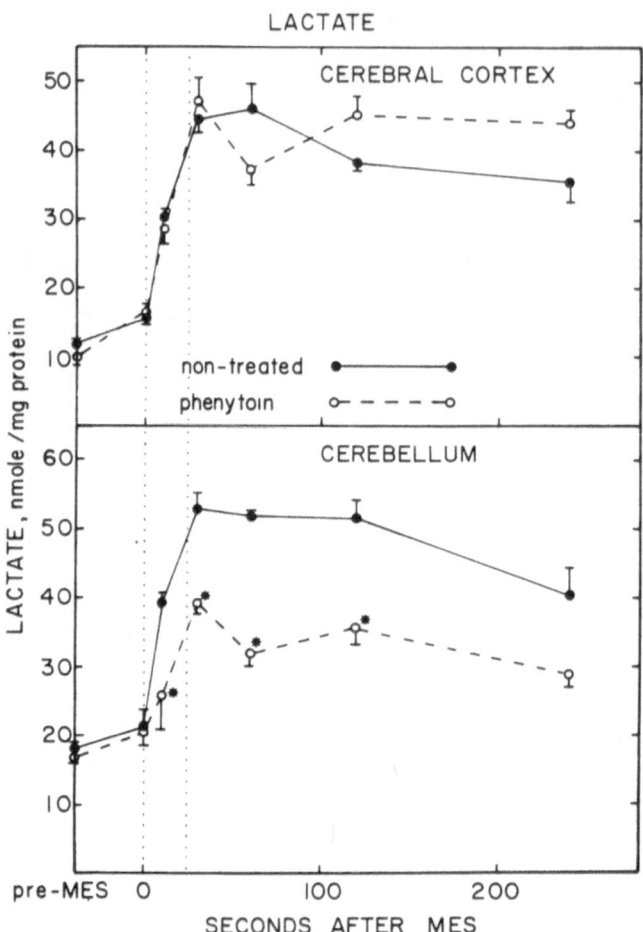

Fig. 2. The concentrations of lactate in the cerebral
        cortex and cerebellum of mouse brain following
        MES; or 25 min after phenytoin treatment (25 mg/
        kg, IP) and MES. A total of 57 mice were used
        with at least 4 in each group. Values signifi-
        cantly different from MES alone are indicated by
        asterisks; p < 0.05.

The 4-fold elevation in the cerebral cortex was unaffected by phenytoin; however, the 3-fold increase in ·the cerebellum was significantly decreased to about 2-fold.

     Cyclic AMP was elevated by MES in both the cortex

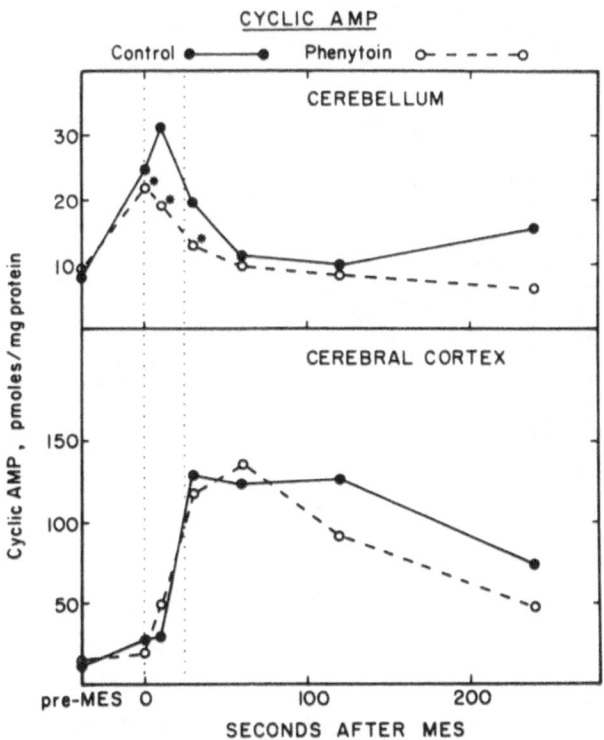

Fig. 3. The concentration of cyclic AMP in the cerebellum
        and cerebral cortex of mouse brain following MES
        in the presence or absence of phenytoin. The
        animals, treatment, and statistics are those of
        Fig. 2.

and cerebellum (Fig. 3). In the presence of phenytoin,
only the cerebellar changes were suppressed.

    Similarly, phenytoin decreased the MES-induced ele-
vation of cyclic GMP in the cerebellum (Fig. 4), but had
no significant effect in the cerebral cortex (data not
given).

    The effects of MES in the presence and absence of
phenytoin pretreatment on cyclic nucleotides in cerebellar layers are shown in Figs. 5 and 6. Cyclic AMP increased in all 3 layers examined. The molecular layer exhibited the greatest increase, and showed a peak value

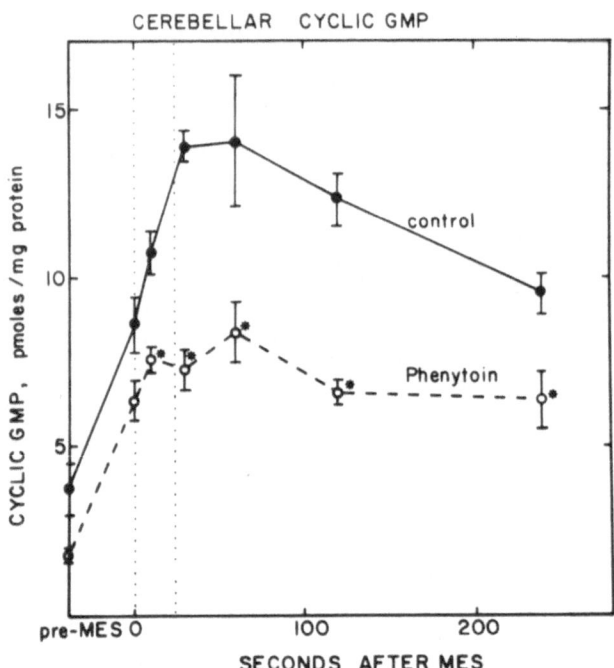

Fig. 4. The concentration of cyclic GMP in the cerebellum
         of mouse brain following MES in the presence or
         absence of phenytoin. The animals, treatment and
         statistics are those of Fig. 2.

at 30 sec post MES. The granular and white layers were
maximally elevated at 10 sec during the excitable phase
of MES and remained high at 30 sec. In the presence of
phenytoin, the increase in cyclic AMP was suppressed at
10 sec only in the granular layer, and was decreased in
all regions at 30 sec. Cyclic GMP increased maximally at
60 sec post MES in all layers, during the postictal depressive phase. Phenytoin markedly suppressed the MES-
induced elevation of cyclic GMP in all layers.

        DISCUSSION

        Experimental seizures induced in mice by MES have a
period of apnea associated with the excitable phase (0-
23 sec), which may account for some of the changes in
brain metabolites and cyclic nucleotides. However, we
have been able in part to dissociate the stimulus from

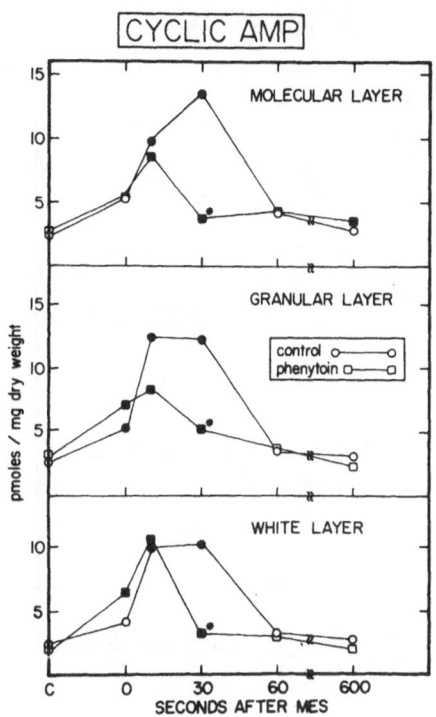

Fig. 5. The concentrations of cyclic AMP in 3 layers of the cerebellar cortex of mouse brain following MES in the presence or absence of phenytoin. The treatment of the animals is described in Materials and Methods. A total of 30 animals were used, and there were at least 6 analyses at each point. Values significantly different from MES alone are indicated by asterisks. $p < 0.05$. solid symbols are significantly different from control; $p < 0.05$.

apnea. When the animals are pretreated with phenytoin, which decreases the apnea due to tonic extension, the increase in lactate concentration in the cerebral cortex is unaffected. However, the lactate increase induced by MES in the cerebellum is diminished in the presence of phenytoin. The contrast between the responses of the two regions argues against anoxia alone since the reversal of such an effect would be expected to be common to both areas. These results provide evidence that the stimulus

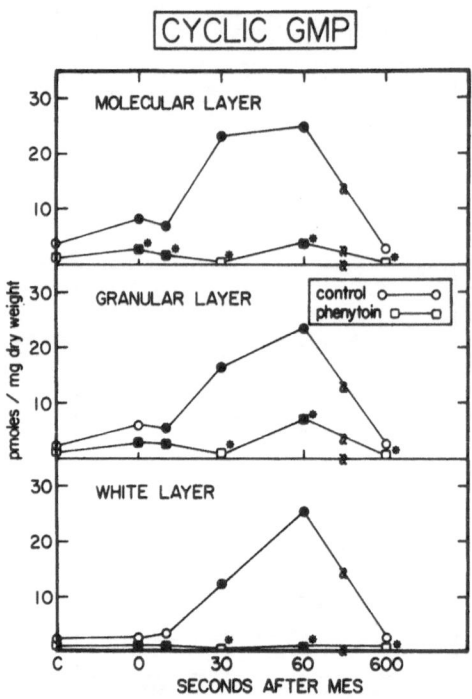

Fig. 6. The concentrations of cyclic GMP in the 3 layers
        of the cerebellar cortex of mouse brain following
        MES in the presence or absence of phenytoin. The
        animals, treatment, and statistics are those of Fig. 5.

of MES creates an increase of energy demand which results
in decrease in ATP and P-creatine and increased lactate,
presumably due to increased glycolysis. The effect of phenytoin, as reflected by changes in energy reserves and
lactate following MES, appears to be largely localized
in the cerebellum, or a pathway to the cerebellum.

     Cyclic AMP and cyclic GMP increased dramatically
following MES. The magnitude of the increase in cyclic
AMP was far greater in the cortex than in the cerebellum
(10-fold and 3-fold respectively). Moreover, the elevated concentrations were sustained in the cerebral cortex
for at least 4 min, but the cerebellar concentrations
of cyclic AMP were near normal by 60 sec. Phenytoin had
no effect on the cortical levels of cyclic AMP induced
by MES, but did diminish the increase in the cerebellum
(Fig.3). It is attractive to relate the MES-induced

changes in cyclic AMP and the effects of phenytoin to
the seizure activity. The output from cerebellum has been
shown to be entirely via the Purkinje cells and to be
inhibitory. Owning to the unique circuitry of the cerebellum, a sustained hyperexcitable state is impossible.
In addition, Bloom (2) and co-workers have shown that
iontophoretically applied cyclic AMP inhibits the firing
of Purkinje cells. The elevated cyclic AMP produced by
MES in the cerebellum could act to reduce the Purkinje
cell firing and have a permissive effect on seizure activity. The diminished increase after MES in the presence
of phenytoin would favor decreased seizure activity. In
the cerebral cortex cyclic AMP has been shown to inhibit
the firing of pyramidal tract neurons (13); therefore
sustained elevation of cyclic AMP would serve to dampen
the seizure state. The lack of effect of phenytoin pretreatment on the cerebral cortical changes in cyclic AMP
would then permit the inhibitory influence of the cyclic
nucleotide to persist.

        The changes in cyclic GMP in the cerebellum occur
later than those of cyclic AMP, and persist during the
period of postictal depression. When seizures are induced
by chemical convulsants, cyclic GMP increases in both the
cortex and cerebellum before and during convulsions whereas cyclic AMP remains unchanged. The increase in cyclic
GMP appears to be common to all seizure states while the
increase in cyclic AMP is unique to MES-induced convulsions. The effect of phenytoin resembles that of a number
of anticonvulsants which reduce the levels of cyclic GMP
in the cerebellum of unstimulated mice (Fig. 4, ref. 1,9).
Furthermore, phenytoin reduced by half the cyclic GMP increase caused by MES. While it is tempting to link the
reduction in cerebellar cyclic GMP to the suppression of
seizures, there are other agents such as chlorpromazine
and ethanol which decrease cyclic GMP but have little or
no anticonvulsant action.

        The diminution of the convulsive and metabolic ef-
fects of MES by phenytoin might be explained by 1) the
attenuation of the electroshock signal, or 2) the suppression of response by a direct effect on the metabolic
pathways. Phenytoin has been shown to prevent the spread
of hyperexcitable foci and possibly may protect the cerebellum from electroshock and consequent massive depolarization. Although high doses of phenytoin have been shown
to decrease the metabolic rate of brain (3), such an effect would not be limited to the cerebellum. Therefore,
a diminished input of the electroshock signal appears to
be more plausible.

The involvement of the cerebellum in seizure activity led to the investigation of the cerebellar layers in
an effort to localize the biochemical events. Cyclic AMP
increased in all 3 layers examined although the increase
was somewhat less in the white than in the molecular and
granular layers (Fig. 5). Phenytoin had the greatest
effect in suppressing the increase in cyclic AMP in the
granular layer, but was effective in all 3 layers 30 sec
after MES. Cyclic GMP also increased in all 3 layers; however, the elevation occurred most rapidly in the molecular
layer. In phenytoin-treated animals, there was little or
no change in cyclic GMP following MES (Fig. 6). These
data indicate that the phenytoin effect may be localized
in the cerebellar cortex. The MES-induced elevation of
cyclic GMP was only reduced 50% by phenytoin in the whole
cerebellum (Fig. 4), whereas in the layers of the cerebellar cortex, phenytoin almost obliterated the changes
in cyclic GMP following MES. However, the effect is not
restricted to a particular layer. Furthermore, the changes
in cyclic GMP cannot be attributed to changes in the
Purkinje cell bodies since the region containing these
cells was not included.

In summary, a major energy deficit occurs during and
directly following MES in which energy production cannot
meet energy demands. The cerebral cortex and cerebellum
differ in the response to MES with respect to change in
cyclic nucleotides, and cyclic GMP increases are subsequent to changes in cyclic AMP. Phenytoin reduces the severity of the seizure, as well as the metabolic changes,
and suppresses the increase in both cyclic nucleotides.
The effect of phenytoin on cyclic GMP appears to be greater in the layers of the cerebellar cortex, than in deeper structures of the cerebellum. Further studies are in
progress to evaluate the effects of other convulsants and
anticonvulsants on both the cerebellar and cortical metabolic pathways.

REFERENCES

1. Biggo, G., Brodie, B.B., Costa, E. and Guidotti, A.
   (1977): Mechanisms by which diazepam, muscimol, and
   other drugs change the content of cGMP in cerebellar
   cortex. Proc. Natl. Acad. Sci. USA, 74: 3592-3596.

2. Bloom, F.E. (1975): The role of cyclic nucleotides in
   central synaptic function. Rev. Physiol. Biochem.
   Pharmacol. 74: 1-103.

3. Broddle, W. and Nelson, S.R. (1968): The effect of
   diphenylhydantoin on energy levels in brain. Fed.
   Proc. 27: 751.

4. Feussner, G.K., McCandless, D.W., Lust, W.D. and
   Passonneau, J.V. (1977): The effect of maximal
   electroshock on energy metabolites and cyclic nucleotides in layers of the cerebellar vermis. Society
   for Neuroscience. 3: 313.

5. Larsell, O. (1952): The morphogenesis and adult pattern of the lobules and fissures of the cerebellum
   of the white rat. J. Comp. Neurol. 97: 281-356.

6. Lowry, O.H. and Passonneau, J.V. (1972): In: A Flexible System of Enzymatic Analysis. Academic Press, New
   York, 151-156.

7. Lowry, O.H., Rosebrough, N.J., Farr, A.L. and Randall
   R.L. (1951): Protein measurement with the folin phenol reagent. J. Biol. Chem. 193: 265-275.

8. Lust, W.D., Goldberg, N.D. and Passonneau, J.V.(1976):
   Cyclic nucleotides in murine brain: The temporal relationship of changes induced in cyclic AMP and cyclic GMP following maximal electroshock or decapitation. J. Neurochem. 26: 5-10.

9. Lust, W.D., Kupferberg, H.J., Yonekawa, W.D., Penry,
   J.K. and Passonneau, J.V. Changes in brain metabolites induced by convulsants or electroshock: Effects
   of anticonvulsant agents. Molecular Pharmacology.
   (in press).

10. Lust, W.D. and Passonneau, J.V. (1976): Cyclic nuc-
    leotides in murine brain: Effect of hypothermia on
    cyclic AMP, glycogen phosphorylase, glycogen synthase
    and metabolites following maximal electroshock or
    decpitation. J. Neurochem. 26: 11-16.

11. McCandless, D.W., Passonneau, J.V. and Lust, W.D.
    (1977): The effect of electroshock on energy meta-
    bolism in cerebellar layers. Anatomical Record. 187:
    649.

12. Steiner, A.L., Wehmann, R.E., Parker, C.W. and Kipnis,
    D.M. (1972): Radioimmunoassay for the measurement of
    cyclic nucleotides. In: Advances in Cyclic Nucleotide
    Research. (Greengard, P. and Robinson, G.A., eds.).
    Raven Press, New York, 2: 51-61.

13. Stone, T.W., Taylor, D.A. and Bloom, F.E. (1975):
    Cyclic AMP and cyclic GMP may mediate opposite
    neuronal responses in the rat cerebral cortex.
    Science. 187: 845-847.

14. Student (Gosset, W.S.) (1907): On the error of
    counting with a hemocytometer. Biometrika. 5: 351-
    360.

BIORHYTHMIC ASPECTS OF EPILEPTIC MANIFESTATIONS

U.J. Jovanović

Neuropsychiatric Clinic, University of Würzburg
D-8700 Würzburg, W. Germany

As early as 1938, Griffiths and Fox (21) found it striking that epileptic manifestations show a certain rhythmicity. In epilepsies on awakening and in epilepsies with combined seizures, biorhythm is affected by numerous external and internal factors. On the other hand, an infradian biorhythm in sleep epilepsies could be demonstrated quite clearly by means of statistical methods (22). Also, in epilepsies with fits on awakening, a relationship between the seizures and the menstrual cycle has been found. In addition, this form of epilepsy shows an annual or a semiannual biorhythm (40). According to our experience with outdoor examinations (39, 40), most fits occur in January/April and September/November. If an epileptic patient seemed to suffer fits irregularly in the course of the year, grand mal seizures accumulated in January/April and September/November, but occurred sporadically or never during the other months.

Psychomotor attacks have a periodic course according to Janz (23); i.e., the episodes accumulate at certain times. The diagnosis could be made only by an exact analysis of these rhythms of epileptic manifestations. The intervals between these series varied between 1 and 6 weeks. The fits phases lasted for 2 to 4 days, while the frequency ranged from 2 to 8 seizures per day. From the biorhythmic point of view, Janz (23) developed the term cycloleptic course (periodically accumulated).

Table I Number, age and sex of epileptic patients investigated during sleep and classification according to the biorhythmic principles

Classification according to biorhythmic principles	Type of epilepsies (type of seizures)	Mean of age	Sex		
			M	F	Total
a) Diffuse epilepsies	1)Combined epilepsies	30.1	27	27	54
b) sleep epilepsies	2)Sleep epilepsies	23.5	18	14	32
	3)Psychomotor epilepsies	27.1	19	23	42
	4)Propulsive petit mal (BNS seizures)?	1.2	3	3	6
c) Epilepsies on awakening	5)Epilepsies on awakening	22.4	13	15	28
	6)Retropulsive petit mal (absences)	18.4	11	13	24
	7)Impulsive petit mal (myoclonic epilepsies)	26.0	1	0	1
All epilepsies	Mean or Total	20.1	92	95	187

Griffiths and Fox (21) gave a number of examples with intervals of 2, 5 and 7 weeks up to 3 months. Helmchen et al. (22) could observe a certain connection between the frequency of the fits and endogenous rhythms, although at first sight this seems not to have been the case in psychomotor epilepsy. Seizures in a monoleptic course occurred rarely (23).

Circadian biorhythms seem to be easier to prove in epilepsies. The highest frequency is, for example, in pycnoleptic seizures [pycnoleptic: daily, according to Janz (23)] between 9:00 and 10:00 h and around 21:00 h. In 777 cases, there was hardly a case of fits between 23:30 and 6:00 h. In one patient with impulsive petit mal, Janz (23) found the highest frequency of seizures around 9:00 and 11:00 h and between 19:00 and 20:00 h. Between 22:00 and 02:00 h no fits were recorded.

In patients in whom there is a relationship between the seizures and sleep, the highest frequency of fits is around 12:00 h (nap at noon) and toward 24:00 h (nocturnal sleep). A second peak of the frequency occurs in the morning hours (7, 29, 33-37). Patients with fits on awakening suffer from their seizures between 7:00 and 8:00 h and 14:00 and 16:00 h (23, 43).

Differences in the sleeping-waking cycle of epileptic patients (7, 8, 14, 19, 51) will be discussed further because of the relationship between the patients' sleep behavior and their epileptic seizures. From 1963 until 1975, 187 patients with epileptic attacks of different manifestations (Table I) were investigated, for a total of 516 investigation nights, by means of polygraphic records during complete sleep; i.e., between evening and spontaneous awakening the next morning.

The method employed has been already described in detail (24-40). In children who had not suffered from epileptic attacks before, but who manifested bilateral-symmetrical spike and wave complexes in the sleep EEG for several seconds each (39, 40), a difference in the sleep profile and in the depth of sleep (Fig. 1) compared to the corresponding parameters in healthy or bed-wetting children was found.

Children with epileptic discharges in the sleeping EEG were delayed in falling asleep and had a higher percentage of wakefulness during the night but a lower percentage of deep sleep (Fig. 1) than the group of the

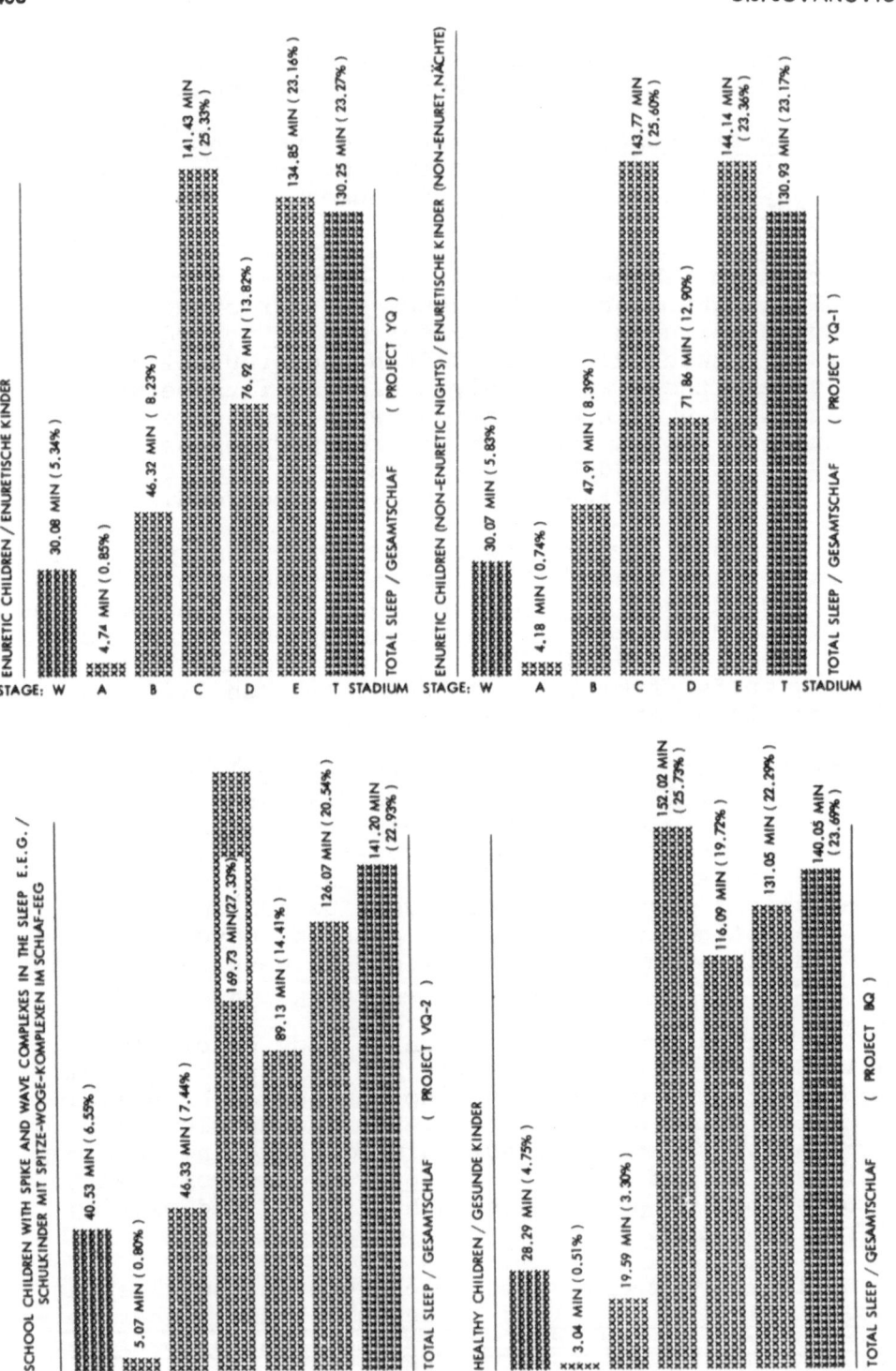

SCHOOL CHILDREN WITH SPIKE AND WAVE COMPLEXES IN THE SLEEP E.E.G. /
SCHULKINDER MIT SPITZE-WOGE-KOMPLEXEN IM SCHLAF-EEG

ENURETIC CHILDREN / ENURETISCHE KINDER

ENURETIC CHILDREN (NON-ENURETIC NIGHTS) / ENURETISCHE KINDER (NON-ENURET. NÄCHTE)

HEALTHY CHILDREN / GESUNDE KINDER

Fig. 1.   Illustration of sleep stages in minutes (and in percentage rates) in 10-year-old children with bilateral-symmetrical spike-and-wave complexes in the sleeping EEG (bottom left) in comparison with the corresponding sleep stages in child bed-wetters (top left and right) and healthy children (bottom right). The children with epileptic discharges in the sleep EEG (Project VQ-2) show a larger percentage of the waking state during the night and a smaller percentage of deep sleep and REM phases (T) in comparison with the sleep of healthy and bed-wetting children.

Stage W = Wakefulness, A = Stage O, B = Stage I, C = Stage II, D = Stage III, E = Stage IV, T = REM phases, dream phases.

Fig. 2.   Top: Percentages of sleep phases W (wakefulness),
          E (deep sleep = Stage IV) and REM (REM phases)
          during 3 investigation nights and as an average
          (framed right) in a 19-year-old female patient
          suffering from a combined form of epilepsy.

          Middle: A 28-year-old female patient suffering
          in past years from epilepsy on awakening.

          Bottom: A female patient 30 years old at the
          time of examination and suffering from sleep
          epilepsy.

          Epilepsies on awakening (middle) show a higher
          percentage of wakefulness during the night in
          comparison with a combined form of epilepsy (top)
          as well as nocturnal epilepsy (bottom), and a
          smaller percentage of deep sleep and REM phases.
          Sleep epilepsies, however, produce a relatively
          high proportion of deep sleep and REM phases.
          These proportions can also be higher in compari-
          son with the corresponding parameters in healthy
          persons, especially during deep sleep.

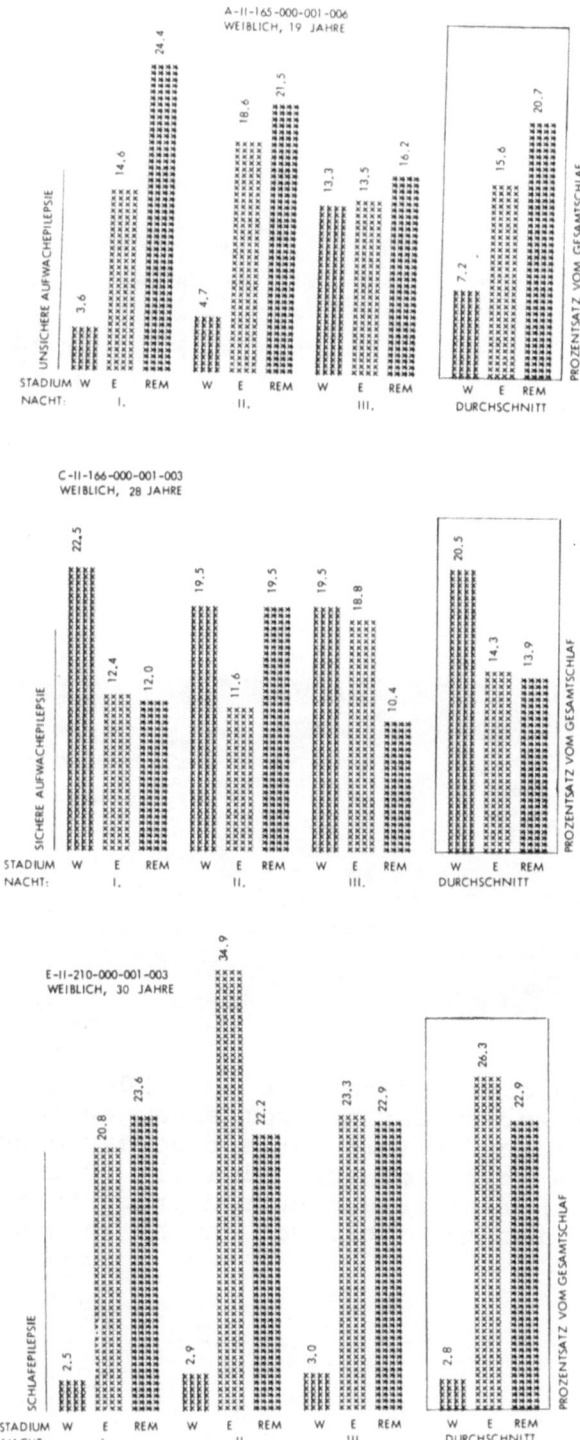

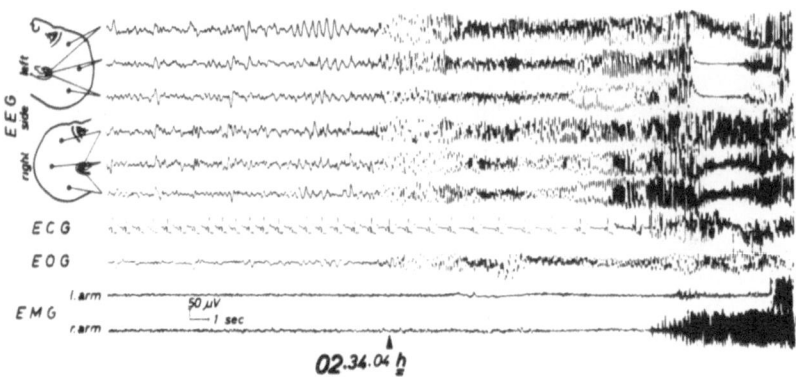

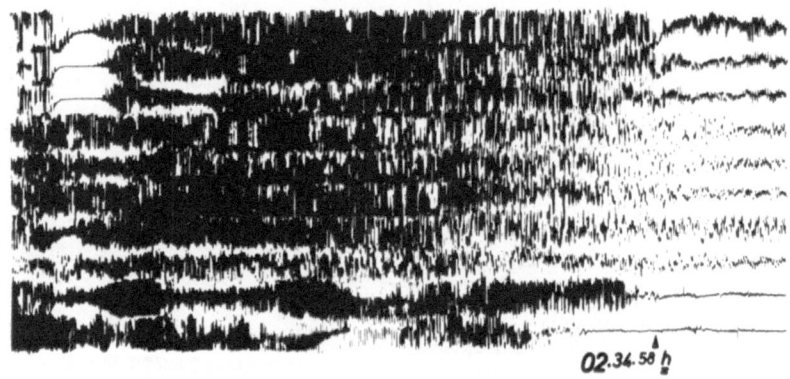

Fig. 3. A generalized seizure in a 20-year-old female
         patient suffering from sleep epilepsy. The
         seizures in this patient occurred during sleep
         in a rhythm of approximately 80 to 85 minutes.
         Another seizure occurred prior to this one,
         and two more followed in the same night.

children with whom they were compared. The percentage of
light sleep (Stage C = Stage II) (Fig. 1) was also relatively increased in children with epileptic discharges in
the sleeping EEG.

In adult patients with epileptic seizures during
wakefulness (epilepsies on awakening, matutinal epilepsies), a relatively very low percentage of deep sleep and
REM phases (Fig. 2) [REM phase = phases of sleep with
rapid eye movements (1-5, 10, 11, 12, 41, 42)] was
recorded as compared to the corresponding parameters of
sleep epilepsies or healthy control persons (Fig. 2).
In sleep epilepsies, on the contrary, surplusses of deep
sleep (25-40) and short periods of wakefulness during the
night were recorded.

Concerning the ultradian biorhythm, a certain con-
stancy has been found by many authors (8, 9, 14-18, 23,
24-40, 47-51). Epileptic equivalents in the EEG seem to
follow a certain rule. In some of our cases (40), the
fits occurred during sleep with a mean of 80 minutes
(Figs. 3 and 4). This ultradian periodicity of 80 to 85
or 90 minutes is almost as long as the ultradian biorhythm in healthy persons (1-5, 10-12, 41, 42, 45, 46,
50).

Our further investigations (36, 37, 39, 40) demonstrated a certain connection between EEG patterns and
the psychic adaptation of the patients to the new sleeping situation. In patients with fits on awakening, for
example, in the first investigation night (unfamiliar
sleeping situation) convulsive potentials in the EEG in
the form of bilateral-symmetrical spike-and-wave complexes exceeded those in the following nights (familiar
sleep situation, adaptation to a new sleep situation)
for 10 minutes (Fig. 5). The deeper the night was, i.e.,
the longer the patient slept, the fewer epileptic potentials were found in the sleeping EEG. Opposite results
were obtained in patients suffering from epileptic seizures during sleep: the better the patient knew the sleep
situation and the more accustomed he was to it, the more
epileptic potentials were recorded in the EEG (40). In
the first group of patients, these elements are therefore
related to fear; in the second group, more to the deep-
ening of sleep. It is well known that the epileptic
equivalent can be provoked more easily by fear and sleep
deprivation (among other factors) in the group with epilepsy during awakening (23) than in the group with epilepsy during sleep (6, 7, 19, 20, 44, 52-54).

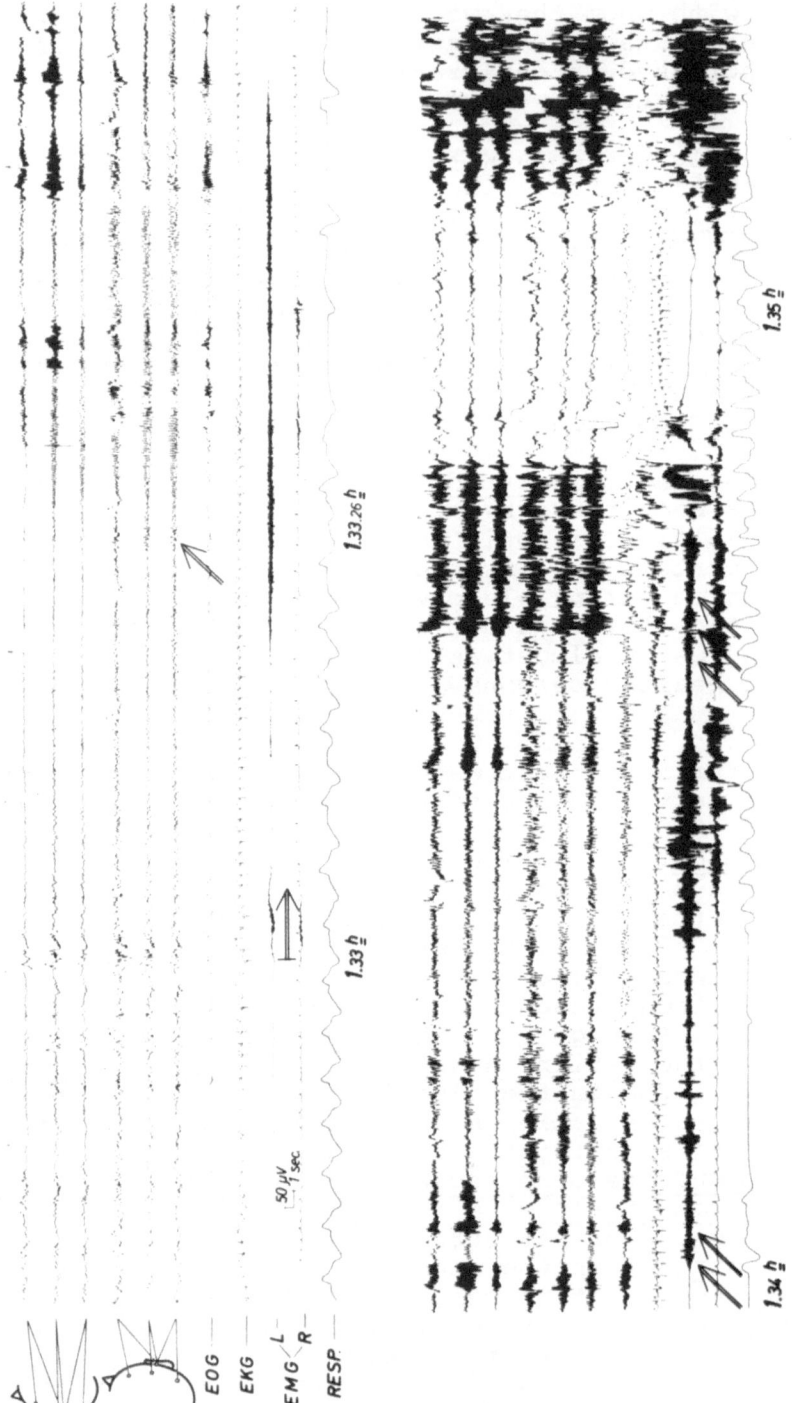

Fig. 4. Somnabulistic psychomotor attack in a 28-year-
         old patient with a somnabulistic psychomotor
         epilepsy. The seizures occurred during sleep
         in a rhythm of approximately 80 to 85 minutes.
         This seizure was followed by 5 other seizures
         during the same night.

Fig. 5.   Left column: Sleep profiles from 3 nights in a
           patient with only 3 anamnestically known epilep-
           tic seizures. Individual spikes were recorded
           (arrows).

           Right column: Sleep profiles from 3 nights in
           a 28-year-old female patient with epilepsy on
           awakening of many years' duration. During the
           first 2 nights, the sleep was restless. Bila-
           tural-symmetrical spike-and-wave-complexes
           occurred periodically and very frequently during
           the first night. During the second night, the
           convulsive potentials in the sleep EEG decreased
           significantly and appeared less and less from
           night till morning. After the patient woke up
           in the morning, a generalized seizure was re-
           corded (circled arrow). During the third night,
           even fewer convulsive potentials were recorded
           in the EEG (adaptation: see the text). In addi-
           tion to long latency times up to falling asleep,
           the course of sleep itself was relatively good.

           The percentage duration of each phase of sleep
           is shown for both patients by means of bar graphs:
           W = Wakefulness; A = Stage O; B = Stage I; C =
           Stage II; D = Stage III; E = Stage IV; T = REM
           phases.

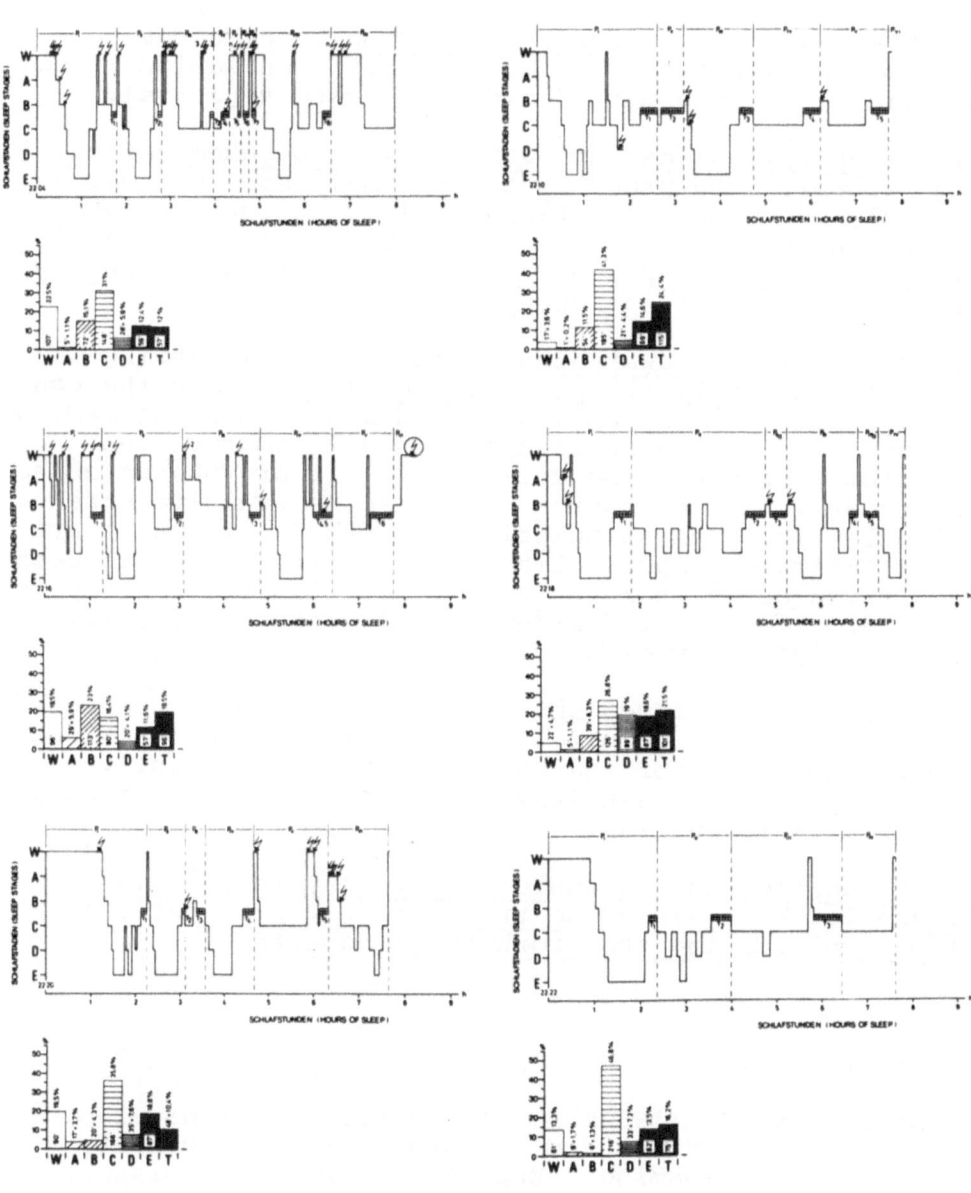

All these findings indicate that the types of epilepsy described in Table I from the biorhythmic aspect
can be systematized into three groups. Therapy should
be selected accordingly, as follows:

(a) The combined (diffuse) type of epilepsy needs combined medication therapy. Further differentiation is not
possible, since one must decide each case separately.

(b) Sleep epilepsies and patients with psychomotor fits
(and probably with propulsive petit mal) can be treated
with only one group of antiepileptics. Stimulants are
also known to have certain antiepileptic effects. Antiepileptics with sleep-inducing effects are contra-indicated in this form of epilepsy.

(c) Epilepsies on awakening (seizures mainly during wakefulness) and nearly all petit mal epilepsies, on the contrary, require an antiepileptic and sleep-inducing medical
therapy. This therapeutic approach will improve the
periodicity of sleep and will not itself provoke attacks.
Furthermore, the medication has an anticonvulsive effect.
Antiepileptics with stimulating effects (stimulants) are
contra-indicated since patients get their attacks at each
stimulation or excitement, and in addition their sleep is
impaired. The impaired sleep itself provokes attacks,
and thus begins a vicious circle.

The differentiated antiepileptic therapy recommended
herein can be applied more successfully to adult patients
than to children, for the child's brain is not fully
matured; it is very sensitive and reacts to every medication in a different way.

In conclusion, we would like to say that it is a
mistake for a physician to treat an epileptic patient
without considering the biorhythmic aspects.

REFERENCES

1. Aserinsky, E. (1953): Ocular motility during sleep
    and its application to the study of rest-activity
    cycles and dreaming. Unpublished Doctoral Disserta-
    tion. University of Chicago.

2. Aserinsky, E. and Kleitman, N. (1953): Eye movements
    during sleep. Fed. Proc. 12: 6-7.

3. Aserinsky, E. and Kleitman, N. (1953): Regularly
    occurring periods of eye motility and concomitant
    phenomena during sleep. Science 118: 273-274.

4. Aserinsky, E. and Kleitman, N. (1955): Two types of
    ocular motility occurring in sleep. J. Appl. Physiol.
    8: 1-10.

5. Aserinsky, E. and Kleitman, N. (1955): A motility
    cycle in sleeping infants as manifested by ocular
    and gross body activity. J. Appl. Physiol. 8: 11-19.

6. Bennet, D.R., Mattson, R.H., Ziter, F.A., Calvery, J.
    R., Liske, E.A. and Pratt, K.L. (1964): Sleep depri-
    vation: neurological and electroencephalographic
    effects. Aerospace Med. 35: 888-890.

7. Beyer, L. and Jovanovic, U.J. (1966): Elektroencepha-
    lographische und klinische Korrelate bei Aufwachepi-
    leptikern mit besonderer Berücksichtigung der thera-
    peutischen Probleme. Nervenarzt 37: 333-336.

8. Christian, W. (1960): Bioelektrische Charakteristik
    tagesperiodische gebundener Verlaufsformen epilepti-
    scher Erkrankungen. Dtsch. Zschr. Nervenheilk. 181:
    413-444.

9. Delmas-Marsalet, P. (1964): Signification prognostique
    des fuseaux alpha parfaits dans l'epilepsie essen-
    tielle. Rev. neurol. 110: 250-254.

10. Dement, W.C. (1965): Studies on the function of rapid
    eye movement (paradoxical sleep) in human subjects.
    In Jouvet, M. (ed): Aspects Anatomo-Fonctionnels de
    la Physiologie du Sommeil. Centre National de la
    Recherche Scientifique, Paris. 571-611.

11. Dement, W.C. and Kleitman, N. (1957): The relation of
    eye movements during sleep and dream activity as
    objective measure for the study of dreaming. J. Exp.
    Psychol. 53: 339-346.

12. Dement, W.C. and Kleitman, N. (1957): Cyclic variations
    in EEG during sleep and their relation to eye movements,
    body motility and dreaming. Electroencephalogr. Clin.
    Neurophysiol. 9: 673-690.

14. Gänshirt, H. and Vetter, K. (1961): Schlafelektroden-
    encephalogramm und Schlaf-Wachperiodik bei Epilepsien.
    Nervenarzt 32: 275-279.

15. Gastaut, H. (1963): Sémiologie et physiopathogénie
     des crises epileptiques généralisées. Helv. med.
     Acta 10: 319-337.

16. Gastaut, H., Broughton, R., Fressey, J. and Tassinarie
     L.A. (1965): Les Phénoménes Episodiques au Cours du
     Sommeil. Masson, Paris.

17. Gastaut, H., Dongier, M., Batini, C. and Rhodes, J.
     (1962): Etude électro-clinique de terreurs nocturnes
     et diurnes concomitantes d'un rêve ou d'une idée
     obsédante chez un névrosé. Rev. neurol. 107: 277-279.

18. Gastaut, H. and Poirier, F. (1964): Experimental or
     "reflex" induction of seizures. Epilepsia 5: 256-270.

19. Gibbs, E.L. and Gibbs, F.A. (1947): Sleep record in
     epilepsy. Res. Publ. Assoc. Res. Nerv. Ment. Dis.
     26: 366-372.

20. Gloor, P., Tsai, C. and Habbad, F. (1958): An assess-
     ment of the value of sleep electroencephalography for
     the diagnosis of temporal lobe epilepsy. Electroen-
     cephalogr. Clin. Neurophysiol. 10: 633-648.

21. Griffiths, G.M. and Fox, I.T. (1938): Rhythm in epi-
     lepsy. Lancet 235: 409-416.

22. Helmchen, H., Künkel, H. and Selbach, H. (1964):
     Periodische Einflüsse auf die individuelle Häufigkeit
     cerebraler Anfälle. Arch. Psychiat. Nervenkr. 206:
     293-308.

23. Janz, D. (1969): Die Epilepsien. Spezielle Pathologie
     und Therapie. G. Thieme. Stuttgart.

24. Jovanović, U.J. (1966): Experimentelle Untersuchungen
     über die Wirkung von 2-Diäthyl-amino-5-Phenyl-Oxazo-
     linon-(4) (Ha 94) auf das Elektroencephalogramm der
     Epileptiker. Ärztl. Forsch. 20: 98-103.

25. Jovanović, U.J. (1966): Die diagnostische Bedeutung
     des Schlaf-Elektroencephalogramms. Dtsch. Med. J.
     17: 121-132.

26. Jovanović, U.J. (1966): Natürlicher Schlaf als beste
     Provokationsmethode bei Epileptikern. Schweiz. Arch.
     Neurol. Psychiat. 98: 244-257.

27. Jovanović, U.J. (1966): Forcierte Normalisierung im
     EEG (Landolt) und forcierte Pathologisierung des
     Schlaf-Elektroencephalogramms. Psychiat. Neurol.
     Basel 152: 370-386.

28. Jovanović, U.J. (1967): Schlaf und Epilepsie. Thera-
     piewoche 95-107.

29. Jovanović, U.J. (1967): Das Schlafelektroencephalo-
     gramm als spezielle diagnostische Methode. Landarzt
     43: 605-615.

30. Jovanović, U.J. (1967): Das Schlafverhalten der Epi-
     leptiker. I. Schlafdauer, Schlaftiefe und Besonder-
     heiten der Schlafperiodik. Dtsch. Zschr. Nervenheilk.
     190: 159-198.

31. Jovanović, U.J. (1967): Das Schlafverhalten der Epi-
     leptiker. II. Elemente des EEG, Vegetativum und Moto-
     rik. Dtsch. Zschr. Nervenheilk. 191: 257-290.

32. Jovanović, U.J. (1967): Neuere Aspekte zur Einteilung
     von Epilepsien auf Grund von Schlafuntersuchungen.
     Jahrestagung der Deutschen Sektion der Internationa-
     len Liga gegen Epilepsie. Tübingen, Juni 1966. Zbl.
     Neurol. Psychiat. 188: 22.

33. Jovanović, U.J. (1968): Die sich aus dem natürlichen
     Schlaf der Epileptiker ergebenden therapeutischen
     Konsequenzen. Nerveanarzt 39: 199-204.

34. Jovanović, U.J. (1968): Das Schlafverhalten der Epi-
     leptiker. III. Epileptische Anfälle und äquivalente.
     Dtsch. Med. J. 19: 76-84.

35. Jovanović, U.J. (1970): Schlafforschung und ihre
     klinischen Aspekte. Nervenarzt 41: 5-23.

36. Jovanović, U.J. (1970): Somnabule psychomotorische
     Epilepsie. Dtsch. Zschr. Nervenheilk. 197: 181-191.

37. Jovanović, U.J. (1974): Psychomotor Epilepsy. A
     Polydimensional Study. Charles C. Thomas, Spring-
     field, Illinois.

38. Jovanović, U.J. (1974): Schlaf und Traum. Physiolo-
     gische und psychologische Grundlagen, Störungen und
     Behandlung. Fischer-Verlag, Stuttgart.

39. Jovanović, U.J. (1977): Zur Methodik der Chronopsy-
     chologie: Psychometrische, polygraphische, klinische,
     und Persönlichkeitsuntersuchungen. Inaugural-Disser-
     tation (Dr. phil.), Universität Würzburg.

40. Jovanović, U.J. (1977): Chronobiologic aspects of
     neurology. Waking and Sleeping. (in preparation).

41. Kleitman, N. (1972): Sleep and Wakefulness. Univer-
     sity of Chicago Press, Chicago, 1939, 1963, 1972 (most
     recent printing).

42. Kleitman, N. and Engelmann, Th. G. (1953): Sleep
     characteristics of infants. J. Appl. Physiol. 6:
     269-282.

43. Langdon-Down, M. and Brain, W.R. (1929): Time of
     day in relation to convulsions in epilepsy. Lancet
     3: 1029.

44. Mattson, R.H., Pratt, L.K. and Calverley, J.R. (1965):
     Electroencephalograms of epilepsies following sleep
     deprivation. Arch. Neurol. 13: 310-315.

45. Ohlmeyer, P. and Brilmayer, H. (1947): Periodische
     Vorgänge im Schlaf. Pflügers Arch. ges. Physiol.
     251: 249-250.

46. Passouant, P. (1965): Olfaction et epilepsie. Rev.
     Laryng. Bordeaux 86: 935-953.

48. Passouant, P. and Cadilhac, J. (1962): EEG and
     clinical study of epilepsy during maturation in man.
     Epilepsia 3: 14-43.

49. Passouant, P., Cadilhac, J., Pternitis, C. and
     Baldy-Moulinier, M. (1967): Epilepsie temporal et
     décharges ammoniques provoquées par l'anoxie oxyprive.
     Rev. neurol. 117: 65-70.

50. Poirel, C. (1974): Some circadian rhythms in experi-
     mental ethology and comparative psychopathology. In:
     Scheving, L.E., Halberg, F. and Pauly, J.E. (eds.):
     Chronobiology. G. Thieme. Stuttgart, 540-543.

51. Popoviciu, L., Badiu, Gh., Corfariu, O., Foisoreanu,
     V., Gáspár, St. and Szabo, L. (1976): Epilepsiile.
     Editura Dacia. Napoca, Cluj (Monogr.).

52. Pratt, K.L., Mattson, R.H., Weikers, N.J. and Williams,
     R. (1968): EEG activation of epileptics following
     sleep deprivation. A prospective study of 114 cases.
     Electroencephalogr. clin. Neurophysiol. 24: 11-15.

53. Scollo-Lavizzari, G., Pralle, E. and Cruz, De L.N.
     (1975): Activation effects of sleep deprivation and
     sleep in seizure patients. An electroencephalographic
     study.   Europ. Neurol. 13: 1-5.

54. Yamamoto, J., Furuya, E., Wakamatsa, H. and Hishikawa,
     Y. (1971): Clinical note: Modification of photosensi-
     tivity in epileptics during sleep. Electroencephalogr.
     Clin. Neurophysiol. 31: 509-513.

REM SLEEP DEPRIVATION: EFFECTS ON INCORPORATION OF INORGANIC SULFATE INTO BRAIN ACID MUCOPOLYSACCHARIDES

M. Rusić, V. Šušić, M. Levental and Lj. Rakić

Department of Neurochemistry
Institute for Biological Research
and
Department of Physiology
Medical School, University of Belgrade
Belgrade, Yugoslavia

INTRODUCTION

The study was a search for sleep-related alterations in brain metabolism through observations on the incorporation of ^{35}S-labeled sulfate into brain acid mucopolysaccharides (AMPS) of REM-sleep-deprived and sleeping rats.

Animals of different ages were used to determine whether developmental stages influence brain metabolism during sleep deprivation and ensuing recovery. Administration of labeled sulfate results in a specific labeling of endogenous AMPS. The global estimate of AMPS turnover in the brain can be made by observing the changes in the specific activity of labeled sulfate.

MATERIALS AND METHODS

Eighty-four female hooded rats (Mill-Hill) of different age (45, 75, 86 and 92 days) weighing from 120 to 180 g were selected from an animal colony and housed individually in plastic cages supplied with food and water ad libitum. The cages were placed in a room with ambient temperature of 22°C with natural light. At the time of experiment, the 84 rats were divided into seven equal groups of 12 rats each: normal controls (group C), stress controls (group S), REM-sleep-deprived animals

animals stressed and then allowed undisturbed sleep for
6 and 24 h (groups RS_6 and RS_{24}), and animals REM-sleep-
deprived and then allowed recuperative sleep for 6 and
24 hr (groups RD_6 and RD_{24}).

Selective REM deprivation for 72 continuous hours
was produced by confining the rats on small inverted
flower pots, measuring 6.5 cm in diameter, in a water
tank with water 1 cm below the top of the pot. This
procedure has been shown by polygraph recordings to suppress REM sleep by 80-100% of the baseline amount, while
slow-wave sleep (SWS) is slightly suppressed (1, 2).

In addition to producing sleep-depriving effect,
this technique may elicit the stress of being immobilized, isolated, cramped and dampened. To determine
what effect REM deprivation had on the variable examined
and to distinguish between the effects caused by sleep
loss and those induced by the stress of the experimental
technique alone, comparisons with a stress control group
were obtained: rats were confined to larger platforms
11.5 cm in diameter, a procedure that was found to cause
hyperplasia of adrenal glands as well as body weight decrease as compared to changes produced by the stress ex-
perienced by animals kept on a small platform. Six animals from the S group and six animals from the D group
were allowed undisturbed sleep for 6 h and 24 h.

All rats were intraperitoneally injected with 30
µC/100 g body weight of carrier-free sodium ^{35}S-sulfate
($Na_2^{35}SO_4$, 640 c/mM, Nen Chemical GmbH) and were returned to their cages for an additional 20 h. The
greatest incorporation of labeled sulfate into the mouse
brain was observed at 12-48 h after injection (3). In
studying the incorporation of ^{35}S-sulfate into various
brain fractions in mature rats, the greatest incorporation into sulfomucopolysaccarides was to occur 2 days
after injection, and decline progressively after that
time.

We decided to sacrifice animals 20 h after injection of labeled sulfate, anticipating that the experimental procedures would change the extent of incorporation. At the end of the experimental procedures the
rats were decapitated and the brains quickly removed
for biochemical analysis. Changes in total brain AMPS
content were investigated by chromatography (5), with
slight modification for the analysis of AMPS in brain
tissue developed in our laboratory. The procedure involved proteolysis of lipid-free, dry brain tissue,

TABLE I

EFFECTS OF REM SLEEP DEPRIVATION ON BRAIN URONIC
ACID IN RATS OF VARIOUS AGES

Groups	45 days	75 days	86 days	92 days
C	2346 ± 212	1167 ± 26	1200 ± 41	2415 ± 328
D	2549 ± 133	1044 ± 26●	943 ± 90●	1738 ± 373
S	1748 ± 133●●	1194 ± 43	1183 ± 91●●	1960 ± 207
RD$_6$	2767 ± 132	1101 ± 47	1595 ± 208	2492 ± 203
RD$_{24}$	2817 ± 113	1551 ± 88●●●		2022 ± 202
RS$_6$	2574 ± 275	1184 ± 78	1126 ± 28	2040 ± 143
RS$_{24}$	1635 ± 129▲	1295 ± 34▲		2039 ± 105

Means $\pm$ S.E.M.
Uronic acid is expressed as µg/g dry weight.
Total MPS are expressed as mg of uronic acid multi-
plied by 25 (assuming there to be 40% hexuronic
acid in the AMPS/per gram of dry tissue).
See Fig. 1 for an explanation of the symbols.

followed by separation and purification of AMPS by chromatography on an Ecteola cellulose column. In the dia-
lyzed eluates, total AMPS content was determined via
uronic acid by the orcinol method (6). Determination
of radioactivity was performed in a Nuclear Chicago
scintillator.

RESULTS

Table I shows that REM sleep deprivation led to a
significant decrease in the amount of uronic acid in
75-, 86- and 92-day-old rats; in the contrast, there
was an increase, although nonsignificant, in young, 45-
day-old rats.

Rats confined on large pots (stress controls) showed
small nonsignificant changes as compared to controls.
However, young, 45-day-old rats showed a significant
decrease.

◪   C   vs.   D

◪ ◪   C   vs.   S

◪ ◪ ◪   C   vs. RD_{24}

▲   C   vs. RS_{24}

▲ ▲   D   vs. RD_6

▲ ▲ ▲   D   vs. RD_{24}

■   S   vs. RS_{24}

■ ■   S   vs. RS_6

■ ■ ■ RD_6 vs. RD_{24}

●   C   vs. RD_6

● ●   D   vs.   S

● ● ●   D   vs. RD_{24}

Fig. 1. Key to symbols in tables I - III.   Student's t-
        test.   The level of significance was set at
        p<0.01.

        Recovery sleep (first 6 h) after REM sleep depriva-
tion resulted in a significant rise in the amount of
uronic acid.

        Recovery sleep (first 6 h) after stress conditions
did not change significantly the amount of uronic acid,
except in 45-day-old rats, in which recovery sleep led
to a significant increase in the uronic acid amount as
compared to the stressed animals.

        Recovery sleep (24 h) after REM sleep deprivation
led to a significant increase of uronic acid in all age
groups, but in 45- and 75-day-old rats the increase was
above control levels.

TABLE II

EFFECTS OF REM SLEEP DEPRIVATION ON THE TOTAL RADIOACTIVITY FOUND IN THE RAT BRAIN 20 HOURS AFTER INTRAPERITONEAL INJECTION OF ^{35}S-SULFATE

Groups	45 days	75 days	86 days	92 days
C	1783±88	3126±268	2263±90	3900±360
D	1366±128◨●●	3300±240◨	1457±219	2768±280◨
S	1581±142◨◨	2780±157●●	1915±107◨◨	3276±133
RD$_6$	1805±157▲▲◨	2570±226▲▲	2060±94●	3420±141●
RD$_{24}$	1110±21◨◨◨ ■■■	2731±336◨◨◨ ▲▲		1948±107
RS$_6$	1820±245	3667±540	1906±170■■	3148±212
RS$_{24}$	1233±121▲■	2478±186■		2674±162■

Values are the means + S.E.M. of groups of 6-12 animals. Cpm/whole brain.

Recovery sleep (24 h) after stress conditions led to a significant decrease in the uronic acid amount as compared to controls in 45- and 92-day-old rats, while there was a significant increase in 75-day-old rats.

In a comparison of REM-sleep-deprived rats and stressed rats, a significant difference in their response to the experimental procedures was apparent, indicating that the changes observed in REM sleep deprivation were not due to the physical stress and discomfort of the experimental environment.

Table II shows the total amount of ^{35}S in the brains of control and experimental groups measured after a relatively short labeling period (20 h after labeled injection). The total amount of radioactivity incorporated showed:

1. Less extensive incorporation in large-pot (stressed) rats as compared to normal controls.

TABLE III

EFFECTS OF REM SLEEP DEPRIVATION ON INCORPORATION
IN VIVO OF ^{35}S-SULFATE INTO CEREBRAL ACID MUCO-
POLYSACCHARIDES IN RATS OF VARIOUS AGES

Groups	45 days	75 days	86 days	92 days
C	3850+406	13,955+1277	9351+570	1820+225
D	2660+133	16,311+1520	7796+1070	2020+360
S	4550+480	12,930+656	7874+1450	1765+102
RD$_6$	3230+332	11,843+1340	5890+1004	1406+409
RD$_{24}$	1850+148	8,777+1070		990+45
RS$_6$	3780+150	15,343+2005	7745+172	1581+129
RS$_{24}$	3660+113	8,695+155		1347+102

Means + S.E.M. Counts /min per mg of uronic acid.

2. Less extensive incorporation in REM-sleep-deprived (small-pot) rats, except in one instance, the
75-day-old rats.

3. That recovery sleep (first 6 h) after D and S
conditions led to an increase in the incorporation of
sulfate as compared with D and S groups.

4. That recovery sleep (24 h) after termination
of D and S conditions led to a sharp drop in the incorporation in all animals as compared with C, D and S
groups.

Table III shows the specific radioactivity of
uronic acid. Data are expressed as specific radioactivity per milligram of uronic acid to enable comparison of different experiments (since the amounts of
^{35}S-sulfate taken up by the brain were different).

The specific activity of uronic acid after REM sleep deprivation was decreased as compared to controls in 45-, 86- and 92-day-old rats, and increased in 75-day-old rats.

In the contrast, stress conditions produced a decrease in all except the young, 45-day-old rats. Changes in the specific activity of uronic acid during recovery sleep as compared to pre and postdeprivational levels were always observed.

DISCUSSION

Shapot (10) conducted an experiment comparing the ^{35}S-methionine uptake into rat brain protein during sleep and wakefulness. He found that the methionine uptake into brain protein was twofold greater in rats allowed to sleep after sleep deprivation than in wakeful animals.

Using the brain biopsy technique, Brodskii et al. (11) measured incorporation of labelled leucine into protein during various physiological states of the animal. They found an increase in protein metabolism during REM sleep and a lesser decrease during slow-wave sleep also.

Reich et al. (8) began an investigation of the incorporation of radioactive phosphorus into brain protein during waking and sleep in rats. Their results suggest that protein synthesis may increase during behavioral sleep.

Mark et al. (9) have investigated the amino acid labeling from ^{14}C-glucose in the whole brain as well as in different brain structures of rats submitted to total sleep and paradoxical sleep deprivation. Their results showed increased specific radioactivity of glutamine in total-sleep-deprived rats, suggesting an increased turnover of this compound.

There are only a few studies of the effects of REM sleep deprivation on the incorporation of labeled amino acids into cerebral proteins of young and adult rats that can be compared. In young rats (24 days old) protein synthesis was studied in vitro (7). A decrease in synthesis in animals deprived for 3 h in comparison with controls was noted; however, an increase in protein synthesis during recuperative sleep of 1 1/2 h was observed. Diminu-

tion of protein synthesis, with no further modifications
during recovery sleep, was recorded in adult rats.

Recently, Bobillier et al. (12) investigated the
incorporation of amino acids into brain proteins in rats
deprived of total and paradoxical sleep. The results
showed that sleep deprivation generally lowered the specific activity of brainstem proteins, and recovery sleep
did result in a return to normal value.

These findings, together with ours, are of special
interest, since they might indicate that sleep serves
the anabolic activity of the nervous systems, thus supporting the rest theory of sleep.

Since this study does not allow any conclusions,
work on estimation of incorporation and turnover rates
and experiments to control variables such as age, size,
and control organ (liver, muscle) is in progress.

The availability of ^{35}S-sulfate in brain tissue
might vary during experimental periods. Although inorganic ^{35}S-sulfate was given in uniform dose based on
body weight, the utilization of ^{35}S-sulfate by organs
other than brain and the excretion of ^{35}S-sulfate may
have altered during the experimental conditions. Accordingly the potential supply of ^{35}S-sulfate to the
brain has probably altered too. It is also possible
that the entry of inorganic ^{35}S-sulfate from the blood
into the central nervous system cells changes according
to the experimental procedures employed. These factors
must be considered in assessing relative rates of ^{35}S-
sulfate incorporation.

AMPS has been considered as brain extracellular
"ground substance." AMPS could provide an aqueous pathway for ion movement between cells. The presence of
such highly charged compounds in the extracellular space,
especially in the case of sulfated AMPS, might be expected to play a role in distribution of cations such
as K^+ and Na^+ in the brain. Such an effect on the distribution of these ions, known to be important in the
conduction of the nervous impulse, may be the basis for
the seizure activity noted after topical (13) or intra-
ventricular (14) administration of hyaluronidase in experimental animals.

It is interesting that after 10 days of REM sleep
deprivation in rats, brain potassium concentrations decreased.

Bowers et al. (15) reported decreased acetylcholine levels in the rat telencephalon.

Decreased AMPS content with REM sleep deprivation was noted, along with decreased potassium content, and the ACh level is suspected to be involved in the increased neural excitability found to be associated with REM sleep deprivation (16, 17).

REFERENCES

1. Duncan, R., Henry, P., Karadžić, V., Mitchell, G., Pivik, T., Cohen, H. and Dement, W. (1968): Manipulation of the sleep-wakefulness cycle in the rat: A longitudinal study. Psychophysiology 4: 379.

2. Morden, B., Mitchell, G. and Dement, W. (1967): Selective REM sleep deprivation and compensation in the rat. Brain Res. 5: 339-349.

3. Ringertz, N. R. (1956): On the sulphate metabolism of the mouse brain. Exp. Cell Res. 10: 230-233.

4. Robinson, J. D., Jr. and Green J. P. (1962): Sulfomucopolysaccharides in brain. Yale J. Biol. Med. 35: 248-256.

5. Stefanovich, V. and Gore, I. J. (1967): A micromethod for the determination of acid mucopolysaccharides in vascular tissue. J. Chromatography 31: 473-478.

6. Boas, N. (1953): Method for the determination of hexosamines in tissues. J. Biol. Chem. 204: 553-563.

7. Bobillier, P., Sakai, K., Seguin, S. and Jouvet, M. (1971): Effects de la privation de sommeil sur l'incorporation d'acides aminés marqués dans les protéines cérébrales du rat. C. R. Soc. Biol. 165: 118-123.

8. Reich, P., Driver, J. and Karnovsky, M. (1967): Sleep: Effects on incorporation of inorganic phosphate into brain fractions. Science 157: 336-338.

9. Mark, J., Godin, Y. and Mandel, P. (1969): Biosynthesis of aspartic, glutamic, -aminobutyric acids

and glutamine in brain of rats deprived of total sleep or of paradoxal sleep. J. Neurochem. 16: 1263-1272.

10. Shapot, V. S. (1957): Brain metabolism in relation to the functional state of the central nervous system. In: Metabolism of the Nervous System. D. Richter ed. Pergamon Press, New York, 257-262.

11. Brodskii, V., Gusatinskii, V. N. Ya, Kogan, A. B., and Nechaeva, N. V. (1974): Variations in the intensity of ^{3}H leucine incorporation into proteins during slow-wave and paradoxical phase of natural sleep in the cat associative cortex. Dokl. Akad. Nauk. SSSR 215: 748-750.

12. Bobillier, P., Sakai, F., Sequin, S. and Jouvet, M. (1974): The effect of sleep deprivation upon in vivo and in vitro incorporation of tritiated amino acids into brain proteins in the rat at three different age levels. J. Neurochem. 22: 23-31.

13. Marovici, N., Stoica, J., Petrescu, A. and Marcovici, G. (1964): Effect of hyaluronidase on normal and experimentally damaged brain. Rev. Roumaine Neurol. I: 37-47.

14. Young, I. J. (1963): Reversible seizures produced by neuronal hyaluronic acid depletion. Exp. Neurol. 8: 195-202.

15. Bowers, M. B., Hartmann, E. L. and Freedman, D. X. (1966): Sleep deprivation and brain acetylcholine. Science 153: 1416-1417.

16. Cohen, H. B. and Dement, W. C. (1965): Sleep: Changes in threshold of ECS in rat after deprivation of the "paradoxical" phase. Science 150: 1318-1319.

17. Cohen, H. B., Thomas, J. and Dement, W. C. (1970): Sleep stages: REM deprivation and electroconvulsive threshold in the cat. Brain Res. 19: 317-321.

REGIONAL CEREBRAL BLOOD FLOW IN MAN DURING

DIFFERENT STAGES OF WAKEFULNESS AND SLEEP

J. S. Meyer, F. Sakai and H. Naritomi

Baylor-Methodist Center for Cerebrovascular
    Research and the
Department of Neurology
Baylor College of Medicine
1200 Moursund, Houston, Texas 77030, U.S.A.

Experience has now been gained with over 1300 measurements of regional cerebral blood flow in normal volunteers
and in patients with neurological disorders using a
modification of the ^{133}Xe inhalation method (1). Since
the method is noninvasive, causes minimal discomfort and,
in our laboratory, is routinely monitored with simultaneous polygraphic recording of electroencephalogram (EEG),
submental electromyography (EMG), extra-ocular movement
(EOG), blood pressure (BP), pulse, body temperature, respiration and end-tidal pCO_2 and pO_2, we have had the
opportunity of observing changes in regional blood flow
during the relaxed, awake state in quiet darkness, during
mild anxiety, during activation and during different stages of spontaneous sleep in normal volunteers and in
patients with narcolepsy.

The subjects reclined comfortably on a padded table,
wearing a light plastic face mask. Sixteen suitably collimated probes were applied over both cerebral hemispheres and the cerebellar and brain stem region. ^{133}Xe gas
was inhaled as a mixture of 5-6 mCi/L of room air for a
one minute interval and the regional desaturation curves
were recorded from the head over the ensuing ten minutes.
Regional flow gray, flow white and their compartmental
weights were calculated by two-compartmental analysis using the method of Obrist et al. (3). Reproducibility of the
method for two serial rCBF measurements for gray matter
in 15 normal volunteers in the resting state is -2.5%.

433

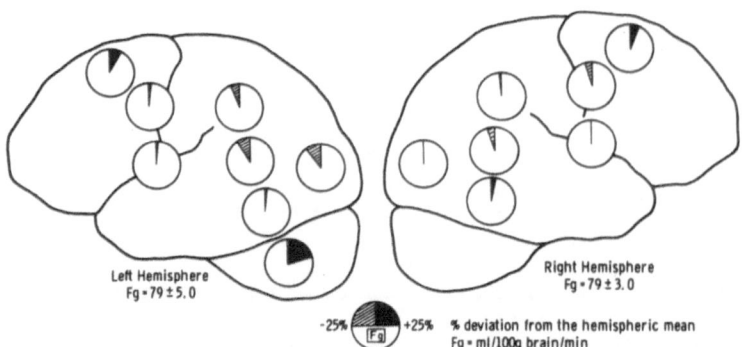

Fig. 1   Regional distribution of Fg at rest in quiet
         darkness (percentage of hemispheric mean values)
         in normal volunteers (n = 14). Age: 40±14 y/o

MEASUREMENTS OF rCBF VALUES AT REST

     At rest, in quiet darkness with the eyes closed
(but without sleep and with alpha activity in the EEG),
Fg values in 14 normal right-handed volunteers were
79.0 + 12.0 ml/100 g brain/min for the left hemisphere
and 79.6 ± 13 for the right hemisphere. The Fg values
in the brain stem-cerebellar regions were significantly
higher than those of the cortex. Fg values for the
cerebral hemispheres were highest in both frontal regions,
being + 10% above the hemispheric mean for the left
frontal and + 4% in the right frontal (Fig. 1).

MEASUREMENTS OF rCBF VALUES DURING ACTIVATION

     The measurements were then repeated 30-40 minutes
later with the lights turned on and during standard multiple psychophysiological activation. This included re-
questing the subject to count silently to himself, to
listen to a brief conversation, and to music, and to
watch figures moving around the laboratory. This cerebral activation increased Fg values by +20% in the brain
stem-cerebellar regions (Fig. 2). The respiration, end-
tidal pCO_2 and blood pressure did not change. Significant
increases were also seen in the left frontal (+14%), pre-
central (11.5%), parietal (+21) and inferior temporal
(+18) regions. The inferior temporal probe is placed so
that midbrain and diencephalic Fg values are also recorded.

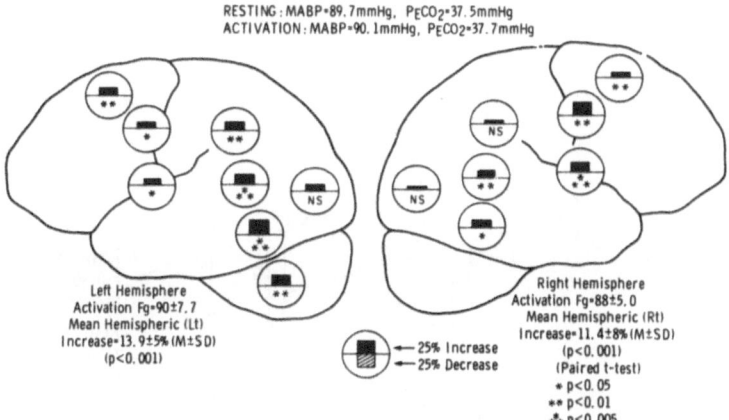

Fig. 2 Regional increases of Fg during cerebral activation, (motor speech and audio-visual stimulation) compared to values at rest in quiet darkness in normal volunteers (n = 14). Age: $40\overset{+}{-}14$ y/o

Although Fg values decrease with advancing age, the increase in Fg during activation was the same in older subjects around 60 years of age as it was in younger subjects (Fig. 3). There was also significant correlation between mean hemispheric Fg values and brain stem-cerebellar Fg values during activation but not at rest, suggesting that the reticular formation influences cortical flow increases (Fig. 4).

In anxious subjects who are poorly acclimated to the procedure, who admitted to feeling tense and anxious during the measurement and who showed beta rather than alpha activity in the EEG together with evidence of sustained muscle contraction in the EMG, the regional Fg values for frontal and brain stem-cerebellar regions were higher than in the relaxed state.

MEASUREMENTS OF rCBF VALUES DURING LIGHT SLEEP
IN NORMAL VOLUNTEERS

Table I shows total mean Fg values during Stage I and II sleep measured in four subjects who spontaneously fell asleep during the rCBF measurements, compared to flow values measured in the same individuals during the steady state when awake in quiet darkness. There was a decrease of -12.6% of total mean Fg values during Stage I & II sleep. The majority of Fg values might have been

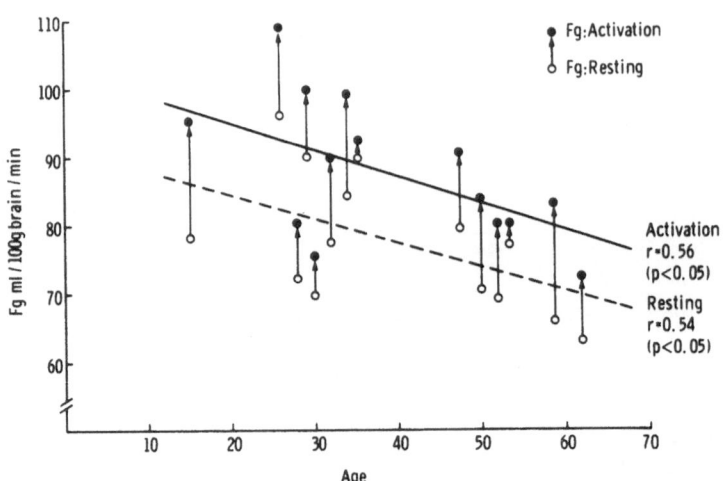

Fig. 3   Correlation of hemispheric mean Fg values with
         advancing age for both resting values (quiet, dark)
         and during multiple psycho-physical activation
         (motor speech and audio-visual stimulation) in
         normal volunteers (N = 14).   Age: 40±14 years

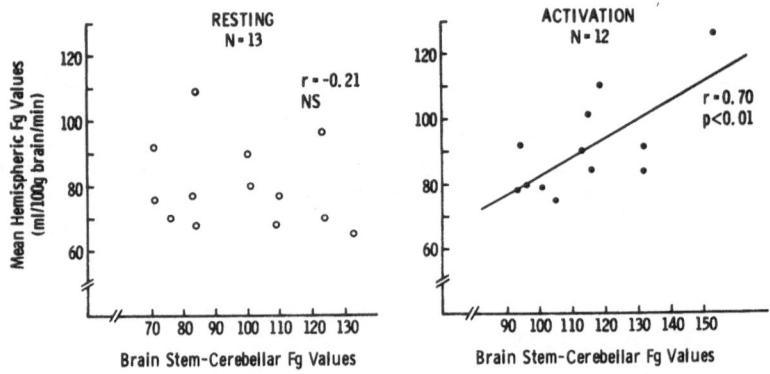

Fig. 4   Correlation between brain stem-cerebellar Fg val-
         ues and mean hemispheric Fg values of both hemi-
         spheres in normal volunteers at rest and during
         activation.   Age: 40±14 y/o

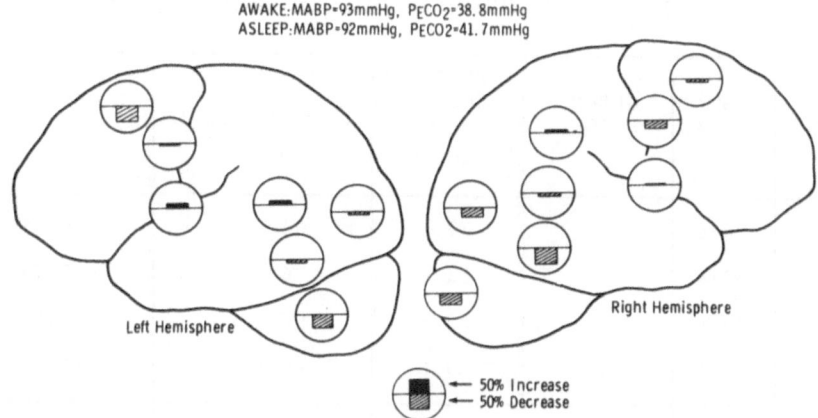

Fig. 5 Regional Fg changes during sleep in a normal vol-
unteer. Values during stage II sleep compared to
values when awake.   33 y/o male

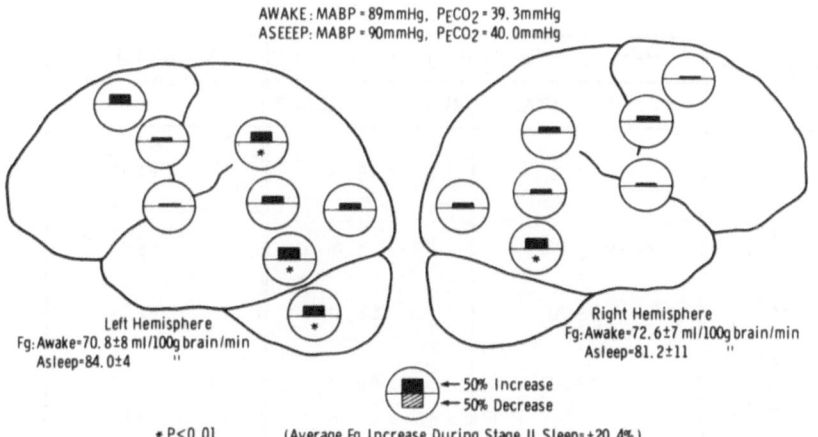

Fig. 6 Regional Fg changes during stage II sleep in narcolepsy compared to values when awake.   (n = 6,
Age 25-59 y/o)

Table I   Fg IN NORMALS AWAKE AND ASLEEP

SUBJECTS	AGE/ SEX	SLEEP STAGE	HEMISPHERIC MEAN Fg (ml/100 g brain/min)		% DECREASE DURING SLEEP
			AWAKE	ASLEEP	
K.Z.	26/M	I	77.5	64.6	-16.6%
N.I.	33/M	I	83.0	76.6	- 7.7%
F.S.	33/M	II	112.5	96.7	-14.0%
D.L.	31/F	I	79.0	69.5	-12.0%
MEAN	30.8		88.0	76.9	-12.6%

Table II Fg IN NARCOLEPTICS AWAKE AND ASLEEP

STAGE OF SLEEP	MEAN Fg (ml/100 g brain/min)		% CHANGE Fg DURING SLEEP
	RUN 1 - AWAKE	RUN 2 - ASLEEP	
REM (N = 2)	77.9	85.5	+10%
II (N = 6)	72.8 ± 6.7	85.5 ± 4	+20% ↑*
III (N = 2)	93.0	82.0	-11%
AWAKE 0 (N = 2)	89.2	88.3	-1%
TOTAL (N =12)	80.5 ± 13		

* p < 0.01

considéred pathologically reduced, if it were known
that the subjects were asleep and that their Fg values
were normal in the waking state.

The regional pattern of flow reduction was characteristic during sleep. Maximum decreases occurred in the
brain stem-cerebellar regions and in the posterior-
inferior temporal regions. Probes placed over the posterior-inferior temporal regions are believed to include
gamma activity derived from the midbrain and diencephalic
regions (Fig. 5). The left frontal region also showed
marked decreases during sleep in these right-handed
volunteers.

MEASUREMENT OF rCBF VALUES DURING SLEEP IN
PATIENTS WITH NARCOLEPSY

Table II summarizes Fg changes during different
stages of sleep and resting wakefulness in 12 patients
with narcolepsy. Two of the narcoleptic patients failed
to fall asleep and in those two individuals there was no
significant change (-1%) in the Fg values between the
two measurements. During rapid eye movement (REM) sleep
in two subjects, there was an increase of +10% in mean
Fg values similar to the results previously reported by
Townsend, Prinz and Obrist (5) and Meyer and Toyoda (2) in
human subjects and Reivich et al. (4) in cats during REM sleep.
During Stage III or slow-wave sleep, there was a decrease
in mean CBF consonant with previous reports (4, 5). However, during the early stages of sleep (Stage I and II),
there was an increase of cerebral blood flow by 20% which
is the opposite of the change seen in normal subjects.
Furthermore, this regional increase in rCBF in narcoleptics was predominantly in the brain stem-cerebellar regions and the posterior-inferior temporal regions. The
probes located in the posterior-inferior temporal regions
are believed to include measurements of regional blood
flow supplying the diencephalic and mesencephalic areas
(Fig. 6).

SUMMARY AND CONCLUSION

Measurements of regional gray matter flow have been
measured in man during the different stages of sleep,
relaxation, wakefulness, anxiety and activation. Maximal
decreases of gray matter flow during slow wave sleep occurred in the brain stem-cerebellar and diencephalic reticular formation in normal healthy subjects and these
same areas showed maximum increases during wakefulness,

activation, anxiety and REM sleep. In general, the cortical Fg values increased or decreased in concert with the
brain stem flow changes but were of less marked degree
except for the left frontal region which showed the
maximum change for all the cortical areas.

Unlike normal volunteers, in patients with narcolepsy, the gray matter flow of the brain stem-cerebellar
regions increased during Stage II sleep which is the
reverse of the pattern seen in normals. Narcolepsy
appears to be a disorder of the reticular activating
system.

ACKNOWLEDGMENTS

This work was supported by grant NS 09287 from U.S.
Public Health Service and the Cooper Laboratories,
Cedar Knolls, New Jersey.

Reprint requests to Dr. John Stirling Meyer, Baylor
College of Medicine, Department of Neurology, 1200 Moursund, Houston, Texas, 77030, U.S:A.

REFERENCES

1. Meyer, J.S., Ishihara,N., Deshumkh, V.D. et al.(1977):
   An improved method for noninvasive measurement of regional cerebral blood flow by ^{133}Xe inhalation. Description of the method and normal values obtained in
   healthy volunteers. Stroke, in press.

2. Meyer, J.S. and Toyoda, M. (1971): Studies of rapid
   changes in cerebral circulation and metabolism during
   arousal and rapid eye movement sleep in human subjects
   with cerebrovascular disease. In Zulch K.J. (Ed.):
   Cerebral Circulation and Stroke. New York, Springer,
   156-163.

3. Obrist, W.D., Thomson, K.H., Jr., Wang, H.S. et al.
   (1975): Regional cerebral blood flow estimated by
   133-Xenon inhalation. Stroke 6: 245-256.

4. Reivich,M., Isaacs,G., Evans,E. and Kety,S. (1968):
   The effect of slow wave sleep and REM sleep in regional cerebral blood flow in cats. J. Neurochem.15: 301 .

5. Townsend, R.E., Prinz, P.N. and Obrist, W.D.(1973):
   Human cerebral blood flow during sleep and waking.
   J. Appl. Physiol. 35: 5: 620-625.

POLYGRAPHICALLY MEASURED EFFECTS OF DIFFERENT ANTIDEPRESSANTS

U. J. Jovanović

Psychiatric Clinic, University of Würzburg
Röntgenring 12, D-8700 Würzburg, W. Germany

ABSTRACT

During the last 15 years more than 100 pharmaceutical compounds have been examined by means of polygraphic recordings, clinical investigations and psychometric tests. Results of the investigations of four antidepressants are presented in detail in this paper. Results with two additional antidepressants are also presented, though in less detail, for comparison.

Twenty patients suffering from endogenous depression (between 25 and 39 years of age, mean 26.7 years, of both sexes: male 9 and female 11 patients) were examined for a total of 5 weeks each. The effects of maprotiline were compared to those of imipramine (both substances in a dosage of 3 x 50 mg a day). Amitriptyline-N-oxide was tested in 15 patients suffering from endogenous depression (24 to 34 years of age, mean 28.5 years). The dosage used was 3 x 20 mg daily. Methylperon was examined in a different way (two series with a dosage of 20 mg and 50 mg, 20 healthy volunteers and 20 depressive patients). The following results were obtained:

(1) The substances investigated had different effects on sleep in endogenous depressive patients. (2) In the first week of therapy imipramine caused a prolongation of nocturnal wakefulness. This could also be observed in the third week of treatment. These effects

were not found in maprotiline and amitriptyline-N-oxide
(ANO). (3) Relative and actual sleep durations were reduced by the increase of wakefulness during the night
under the effects of imipramine and chlorimipramine in
the first week of therapy. There were no important differences concerning the effects of imipramine after the
first week of therapy compared to the third week of therapy with the same substance. These effects were not observed when maprotiline or ANO was administered. (4)
Duration of wakefulness shifted from the first to the
last third of the night after administration of impra-
mine, which was not the case after maprotiline and ANO.
(5) The ultradian sleep periods (PC) changed in patients
under the effects of imipramine and chlorimipramine, especially in the first PC. (6) In the first and third
weeks of therapy imipramine and chlorimipramine (chloramine) caused a reduction of REM phases in patients; this
reduction was either not found at all with maprotiline
or it was very small compared to that with imipramine.
ANO caused no REM suppression. (7) Further, imipramine
had a stronger influence on the duration of sleep stages
within the sleep periods in the first and third weeks of
therapy compared to the other substances. Wakefulness
and the superficial sleep (Stage II) were much longer
in the first PC under the effects of imipramine and
chlopramine. (8) An assimilation of the effects of the
substances investigated was found only in the morning
sleep hours. (9) After the fourth week of therapy, the
effects of imipramine, maprotiline and chloramine were
similar. In the fifth week of therapy there were no differences in the effects of the four substances examined.
(10) According to the findings obtained by polygraphic
sleep recordings and by clinical examinations, imipramine
and chlopramine had stronger effects on the depressive
symptoms already in the first and especially in the second week of therapy compared with ANO. These effects
assimilated at the end of the fourth week of therapy.
(11) According to these investigations, amitriptyline-N-
oxide and maprotiline are antidepressants; however, they
have stronger relaxing effects than imipramine. On the
other hand, imipramine has stronger antidepressive effects, but in the first three weeks of therapy, nocturnal
wakefulness is prolonged, sleep becomes more superficial,
latency times up to the first REM phase are increased
and REM phases are suppressed more strongly than after
administration of maprotiline. (12) Methylperon effected
a prolongation of the REM phases.

INTRODUCTION AND METHODS

During the last 15 years we have examined more than 100 chemical (pharmaceutical) compounds in man by means of polygraphic recordings of bioelectrical potentials (EEG, EMG, EOG, ECG, GDS, EDG) and by clinical investigations and questionnaires (3, 4, 6-22). In this paper we will present only some results of our investigations of five antidepressants: imipramine (N-/gamma-dimethyl-amino-propyl/-imino-dibenzyl-chloride), maprotiline (1-/3-methylamino-propyl/-dibenzo/b,e/-bicyclo-/2.2.2/-octa-diene), amitriptyline-N-oxide (5-/gamma-dimethyl-amino-propyliden/-5H-dibenzo-/a,d//1,4/-cycloheptan-N-oxide), clorimipramine (monochlorimipramine = chlopramine: 3-chloro-5-/3-dimethylamino-propyl/-10,11-dihydro-5H-di-benzo-/b,f/-azepine hydrochloride), methylperon (4'-fluor-4-/4-methyl-piperidino/-butyrophenon-hydrochloride). In addition some points concerning lithium carbonate will be mentioned.

The methods of our investigations are described in more detail by Jovanovic (15, 17, 19, 20) and Jovanovic and Schulte (22). Imipramine and maprotiline were tested in a double-blind trial in 60 patients suffering from endogenous depression. During a period of 4 weeks poly-graphic sleep recordings were made from going to bed in the evening until waking up next morning. In addition the Hamilton Rating Scale for Depression, Hamilton Anxiety Rating Scale, Taylor Manifesting Anxiety Scale, clinical investigations, and self-judgment questionnaires were applied (see also refs. 1-5, 22, 23). The 4-week tests were distributed as follows:

From 3 to 5 recording nights and investigation days from each week were confirmed for evaluation. Until the end of the 4th week only 20 (9 men and 11 women aged 25 to 39 years, mean 26.7 years) of 60 patients were good for evaluation, because the rest did not fulfill the criteria for a double-blind trial. Single results from this trial series have already been published (17-20). The dosage of both preparations (imipramine and maprotiline) was 3 x 50 mg (total 150 mg) per os daily.

In the same laboratory and under the same conditions, but differentiated from the two substances mentioned before, amiproptyline-N-oxide was tested in 15 patients suffering from depression (7 men and 8 women) between 24 and 34 years of age (mean 28.5 years). The daily dosage was 3 x 20 mg (total: 60 mg/day). It was considered as

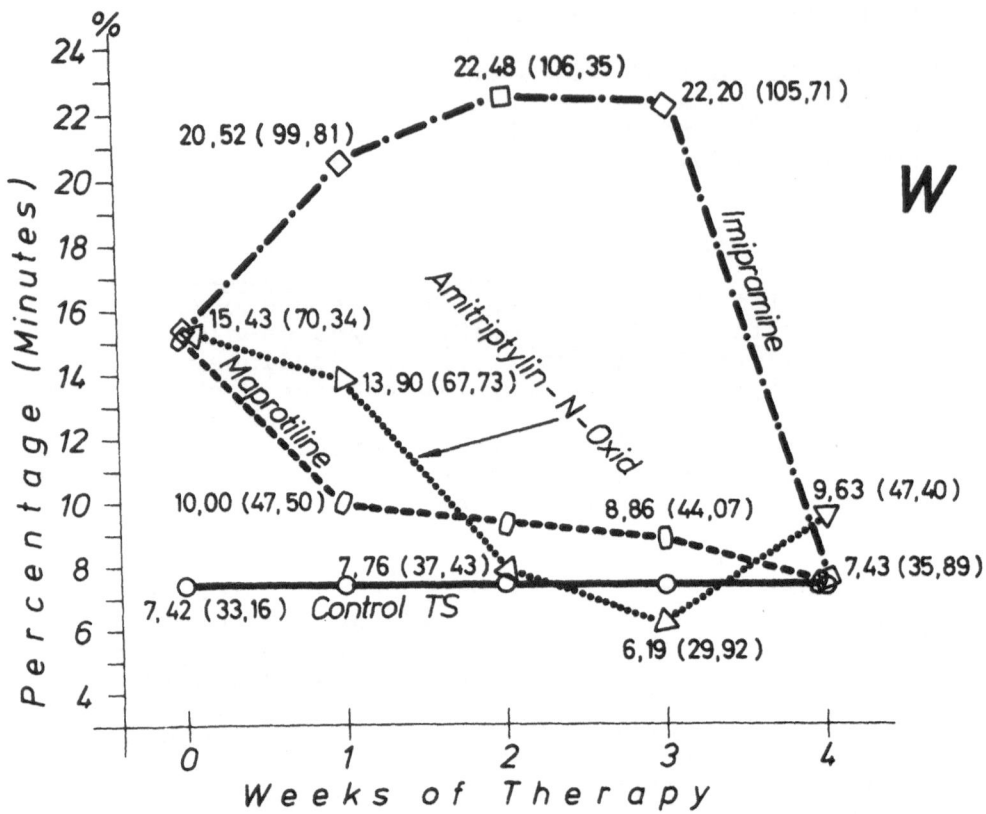

Fig. 1. Effect of imipramine, maprotiline and amitrip-
         tyline-N-oxide on sleep in endogenous depressive
         patients. Imipramine prolongs the waking state
         which already existed previously during the night.
         This prolongation continues to the end of the
         third week of therapy and undergoes a correction
         not before the fourth week of therapy. maprote-
         line and amitriptyline-N-oxide shorten the waking
         state during the night already during the first
         week of therapy.

an equivalent to the dosage of 150 mg of imipramine and
maprotiline (see also refs. 22,23).

     Methylperon was tested in a different way. During
one phase 20 healthy volunteers were investigated continuously during 20 nights by means of psychological tests
and electroencephalographic spectral analysis (ESAP)

(P = program). In a double-blind trial, we also administered placebo as a crossover study. A group of 10 volunteers received a dose of 20 mg each on nights 5, 6 and 7 and a dose of 50 mg each on nights 14, 15 and 16. On the remaining nights, placebo was administered. The other group received the mentioned dosage in a reverse sequence. In another series, 20 patients suffering from endogenous depression, aged between 21 and 54 years, were examined similar to the first group. In addition, four times a day reaction times (RT) and ESAP were undertaken.

The fifth of substance mentioned is chlorimipramine (chlopramine), which was tested in different groups (15, 16). It is given here only as an example, without any details. Lithium carbonate (3 x 450 mg/day is presented for the same reason.

RESULTS

In this study, we wished to establish a direct comparison only among imipramine, maprotiline and amitriptyline-N-oxide (ANO) (Figs. 1-5) (results of the control healthy persons are also given).

Concerning the effects of these substances on sleep patterns in endogenous depressive patients, considerable differences were found. The sleep profile of patients treated with imipramine did not improve in the first week of therapy. ANO effected a slight but continuous improvement. In the first week of therapy there were no differences among the groups treated with imipramine, maprotiline and ANO concerning total sleep duration, (TSD). When relative sleep duration (RSD) was taken into consideration, differences were observed, since wakefulness (W) lasted relatively long in patients treated with imipramine for a week (Fig. 1): sleep duration decreased from 483.3 min (total) to 383.5 min (relative). In the first week of therapy RSD amounted to 429.5 min in the patients treated with maprotiline. Actual sleep duration (ASD) changed accordingly and was 375.5 min in the patients treated with imipramine and 421.7 min in the maprotiline group. In the first third of the night the patients treated with imipramine were awake for 48.3 min; the corresponding time was 27.8 min for the patients treated with maprotiline (p<0.02). There was no difference concerning Stage A. Stage B (Stage I) was shorter in the patients treated with imipramine than in those treated with maprotiline (p<0.2). Further, the duration of Stage C (Stage II) was considerably shorter in the group of patients treated with imipramine (p<0.05).

In the first third of the night, the duration of REM
phases was longer in patients treated with maprotiline
than in the imipramine group. In the second third of the
night, the duration of W was three times as long in the
patients treated with imipramine as in the maprotiline
group (p<0.02). Stages B and C were of considerably
shorter duration in the patients treated with imipramine
than in the maprotiline group (p<0.2 and p<0.005). However, the patients treated with imipramine spent more
time in Stage D (Stage III) and E (Stage IV) (p<0.01 and
p<0.2). In the last third of the night, sleep of patients treated with imipramine was different from that
of the maprotiline group. The imipramine-treated patients stayed awake for 29.2 min compared to the 13.4
min recorded in the maprotiline group (p<0.05). The time
spent in Stage B was significantly shorter in patients
treated with imipramine than in those treated with maprotiline (p<0.05). On the other hand, more time was
spent in Stage D and E by the imipramine group (p<0.05
and p<0.1). In the last third of the night, the duration
of REM phases was shorter in the patients treated with
imipramine (p<0.2). The effects of ANO on sleep in the
three thirds of the night are discussed in Jovanovic and
Schulte (22) as well as in Schulte (23).

From these results, we draw the conclusion that the
patients treated with imipramine for one week stayed a-
wake for a relatively longer period of time compared to
the patients who had been given maprotiline or amitriptyline-N-oxide (ANO). According to these findings, after
treatment with imipramine for one week, the patients by-
passed the superficial sleep stages and slipped into
wakefulness. A different situation was noted with maprotiline and amitriptyline-N-oxide (Fig. 1).

Patients treated with imipramine showed a delay in
falling asleep. For this reason, the first ultradian
sleep period (sleep cycle; PC) lasted quite long (19).

The first ultradian PC was 176.9 min in the patients
treated with imipramine and 136.5 min in the maprotiline
group (p<0.05). The second PC was slightly shorter
(p<0.1) and the fourth PC considerably shorter (p<0.05)
in the imipramine patients. No differences in the remaining PCs were found.

When the duration of sleep stages in minutes within
the PCs were compared, differences among the patient
groups were again recorded. These differences, which
were found in only a very few stages, were very important.

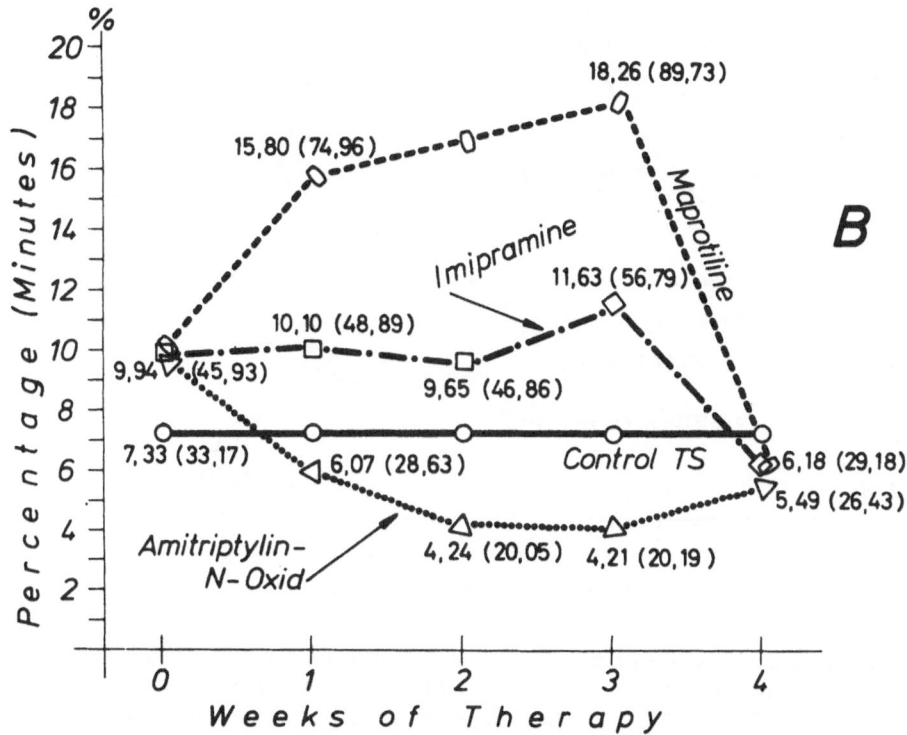

Fig. 2. Effect of imipramine, maprotiline and amitriptyline-N-oxide on Stage I (B) sleep. While the
waking state was prolonged by imipramine during
the night (see Fig. 1), it can be seen here that
Stage I (B) is increased by maprotiline during
the first three weeks of therapy. The effect of
imipramine on Stage B (Stage I) is different from
that of maprotiline and amitriptyline-N-oxide.

In the first PC, wakefulness, for example, lasted
for 67.97 min in the patients treated with imipramine,
which was more than twice the time the maprotiline patients stayed awake (29.1 min) (p<0.01). Findings similar to those in the maprotiline group were seen after
amitriptyline-N-oxide. After treatment with imipramine,

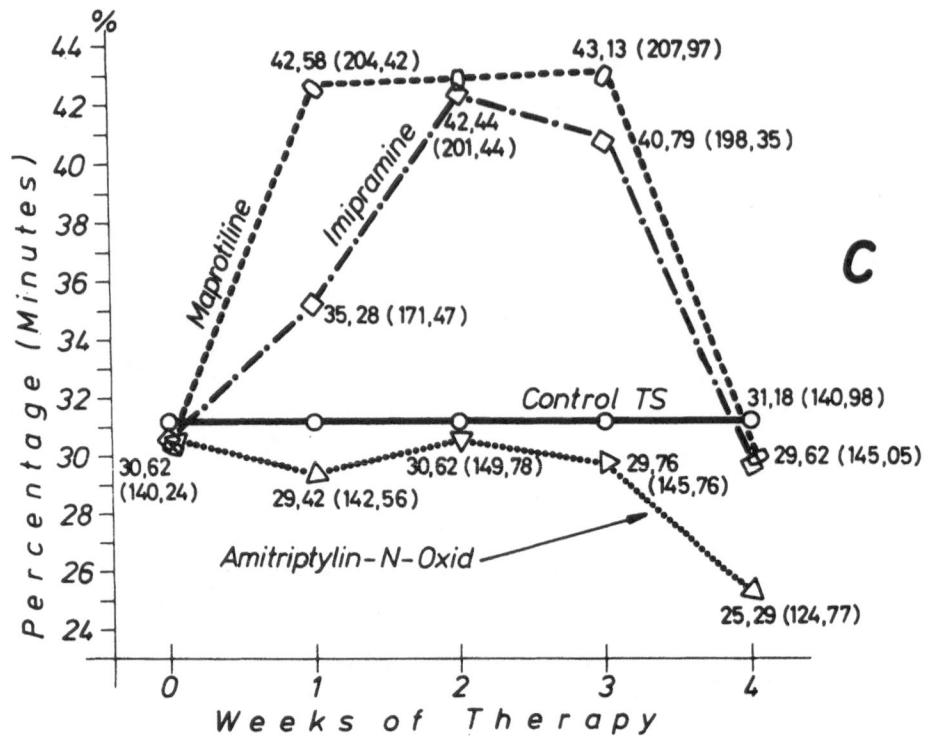

Fig. 3. While the effects of imipramine and maprotiline
on the waking state during the night and on Stage
I differ (cf. Figs. 1 and 2), it can be seen here
that these two substances have an almost parallel
effect on Stage C (Stage II = superficial sleep).
Amitriptyline-N-oxide is different from the other
two substances mentioned in regard to its effects
on all stages.

the time spent in wakefulness in the second PC was ten
times longer (10.6 min) than with maprotiline (1.3 min)
(p<0.02). Superficial sleep stages were considerably
shorter in the patients treated with imipramine (p<0.005).
On the other hand, in the second PC, Stage E (Stage IV)
lasted longer in the imipramine group. In the fifth PC,
further differences were found: the patients treated with
imipramine were awake double the time the maprotiline
group spent in wakefulness (p<0.2). The ANO group was
awake 6.77 min.

In the course of the total night (Fig. 1), patients treated with imipramine were awake for 99.8 min, while the maprotiline group spent 47.5 min in wakefulness (p<0.005). Stage B was considerably shorter in the imipramine group (48.89 min) than in the maprotiline group (74.96 min.) (p<0.05). The shortest Stage B was found after ANO (28.6 min) Fig. 2. Stage C was also shorter in the patients treated with imipramine (171.5 min) than in the maprotiline group (204.4 min) (p<0.05). Again, the shortest Stage C was found after ANO (Fig. 3). Stages D and E lasted longer in the imipramine group (p<0.02 and p<0.2) than in the maprotiline group. The longest Stage E was found in the ANO group (Fig. 4). Despite increased wakefulness in the patients treated with imipramine, there was no difference in duration of the REM phases (Fig. 5). The duration of REM phases was shorter by 11.5 min in the imipramine group. However, this difference was not significant. The longest REM phase was found after ANO. The difference between the ANO and the imiimipramine group was significant.

Usually, latency times were longer in the patients treated with imipramine than in the maprotiline and ANO groups. While the imipramine patients were awake for 32.2 min until they reached Stage A, the corresponding time was 21.1 min in the maprotiline group (p<0.05). Patients treated with imipramine reached Stage B after 55.8 min, compared to 30.0 min for the maprotiline patients (p<0.005). Deep sleep (Stage E, Stage IV) was found 91.1 min after the onset of the recordings in patients treated with imipramine; the maprotiline group reached this sleep stage faster. Latency times after ANO revealed wide variations.

Since, after treatment for one week, maprotiline had shown better effects than imipramine, the second week of therapy for the imipramine patients was evaluated. The second week of treatment with imipramine did not cause an improvement of sleep. On the contrary, in some stages a further deterioration (Figs. 1-5) was observed. For the results of the effects after ANO, see Figs. 1-5.

Now let us compare the groups after treatment for three weeks. There was no difference in total sleep duration in the patients treated with imipramine compared to those treated with maprotiline and ANO. However, there was an enormous difference in relative sleep duration, which amounted to 380.5 min in the patients

treated with imipramine and to 441.5 min in the maprotiline group (p<0.05). Actual sleep duration showed even
more significant differences. Imipramine patients slept
for 369.9 min compared to 436.5 min in the maprotiline
group (p<0.05). According to these criteria, treatment
with imipramine did not show the same quantitative positive results in patients suffering from endogenous depression. The third week of therapy with maprotiline
caused an improvement or tendency toward improvement of
sleep, which was already recorded after the first week
of treatment. In several stages, normalization seems
to have been reached. ANO continued in its tendency toward normalization of sleep (Figs. 1-5).

     In the first third of the night, imipramine patients
were awake for twice as long as maprotiline patients
(p<0.005). Stage B was three times longer in the maprotiline patients (p<0.2). Concerning REM phases, a
considerable deficiency in the imipramine patients in
the first third of the night was recorded. Patients
treated with imipramine spent 2.96 min in REM phases compared to the 15.7 min being recorded in the maprotiline
group (p<0.005).

     In the second third of the night, wakefulness a-
mounted to 18.3 min in the imipramine patients compared
to 7.2 min in the maprotiline group (p<0.05). Stage A
was twenty times longer in the patients treated with
imipramine (p<0.05). In the last third of the night,
patients treated with imipramine stayed awake three
times longer than the maprotiline patients (p<0.005).
Stage A was also longer in the imipramine patients
(p<0.1). In this third of the night, Stage C was shorter
in the patients treated with imipramine (p<0.005). For
a discussion of the thirds of the night after ANO, see
Jovanović and Schulte (23).

Fig. 4. Effects of imipramine, maprotiline and amitrip-
        tyline-N-oxide on deep sleep (Stage E = Stage IV)
        in endogenous depressive patients in the course
        of therapy with a duration of four weeks. Imi-
        pramine and maprotiline shortened the deep sleep
        of patients during the first three weeks of ther-
        apy. Amitriptyline-N-oxide, however, prolonged
        deep sleep continuously from one week to the
        other. At the end of the fourth week, the ef-
        fects of the substances mentioned are similar. →

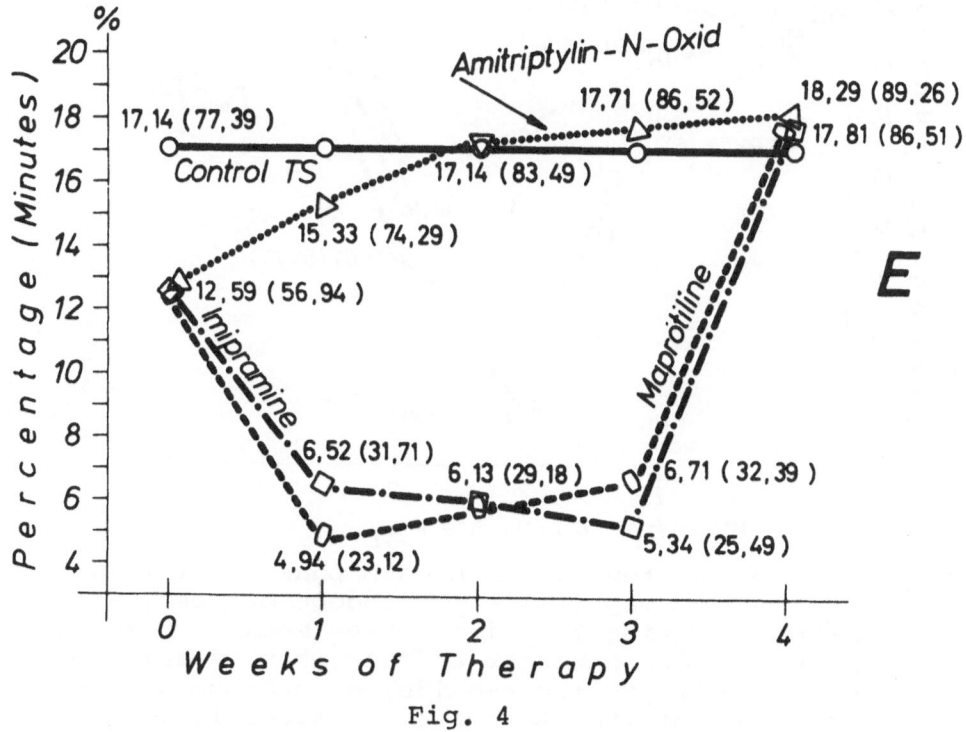

Fig. 4

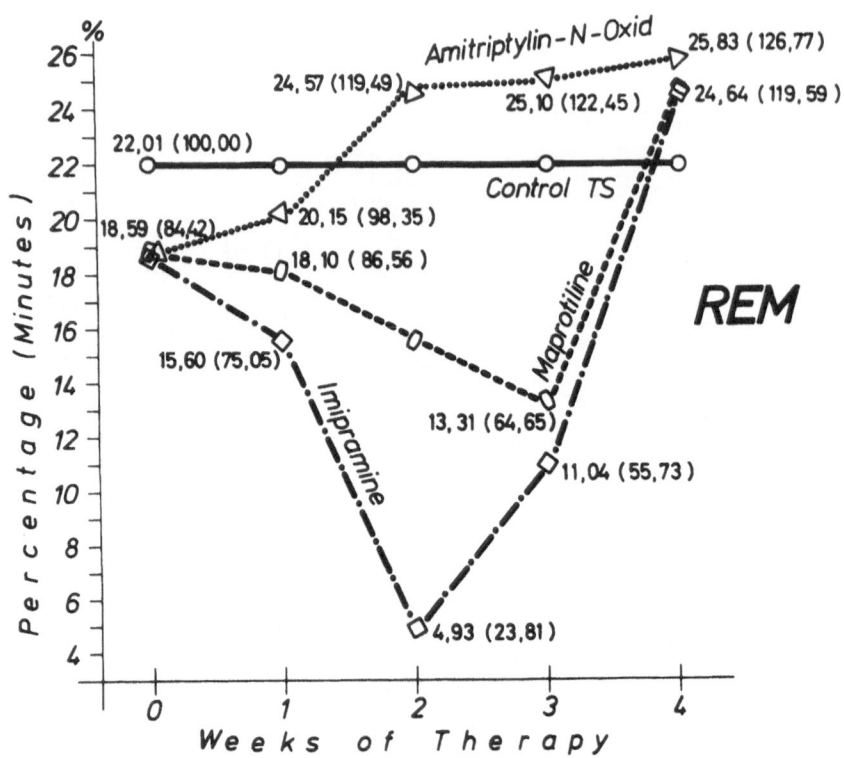

Fig. 5. Evident shortening of the REM phases after imipramine and maprotiline in endogenous depressive
patients during the first three weeks. Here imipramine shows stronger effects than maprotiline.
After amitriptyline-N-oxide, a continuous prolongation of the REM phases of sleep from one
week of therapy to the other can be seen.

The ultradian PCs were different in all stages. The
first PC was of enormously long duration in the imipramine group, probably because of the lengthening of the
latency times. This PC lasted for 331.4 min in the imipramine patients, i.e., 5 1/2 hours. The duration of the
first PC was 150.3 min in the patients treated with maprotiline (p < 0.005). The second PC lasted longer in the

maprotiline patients (p < 0.2), and the third and fourth
PCs lasted four times longer in the maprotiline group
than in the patients treated with imipramine (p < 0.005
for both). Corresponding results were found in the remaining two PCs (p < 0.005 and p < 0.05).

When the three antidepressants imipramine, maprotiline and ANO were compared, three weeks of treatment
showed different effects. When treatment was stopped,
i.e., after the fourth week of therapy, the effects of
the three preparations had assimilated (except Stage II)
(see also ref. 20).

Chlorimipramine has effects similar to those of imipramine (Fig. 6). On the other hand the relatively new
substance methylperon has quite a different effect (Fig.
7). The dosage of 20 mg administered in the evening
prolonged the deep sleep and REM phases simultaneously
(p < 0.05) and to a considerable extent in healthy subjects. The dosage of 50 mg prolonged deep sleep, only
slightly, but with regard to the REM phases, a very pronounced effect was to be seen (see Fig. 7). In depressive patients, a similar situation was noted.

Treatment with lithium carbonate completely destroys
the already altered course of sleep in manic patients.
The sleep becomes very unquiet. Patients many often
shift from one stage to another in the phases of superficial sleep and are unable either to wake up or to fall
into deeper sleep (Fig. 8). We were unable to find any
effect similar to this in any of the 100 psychotropic
or antidepressive drugs we tested.

DISCUSSION

Our investigations have indicated that the effects
of antidepressants differ considerably. As we were able
to demonstrate in an earlier report (19), there are pronounced differences in regard to the types of endogenous
depression. All these results reveal the manifold characteristics of the depressions and emphasize that the
therapy of this disease must be a differentiated one.
According to these findings, the antidepressants have a
wide spectrum of effects: from a suppression to a prolongation of the REM phases; from an antidepressive effect varying in time of onset from only after a period
of four weeks to just after the first dosage is administered. Further studies on this subject will supply
more details with which to approach a resolution of the
problem, which now is seen only relatively.

Fig. 6. Sleep profiles (ordinate: sleep-stages; abscissa:
duration of sleep) of three nights from a 19-
year-old female patient suffering from enuresis.
Top: Sleep profile at night before administration
of 50 mg chlorimipramine. The sleep periodicity
is relatively regular; the dream phases occur
four times and last relatively long. The first
REM-phase (dream-phase = T) is relatively long,
as is usually the case in adult bed-wetters.
Middle: Effect of 50 mg chlorimipramine, given in
the evening before going to bed. The REM phases
were almost totally suppressed. Only in the
morning sleeping hours, shortly before the pat-
tient arose, could a single REM phase (hatched)
with a duration of 10 minutes be recorded. Bottom: Rebound phenomenon after the withdrawal of
chlorimipramine. Now the REM phases occur more
often, and the individual REM phases are relatively long. The effect of chlorimipramine is
similar to that of imipramine (see text).

AL -♀- 051251 - 110870 - 214 - 07 - 000

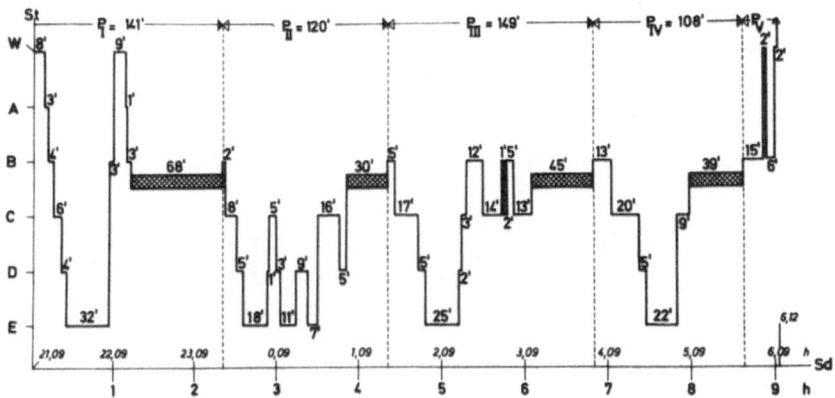

AL -♀- 051251 - 140870 - 214 - 10 - CHLORIMIPRAMIN 50 mg

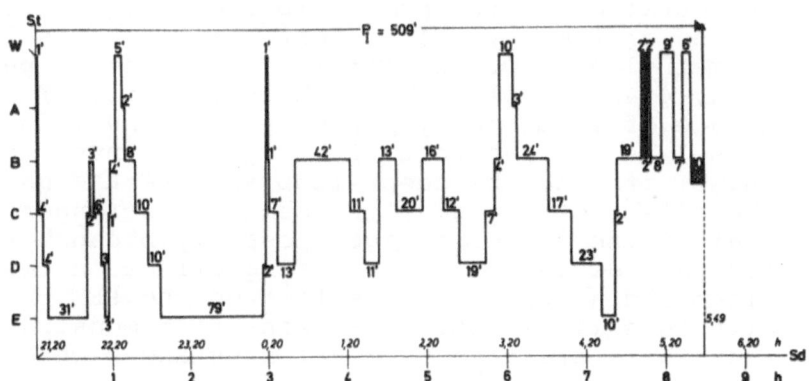

AL -♀- 051251 - 180870 - 214 - 14 - 000

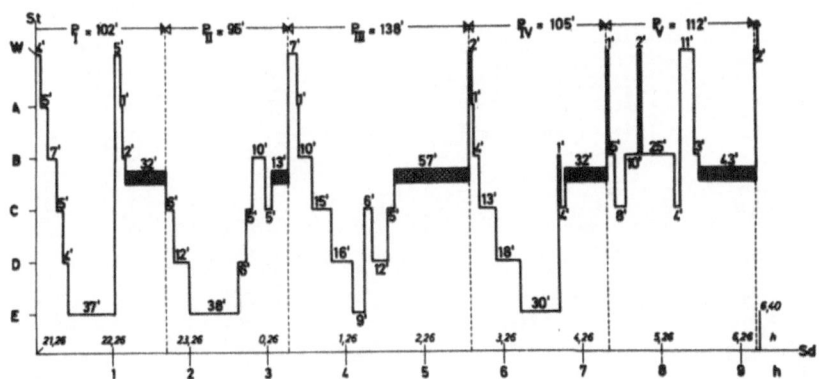

Fig. 7. Effect of methylperon (see the formula in the
text) on deep sleep (Stage E = IV) (top) and on
the duration of REM phases (bottom) after various
phases of therapy. In investigation phases 2 and
5, methylperon was given; in investigation phases
1, 3, 4 and 6, placebo (double-blind trial). The
dosage of 20 mg methylperon (second investigation
phase) as well as that of 50 mg (5th investigation
phase) prolong the deep sleep and the REM phases.
The effect of the smaller dosage is stronger than
that of the larger dosage, possibly because the
stronger dosage has an arousing effect on deep
sleep and a suppressing effect on the REM phases.
With this illustration, we want to demonstrate
that there are also chemical substances which are
able to prolong the REM phases.

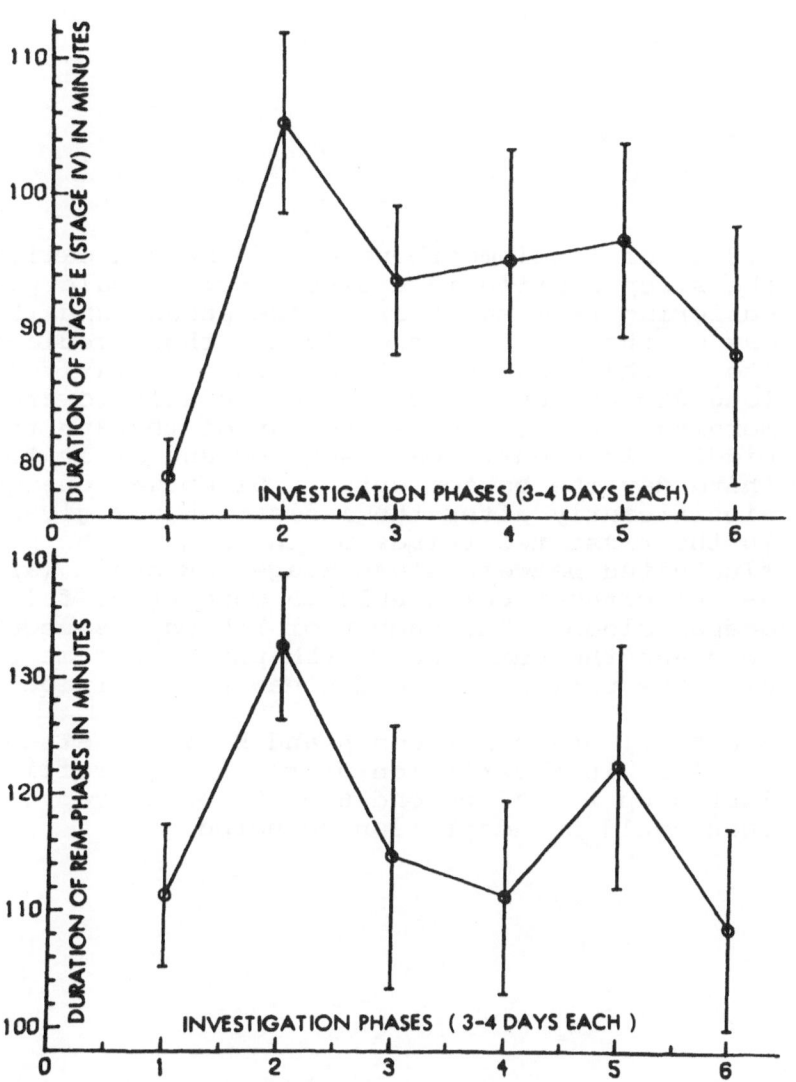

Fig. 8. Effect of lithiumcarbonate (450 mg 3 x daily) on
        the sleep profile in a 34-year-old female patient
        suffering from mania after the second and third
        day of therapy. In the night without medication
        (top), the sleep profile evidences relatively
        long REM phases (hatched), especially toward
        morning. After the second day of therapy the
        middle sleep curve was recorded and after the
        third day the bottom curve. Sleep was changed
        significantly after the medication was given.
        In the first two thirds of the night, the patient
        fluctuated between sleep Stages AB and C (as far
        as CD) without being able to wake up or fall into
        deeper sleep. The number of REM phases became
        less and the duration of REM phases became short-
        er. The patient changed sleep stages frequently.

        Stage AB: Stage between A and B (i.e., Stages O
        and I). This variation of the sleep profile
        lasted up to and beyond 6 weeks of therapy. Only
        then could an adaptation be noted.

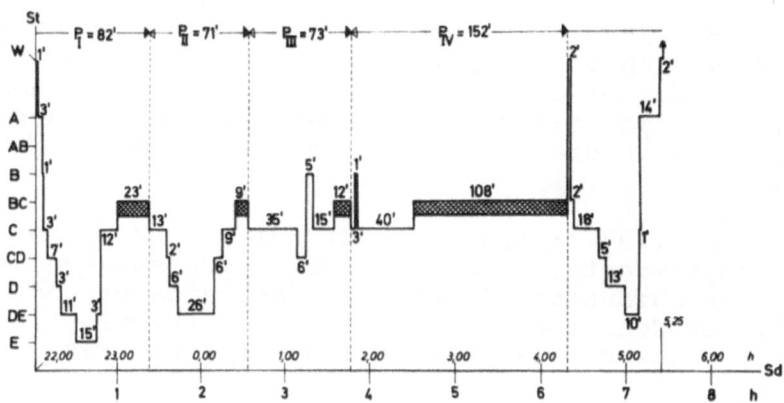

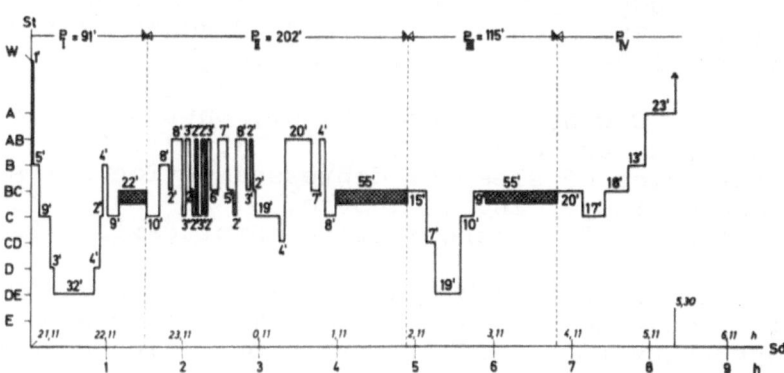

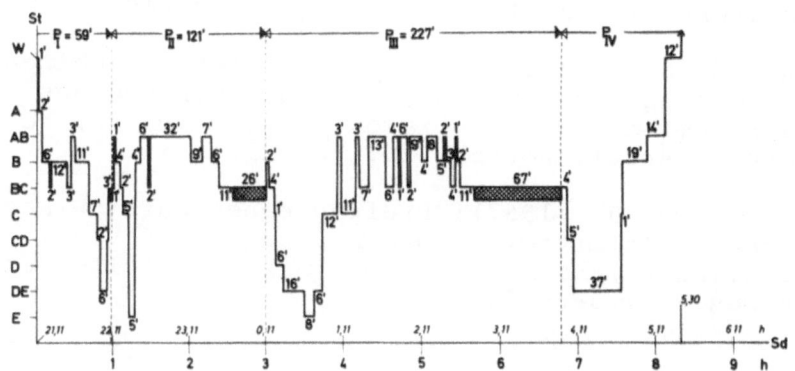

REFERENCES

1. Brunner, M. (1977): Zum Problem der Störungen der Bio-
   rhythmik im Schlaf bei Patienten mit endogener Depres-
   sion: Schlafdauer und ultradiane Schlafperiodik nach
   Placebo und nach einer einwochigen thymoleptischen
   Therapie. Inaugural-Dissertation, Universität
   Würzburg.

2. Brunner, W. (1977): Ein weiterer Beitrag zum Problem
   der Biorhythmusstorungen bei endogen-depressiven Patienten: Verlaufsuntersuchungen der Schlafdauer und der
   ultradianen Schlafperiodik wahrend der thymoleptischen
   Therapie. Inaugural-Dissertation, Universität
   Würzburg.

3. Dürrigl, V., Rogina, V., Stojanović, V., Hajnšek, F.,
   Gubarev, N. and Jovanović, U. J. (1973): Sleep of de-
   pressed patients under the influence of antidepressive
   drugs - A study of two substances. In: Jovanović,
   U. J. (Ed.) "The Nature of Sleep." Fischer-Verlag,
   Stuttgart. 197-202.

4. Hajnšek, F., Dogan, S., Gubarev, N., Dürrigl, V.,
   Stojanović, V. and Jovanović, U. J. (1973): Some
   characteristics of sleep of depressed patients - A
   polygraphic study. In: Jovanović, U. J. (Ed.) "The
   Nature of Sleep." Fischer-Verlag, Stuttgart. 197-202.

5. Hofmann, E. W. (1977): Schlafprofil und ultradiane
   Schlafperiodik beim Menschen unter Wirkung eines
   Butyrophenonhydrochlorid (Methylperon). Inaugural-
   Dissertation. Universität Würzburg.

6. Jovanović, U. J. (1965): Das Elektroencephalogramm des
   Menschen unter Wirkung von 2-Diathylamino-5-Phenyl-
   Oxazolinon-(4). Ärztl. Forsch. 12: 640-648.

7. Jovanović, U. J. (1966): Elektroencephalographische
   und klinische Betrachtungen über die Wirkung des
   2-Diäthylamino-5-Phenyl-Oxazolinon-(4) (Ha 94) beim
   Menschen. Ärztl. Forsch. 20: 206-216.

8. Jovanović, U. J. (1967): Prufung eines barbitursäure-
   haltigen Schlafmittels. Die Wirkung von 5-Vinyl-5-
   (1-methylbutyl)-barbitursäure. Arzneimittelforsch.
   (Drug Res.) 17:365-376.

9. Jovanović, U. J. (1972): Dreifacher Blindversuch im klinischen Experiment. Vergleich der therapeutischen Resultate nach Amitryptylin, Imipramin und Maprotilin (unpublished results).

10. Jovanović, U. J. (1972): Polygraphische Registrierung nach parenteraler Applikation von 3,7-Dimethyl-1-(5-oxo-hexyl)-xathin (BL 191). Arzneimittelforsch. (Drug Res.) 22: 994-999.

11. Jovanović, U. J. (1973): Suggestions for the Treatment of Sleep Disturbances and Conclusions. 1st European Sleep Research Society, October 1972, Basel. In Koella, W. P. and Levin, P. (Eds.): Sleep - Physiology Biochemistry, Psychology, Pharmacology and Clinical Implications. S. Karger, Basel, 145-160.

12. Jovanović, U. J. (1973): Der Effekt einer Kombination von Methaqualon und Diphenhydramin auf den Schlaf gesunder Menschen. Med. Klin. 68: 334-339.

13. Jovanović, U. J. (1974): Zum Problem der Effekte zentral-wirkender Pharmaka auf den REM-Anteil des Schlafes. Untersuchungen mit höheren Dosen von Chloraldurat 500. Fortschr. Med. 92: 1090-1094.

14. Jovanović, U. J. (1974): Rating scale for drug effects on sleep electroencephalograms. Ten categories of psychopharmacological agents. In: Itil, T. M. (Ed.): Psychotropic Drugs and the Human EEG. Mod. Prob. Pharmacopsychiatrie. Karger, Basel, 8: 158-181.

15. Jovanović, U. J. (1974): Categories of psychotropic drug effects on sleep EEG and EOG. 16th Psychopharmacological Meeting of the Czechoslovakian Psychopharmacological Society, Jesenik Spa., Arch. nerv. sup. (Praha) 16: 209-210.

16. Jovanović, U. J. (1975): Der Einfluss von Psychopharmaka auf das Elektroencephalogramm und das Elektrookulogramm. Verhandlungen der Ges. Neurootol. Aequilibriometr., Bad Kissingen. In Clausen, F. (Hsg.): Therapie bei Schwindel. Edition Medizin und Pharmazie Frankfurt/M, 4: 269-285.

17. Jovanović, U. J. (1976): Sleep disturbances in neuropsychiatric patients. Waking and Sleeping 1: 67-88.

18. Jovanović, U. J. (1977): The sleep profile in manic-depressive patients in the depressive phase: Studies to compare these patients with healthy human subjects. Waking and Sleeping 1: 199-210.

19. Jovanović, U. J. (1977): Sleep profile and ultradian sleep periodicity in endogenous depressive patients: Studies during antidepressive therapy. Waking and Sleeping. 1: 297-306.

20. Jovanović, U. J. (1977): Sleep profile and ultradian sleep periodicity in patients suffering from endogenous depression during the treatment with imipramine and maprotiline. Waking and Sleeping 1: 415-421.

21. Jovanović, U. J. (1977): Chronobiologic aspects of psychiatry. Waking and Sleeping 1: 335-341.

22. Jovanović, U. J. und Schulte, W. (1976): Polygraphische Registrierungen des Schlafes bei endogen-depressiven Patienten vor und nach der Behandlung mit Amitriptylin-N-Oxid. Arzneim. Forsch. (Drug Res.) 26: 2106-2113.

23. Schulte, W. (1976): Der polygraphisch registrierte Schlaf endogen-depressiver Patienten unter Wirkung von Amitriptylin-N-Oxid. Inaugural-Dissertation, Universitat Wurzburg.

PARTICIPANTS

R.K. Andjus, Institute of Physiology, Faculty of
    Science, University of Belgrade, 11000 Bel-
    grade, Yugoslavia.

A. Baethmann, Institute for Surgical Research,
    Department of Surgey, University of Munich,
    Nussbaumstrasse 20. 8000 München, West Ger-
    many.

W.F. Caveness, Laboratory of Experimental Neuro-
    logy, National Institute of Neurological
    and Communicative Disorders and Stroke,
    National Institutes of Health, Bethesda,
    Maryland 20014, U.S.A.

R. Cohn, Department of Neurology, Howard Univer-
    sity College of Medicine, 2041 Georgia
    Avenue, N.W., Washington, D.C., 20060, U.
    S.A.

B.M. Djuričić, Laboratory for Neurochemistry,
    Institute of Biochemistry, Faculty of
    Medicine, 11000 Belgrade, Yugoslavia.

J.H. Garcia, Division of Neuropathology, Univer-
    sity of Maryland, School of Medicine,
    Baltimore, Md 21201, U.S.A.

K.-A. Hossmann, Max-Planck-Institute for Brain
    Research, 5000 Köln (Merheim) 91, West
    Germany.

F. Joó, Institute of Biophysics, Biological Re-
    search Center, Hungarian Academy of Sci-
    ences, Szeged, Hungary.

U.J. Jovanović, Universitäts-Nervenklinik, Rönt-
      genring 12, D-8700 Würzburg, West Germany.

I. Klatzo, Laboratory for Neuropathology and
      Neuroanatomical Sciences, National Insti-
      tute of Neurological and Communicative
      Disorders and Stroke, National Institutes
      of Health, Bethesda, Maryland 20014,
      U.S.A.

W.D. Lust, Laboratory for Neurochemistry, Nat-
      ional Institutes of Neurological and
      Communicative Disorders and Stroke, Nat-
      ional Institutes of Health, Bethesda,
      Maryland 20014, U.S.A.

J.S. Meyer, Department of Neurology, Baylor
      College of Medicine, Texas Medical Center,
      Houston, Texas 77025, U.S.A.

B.B. Mršulja, Laboratory for Neurochemistry, In-
      stitute of Biochemistry, Faculty of Medi-
      cine, University of Belgrade, 11000 Bel-
      grade, Yugoslavia.

B.J. Mršulja, Department of Neurophysiology and
      Neurochemistry, Institute for Biological
      Research, Belgrade, Yugoslavia.

V.M. Okujava, Academy of Sciences Georgian SSR,
      Tbilisi, U.S.S.R.

H. Pappius, Laboratory for Neurochemistry, Mon-
      treal Neurological Institute, McGill Uni-
      versity, Quebec, Canada.

LJ.M. Rakić, Institute of Biochemistry, Faculty
      of Medicine, University of Belgrade, 11000
      Belgrade, Yugoslavia.

P. Riederer, Ludwig Bolzman Institut für Klinis-
      che Neurobiologie, A-1130 Wien, Austria.

J. Ristić, Department of Neurology, University
      Clinic of Neurology and Psychiatry, Uni-
      versity of Belgrade, 11000 Belgrade, Yugo-
      slavia.

M. Spatz, Laboratory for Neuropathology and Neuro-
    anatomical Sciences, National Institute of
    Neurological and Communicative Disorders and
    Stroke, National Institutes of Health, Be-
    thesda, Maryland 20014, U.S.A.

V. Šušić, Institute of Physiology, Faculty of Medi-
    cine, University of Belgrade, 11000 Belgrade,
    Yugoslavia.

K.M.A. Welch, Department of Neurology, Baylor Col-
    lege of Medicine, Texas Medical Center, Hous-
    ton, Texas 77025, U.S.A.

D.M. Woodbury, University of Utah, College of Medi-
    cine, Salt Lake City, Utah 84112, U.S.A.

INDEX